Places of Quiet Beauty

The American Land and Life Series

Edited by WAYNE FRANKLIN

Places of Quiet Beauty

PARKS, PRESERVES, AND ENVIRONMENTALISM

by Rebecca Conard

University of Iowa Press Ψ Iowa City

University of Iowa Press, Iowa City 52242

Printed in the United States of America

Portions of this work were published as "Hot Kitchens in Places of Quiet Beauty: Iowa State Parks and the Transformation of Conservation Goals," *Annals of Iowa* 51 (1992), 441–479. Adapted by permission of the publisher.

Publication of this book was supported in part by the Iowa Department of Natural Resources.

Printed on acid-free paper

Library of Congress Cataloging-in-Publication Data
Conard, Rebecca.
Places of quiet beauty: parks, preserves, and environmentalism / by Rebecca Conard.
p. cm.—(The American land and life series)
Includes bibliographical references (p.) and index.
ISBN 0-87745-558-9 (pbk.)
1. Nature conservation—Iowa—History. 2. Parks—Iowa—History. 3. Natural areas—Iowa—History. 4. Environmentalism—Iowa—History. I. Title. II. Series.
QH76.5.I8C66 1997
333.78′216′09777—dc20 96-27207
CIP

02 01 00 99 98 97 P 5 4 3 2 1

Title page: Springbrook State Park in the 1930s. Courtesy Iowa Department of Transportation.

For Lowell Soike, who opened the path

Contents

Acknowledgments

The initial research for this book was funded under the provisions of the National Historic Preservation Act of 1966, as amended, from the U.S. Department of Interior and the State Historical Society of Iowa. A travel to collections grant from the National Endowment for the Humanities and a faculty summer research grant from Wichita State University assisted additional research.

My sincere appreciation to Larry Wilson and the staff at the Iowa Department of Natural Resources, who gave me unlimited access to agency records and made an office available to me whenever I requested work space. This was not a history the agency initially sought; however, my research and writing coincided with the approach of the Iowa state park system's seventy-fifth anniversary in 1995. I happened to be in the water when the tide of interest began to rise. While I may have contributed to the reflective spirit that milestone generated among staff at headquarters and in the field, I was more the beneficiary of it. Michael Carrier, administrator of the Parks, Recreation and Preserves Division, took the event as an opportunity to reflect on the past in relation to the present as the agency heads toward a new century. Accordingly, in April 1995, I was privileged to be part of a symposium devoted to examining the meaning of state parks and preserves and the mission of those who manage them. That Mike and others became genuinely interested in this study of the state park system is reflected in the department's offer to underwrite, in part, its publication.

Writing a book demands long stretches of solitude, but thankfully it is not a solitary endeavor. I owe thanks to many who have contributed in ways great and small to the final product. To the many people at DNR who helped me find what I was looking for, answered questions, and relieved the tedium of reading through sixty-five years of commission minutes, my deepest gratitude. Thanks especially to Angela Corio, Nancy Exline-Downing, John Fleckenstein, Ken Formanek, Daryl Howell, John Pearson, Jim Scheffler, and Arnie Sohn.

Research assistants never get enough credit for helping to lay the foundation, organize the data, and apply the polish. For this, my thanks to Stacie Bromley Stoffregan, Tracy Cunning Chadderdon, Jennifer Hudson, Ellen Keegan-Palacio, Marie Neubauer, Vicki Schipul, and Debbie Taylor. Virginia Cunning, Leah Rogers, and Dan Towers also assisted me with research that ended up in this book. Thanks also to the staff at the State Historical Society of Iowa, both Iowa City and Des Moines, the University of Iowa Special Collections Department, Iowa State University Special Collections Department, the Iowa Federation of Women's Clubs, the General Federation of Women's Clubs in Washington, D.C., the Sioux City Public Library, the Des Moines Public Library, and the interlibrary loan departments at Wichita State University as well as UC Santa Barbara. Those who shared information in taped interviews, telephone conversations, or informal discussions are acknowledged in the bibliography. In addition to these people, Whitley Hemingway graciously allowed me to look through materials belonging to his grandmother, Cora Call Whitley, and then donated the collection to the Iowa Women's Archives. Special thanks to Patricia Wiegel, my mother, for transcribing several lengthy interviews.

Susan Flader's critical reading of the draft manuscript gave me both direction and inspiration as I began the revisions. Michael Carrier, David Glenn-Lewin, Jone Miller, Jan Nash, Michel Nellis, Galen Smith, and Lowell Soike read parts or all of the work in progress, and the resulting work is far better because of their honest appraisals and impromptu marginalia. My companion, Steve Kettering, read the first draft of every chapter, an act of transcendent kindness, and provided constant support, especially when the days were long and the word well ran dry.

Finally, thank you, Holly, Mary, and Bob for keeping faith.

The merits of this work I attribute to good counsel and good friendship. Its shortcomings I am left to claim as my own.

Foreword

Studies of landscape work best when they combine a sense of the concrete places in which we all live, work, and play with a larger inquiry about the values and ideas, the forces and influences, that shape those places and our continuing experience of them. In this informative and compelling account of the fortunes of the modern park movement in the single state of Iowa, Rebecca Conard succeeds splendidly on both fronts. Her narrative of the institutional context of the Iowa story is impeccably and patiently developed, while at the same time the larger national scene within which that one story emerged is always before us. We do not tour the particular parks of the Hawkeye State, nor are we overwhelmed with facts about this site or that. The focus instead aims steadily at the play of ideas, interests, and intentions in a state that devoted little space to public purposes in 1915 but that by the eve of the Great Depression had several dozen diverse parcels actively in use for recreation, preservation, and scientific research.

This institutional story is fascinating and significant. It provides insight not only into the particular history of public land creation and management in Iowa but also into a wide array of larger questions, to which I want to return later. Yet it remains true that the individual parcels about which Conard writes at one remove, as it were, underpin her narrative, and anyone who has spent much time in Iowa will easily supply the appropriate images. For me, her story inevitably called to mind the very first moment I saw the Hawkeye State almost three decades ago. Returning to Pennsylvania from a camping trip in the

West, I was crossing the Missouri River from Nebraska City when the gorgeous Loess Hills abruptly rising on the Iowa side of the Missouri grabbed my eye. Marking where glacial drift, blown eastward thousands of years before by the wind, fell to earth and slowly built bluffs along the river, the hills are peculiarly undulating and remarkable for their fragility. The hills stood out with a green unreality as I came over the bridge in 1970, slowly crossed the alluvial flats separating them from the river, and then followed the twists of Highway 2 up onto their wooded heights. I read later that these are very special places, unparalleled except in the valley of the Yellow River in China. Mimic mountains as you approach them, they form a separate biological region above the monoculture that dominates the surrounding human landscape.

From a vantage point three hundred feet above the Missouri flats in 1970, it seemed as if half the world were visible. In such a rolling landscape, devoid as it is of real mountains, a modest rise in elevation yields surprising gains of prospect. Hills, as opposed to mountains, have rarely received their due. On a continent where so many sizable peaks cluster, especially beyond the 100th meridian, we too often overlook—literally and figuratively—the more modest lift and the more subtle difference that the uplands of every region offer, from the hills of New England or Kentucky to the blond sculpture of the Palouse in Washington and Idaho when the wheat is ripe. I have been astonished by how much can change as one gains only a few hundred feet in altitude. In the desert Southwest, a microshift in vertical position—three hundred feet—will move you from sage brush to pinyon pines, from bluish scrub to a small but deep green forest. Iowa, the heart of the heart of the country, lies in the national imagination like a flat, almost geometric space, but it is not flat even in its hill-less regions, and the landscape as Grant Wood painted it flexes with the supple beauty of the place. Poets of the last century, who instinctively compared the prairie to the sea, knew how much motion there was in it—in the surf of the grass but also in the groundswell of the earth underneath. I could have had no better introduction to the state than the Loess Hills provided on that warm summer evening in 1970.

As it happened, there was an Iowa state park crowning the ravine-cut hills where I camped that night, sung to sleep by whippoorwills echoing in the deep deciduous woods all around. Waubonsie State Park, as those who read Conard's narrative will discover, was among the first of the many sites set aside by this prairie state during its progressive years. As early as 1926, when Iowa was still struggling out of the "gumbo" of its first automobile roads, Waubonsie had been preserved. Because of the rough terrain there, the hills had been spared the intensive use visited on the alluvial grounds below them, and to judge from the rich population of wildflowers in the park, they must have been used little, if at all, for pasture in the years prior to 1926. The establishment of the park there in that year helped prevent further deterioration, and from then until the present—seventy years now, almost half of Iowa's history as a state—this one piece of earth has been removed from the ordinary economy.

Iowa entered the field of park-making by dedicating a handful of sites in 1919 and soon emerged as a leader in developing parkland. In 1925, it was fourth in the nation in the number of parks, behind New York, Michigan, and Texas, but ahead of many other more populous and wealthier states. By 1931, not a decade and a half after the effort commenced (the enabling legislation had been passed in 1917), Iowa had established forty state parks. They included a variety of sites, from those such as Waubonsie, which were deemed worthy for their scenic value, to those such as Backbone, on the Maquoketa River north of Cedar Rapids, which were seen as sites of geological and biological significance worthy of preservation for their own sake, whatever their immediate human appeal. Later came historic sites and prairie parcels, postage-stamp in size by comparison with most modest farms but evocative in their coat of once abundant but now rare foliage.

It's a false impression, of course, and one that Conard effectively counters, to think of such places as frozen in some premodern or prehuman condition. Waubonsie itself has evolved over time, so that its original scenic purpose has yielded more recently to the broader purpose of preserving the unique ecosystem of the wider Loess Hills region. Other changes have altered the face of Waubonsie and its uses as well as its larger social rationale. As late as the eve of World War II, according to the Iowa volume in the WPA state guide series published in 1938, Waubonsie had picnic tables but no overnight facilities. Other Iowa parks already did offer the latter, and the way in which the park system more generally accommodated campers offers interesting sidelights on the shifting relation of Americans to their parks. Recreation in the Victorian era tended to be more institutionally cushioned—hence the big mountain houses of the East and West or hotels such as the Grand on Mackinac Island. With the new century came a scaling down and back, as well as a broadening of participation in tourism. Camping, particularly auto-camping, came into vogue in the teens and twenties, when the "tin-can tourists" poured out of cities and towns in their newly affordable cars, took to the expanding road system, and proved by their numbers how thoroughly the cheap car would transform the older rural landscape. Their popular nickname denoted both the cheap vehicles they drove and the litter with which they decorated the makeshift camping grounds they occupied along the pavement, often on private land. Few groves within easy reach eluded them, and farmers soon found new creatures occupying their pastures.

An aspiring Chicago newspaper cartoonist by the name of J. Fergus O'Ryan caught the farmer's view of this development in a drawing he submitted to a competition in the early 1920s. Showing a pair of auto tramps in the foreground, parked along a country road and cooking their dinner over a fire in the ditch, he crowded the background with loafers and trespassers from other cars, as well as a half-dozen large makeshift signs—posted, obviously, by the farmer beset by such interlopers—declaring, "BEWARE: DANGEROUS DOGS," "BEAR TRAPS HERE," "GROUNDS COVERED WITH POISON IVY," and the like. His title for the bit of social commentary, playing on the loftier ideals of America's recent involvement in the Great War, suggested a more realistic domestic truth: "Making the World Safe FROM Democracy."

Nationally, state parks aimed in part at doing just that: they diverted the wandering hordes into specially prepared enclaves. The purposes of such enclaves, though, were quite varied, and they were defined, refined, and realized via a complex social process that is rarely evidenced in the sites themselves. Rebecca Conard helps us see precisely how institutional the story of the founding and development of parks is. Those who use them today—like myself in 1970—are prone to see them as self-evidently placed where they belong: Waubonsie certainly is brilliantly situated, so that on first encounter it is hard to imagine that it hasn't "always" been there, as if no other use for such a place ever occurred to anybody. As a matter of course, park visitors do not see the park's history or the changes, material as well as immaterial, that have been witnessed in it over time. Rebecca Conard's great gift, aside from the patience that sustained her as she camped out in the forest of original sources she has used, is an ability to see human values and human purposes at work in complex institutional settings. She never reduces even in the more political episodes to mere politics; nor does she lose track of the individuals, from botanist Thomas Macbride to cartoonist (and biologist) "Ding" Darling or ecologist Aldo Leopold, who figured prominently in the story of Iowa's, and America's, public lands.

Iowa's story in this regard is, on one level, barely a fiftieth-part of the whole. Yet in other ways it looms larger. First, there is the prominence the state quickly established (and in more recent decades has lost), a prominence typified by such figures as Darling (who served under Franklin Roosevelt as chief of the Bureau of Biological Survey in 1934–1935) or Leopold, also an Iowan native. Iowa, with its early record of accomplishments and its in-place master plan for state parks as of 1931, was perfectly positioned for the new initiatives of Roosevelt, including the CCC. Roosevelt, though perhaps his generosity to the recently thrashed Herbert Hoover's state had an element of crowing about it, did underscore Iowa's leadership in the parks movement nationally when he ordered his subordinates, in the spring of 1933, to "give Iowa all it wants" for the new program. Iowa *was* in the lead here and on other fronts. The pertinence of Rebecca Conard's narrative of the Iowa park experience, though, goes beyond mere precedence. If Iowa in some ways was qualitatively a leader, in other ways it offers a quite typical example of how the competing issues of preservation, recreation, conservation, and resource management worked themselves out at the local level. If all politics is local, this is doubly true of politics that affects the immovable locales of life, the landscapes of home, of civic space, of farm and factory—and parkland. While the parks movement was national in inspiration, and the federal government obviously played a strong role in articulating goals, setting agendas, and providing models, we need to recall, as Rebecca Conard helps us to do with careful detail set in a revealingly broad context, precisely how much was done in these years by citizens acting in and for their immediate environs.

Places of Quiet Beauty also raises profound questions about the ultimate ethics of the parks movement nationally. Did Americans, by institutionalizing such sacred spaces as Waubonsie or Backbone, Yosemite or the great Adirondack Park of New York, suggest a deep split in their values? Were they willing to allow the devastation of much of the continent in the search of gain or pleasure or power as long as such enclaves, even with their crowds of tin-can tourists, were symbolically removed from the fray? Whereas we may tend to tell ourselves that we have to "save as much as we can" from wrongful use, of course in practice we set very little wholly aside. And even what we set aside is never wholly set aside anyway: that it is at our disposal in the first place, be it "wilderness" or not, suggests that what passes as "nature" is always an artifact of our own contriving. If the parks movement shows how complexly entangled in social life the environment is, it also reveals, as Rebecca Conard suggests with such perceptive analysis, a kind of theology of atonement at work here, salving our conscience for the large scale and complex environmental changes we create elsewhere. Hence it is that reading her wonderful book has given me a new understanding not only of what I saw there in the Missouri Valley in 1970 but also of how it came to be there—of who made it and what their motives may have been. As with the best books on the environment, *Places of Quiet Beauty* reveals how much we may have missed in our routine experience of the landscapes of everyday life.

WAYNE FRANKLIN

Prologue

Ladies and Gentlemen, friends of the living world, lovers of Iowa and her beauty;—successfully administered, this park-movement shall indeed go far. Well administered, parks become much more than havens for birds and flowers; much more than game-preserves, a refuge for wild life of every sort; even something more than a play-ground for all the people . . .—something more than all these, the well administered park shall show. It shall show us real democracy.

—THOMAS H. MACBRIDE[1]

State parks are places that augment our concepts of state and local identity. Yet they are increasingly distinct from their surroundings. As cities expand at the periphery and decay at the center, as small towns wane, as highways widen, and as farmsteads disappear, state parks and preserves exist in ever-greater contrast to the reworked landscape around them. The image of state parks as oases of solitude and unspoiled nature is an old one. Still, it is an image that bears reexamination within a context that goes beyond the niche of sentimental preservationist motives to which the origins of state parks are typically assigned.

In one form or another, the following questions have been rolling over in my mind for the past few years: Will we become a nation that is remembered for having squandered its resources while its people, in a vain attempt to ignore the environmental consequences, set aside a tiny portion of their landed wealth in parks and preserves? Do parks, protected as they are from the encroachments of our technological selves and our grossest economic machinations, represent a sort of collective atonement for overexploiting resources? This study does not contain florid descriptions celebrating the history as well as the splendid natural and cultural features of this park and that park and so on; in other words, this is not a historical guide to Iowa's parks and preserves.

Rather, it is an attempt to understand the meaning of Iowa's state park system within the progress of environmentalism in the twentieth century.

In part, this is a story of park politics. It is also a story about people and their motives, about men and women who, for one reason or another, were compelled to a mission they defined as creating a legacy for future generations. Inasmuch as the history of creating and reshaping the park system is reflected in the parks and preserves themselves, it is also a story about places. Just as every park and every preserve has a history of land use before it entered the system, so too do these places embody the history of the system itself.

Chapter 1 opens with a look at botanist Thomas Macbride, considered by many to be the "father" of conservation in Iowa because he sent out the call for parks in 1895. Macbride and other academic scientists succeeded in nurturing a broad-based coalition that gave birth to the 1917 State Park Act and a new state agency, the Board of Conservation. The nature of this coalition, which mixed academicians and clubwomen, preservationists and conservationists, is examined. While it is well known that scientists helped to shape federal conservation policy during the early twentieth century, their role at the state level has been less clear. Likewise, historians have always understood that women were a vital force in the conservation movement because of their numbers, but their relatively minor role in politics at the national level has overshadowed whatever they accomplished at the state and local levels.

Botanist Louis Pammel commands center stage in chapters 2 and 3, which examine the formative years of the state park system and the reshaping of conservation goals during the 1920s. As the first chairman of the Board of Conservation, Pammel energetically directed the assembly of a park system that reflected Iowa's environmental diversity as well as those cultural antiquities that were considered important landmarks of Iowa's prehistory and history. From 1920, when the first state parks were dedicated, to 1927, when Pammel retired from the board, the vision of state parks that scientists held was constantly challenged by a variety of special interests and social forces. Pammel and his cohorts attempted, without much success, to expand the board's purview to include scientific forestry and water pollution control. However, through the Board of Conservation, Pammel saw to it that women continued to have a voice in policymaking.

Chapter 4 examines the contradictory impulses of the 1930s. On the one hand, federal conservation and relief programs funded state park development for recreational purposes, producing a shift toward aesthetics and amenities. On the other hand, Jay N. Darling, a forceful conservationist, steered the state back toward centralized resource management. While state parks were beautified, attracting greater and greater use, they also were enfolded in the mandate of a new State Conservation Commission, which merged the functions of the Board of Conservation with the Fish and Game Commission. The *Twenty-five Year Conservation Plan*, completed in 1933 on the eve of the New Deal, reaffirmed the role of the state park system in resource protection and conservation. It also provided a roadmap for developing a comprehensive resources agency.

World War II effectively ended federal funding for park development, but the statewide conservation plan remained a powerful policy influence during the wartime years and into the postwar period. Chapter 5 examines renewed and reinvigorated emphases on prairie preservation, historic preservation, and wildlife conservation during the 1940s and early 1950s. Interest-group politics were relatively absent during this decade, allowing the commission to work toward goals set forth in the *Twenty-five Year Conservation Plan*. Botanists, notably Ada Hayden, and historic preservationists found a strong ally in Commissioner Louise Lange Parker, who demonstrated remarkable leadership skills. At the same time, federal initiatives to encourage wildlife habitat protection and to control devastating floods pointed to a new era of intergovernmental relations.

Chapter 6 examines a period of neglect that set in during the 1950s. Federal flood control projects and the outdoor recreation boom fueled park expansion, particularly in association with artificial lake construction and flood control reservoirs. However, while the system continued to grow, budgets for ongoing maintenance and operations increasingly lagged behind park use. Park neglect could have been ascribed to growing pains except that it coincided with a decided shift in environmental values. Fish and game interests dominated the State Conservation Commission during the 1950s and into the 1960s. This factor, which contributed to unexamined growth, transformed the state park system once again, giving primacy to large scenic areas that could accommodate many outdoor recreational uses.

Passage of the 1965 State Preserves Act signaled the beginning of the environmental movement in Iowa, which brought a renewed concern for protecting fragile and increasingly threatened natural and cultural resources. Chapter 7 covers the controversies that surrounded construction of Saylorville Dam and the environmental threats this posed to Ledges State Park, as well as the interest group conflicts that postponed and ultimately recast the development of Brushy Creek State Park. This chapter also examines the state government reorganization effort that, in 1986, merged the State Conservation Commission with the Department of Water, Air, and Waste Management, which had developed as the state's environmental regulatory arm in response to federal environmental legislation. The environmentalism of the early 1900s reverberated in this merger, which produced the Iowa Department of Natural Resources. However, the complexity of resource management and protection issues continues to challenge, and may ultimately undermine, the agency's effectiveness.

Fundamentally, the vision of early state park advocates was realized in the Department of Natural Resources, but the role of state parks has been redefined more than once, and their presence within the state lands system has gradually decreased over the decades. Nonetheless, state parks remain important repositories of Iowa's past. An epilogue brings this study up to the present, briefly examining the direction state park policy has taken since 1986.

After I began this study in earnest, many people asked, "Why Iowa?" or "Why study just one state?" The answer to the first question is that one begins with what one knows. In 1990 I set out to research and write a brief historical context for identifying important historic places in Iowa that were associated with the conservation and parks movements, a study sponsored by what was then the Iowa Bureau of Historic Preservation and one in which the Iowa Department of Natural Resources cooperated. The initial research questions, set by the bureau's historian, Lowell Soike, called for treating the conservation movement and the parks movement under one rubric, a research agenda I questioned at first. Despite the long-term existence of an organization that recognizes the historic intersection of these two movements, the National Parks and Conservation Association, founded in 1919, the literature nonetheless conditions one to associate the term "conservation," historically speaking, with forestry, water control, wildlife management, soil erosion control, and the scientific management of natural resources in general. Conversely, parks have been associated chiefly with preservationist and aesthetic values, notably the desire to preserve natural scenery and to promote outdoor recreation.[2] Moreover, because many state park systems were established in the wake of the 1916 National Park Act, the common perception is that state parks are somehow lesser clones of our national parks. The implication, of course, is that the history of the national parks is, by extension, the history of the state parks.

In any case, I may have questioned the initial research agenda, but I was nonetheless contractually bound to follow it. What I discovered is that the project director knew more than I did. In the course of my investigations, I began to understand that Iowa's state park system was born from a hope of centralizing control over resource use. Scenic preservation and outdoor recreation were among the motives, to be sure, but they were not the driving force. In large part, those who framed Iowa's 1917 State Park Act viewed parks as a wedge for creating a state agency that would oversee the use, protection, and conservation of many natural and some cultural resources. The implications of this discovery intrigued me and caused me to reconsider my ideas about the meaning of state parks in environmental history.

If this was Iowa's story, then why not compare the origins of its park system with those of other states. Are the origins of Iowa's park system anomalous or part of a larger and perhaps important pattern? The merits of this question are evident in the fact that, between 1900 and 1930, about a dozen states established state park systems *and* placed them under the jurisdiction of a unified department of conservation or natural resources. Some of these states, such as Wisconsin, New York, and Michigan, were, like Iowa, politically progressive during these decades. Others, such as Alabama, Kentucky, and North Carolina, did not fit the progressive model.[3] Thus, there is no single answer to the question of what links the histories of these state park systems. Questions concerning shared patterns of origin and development also seem pertinent, considering that state park systems today tend to be administered variously: by an environmental superagency, as part of a smaller state agency division, or by a separate parks department. Ultimately, the question of common bonds held less interest for me than did the question of what happened to Iowa's state parks in the post–World War II decades.

Part of my desire to probe questions about one system for the span of nearly a century sprang from the wealth of research material available. The volume and quality of records for the early decades, particularly the papers of botanist Louis Pammel and conservationist J. N. "Ding" Darling, leave one with the distinct impression that the people involved in creating the state's resource conservation agency, and then in reshaping its mandate, understood the historical importance of their work.[4] What then became of their vision— or visions? My interest was fueled by the realization that, having grown up near a state park, I thought of them as extensions of fish-and-game work. Since this personal knowledge did not square with their historical origins, I wanted to know why. During the 1920s and 1930s, Iowa was considered a leader in the state park movement. What changed between then and the postwar era?

More important, I realized that as environmental values and issues recycled through the twentieth century, the state conservation agency was the only entity in which all of these transformations would be reflected. Here then was an opportunity to study the complexity of American environmentalism and its roots as played out in one locale. By "environmentalism" I mean issues concerning human relationships to the land that have been expressed as interest-group politics and that, from time to time, have achieved the force of a social movement with the power to change public policy, if not public attitudes.[5]

Since state parks were the rationale for creating Iowa's resources agency, the state park system, including preserves, became the logical place to trace the evolution of environmentalism in Iowa. Besides all this, I wanted to examine some interesting personalities I had come to know through the historical record, and I wanted a history of parks that could be linked through place, not abstract policy or pedantic administrative detail.

The history of Iowa's state park system set within the context of environmentalism forces one to see parks and preserves, even forests and wildlife refuges—all those special places that the term "park" conjures up—as measures of our own commitment to caring for the environment. Each of us will judge the distance we have come by different standards, it is true, and my purpose is not to suggest that the number of parks and preserves or the acreage of state-managed lands can or should be used as a standard by which to measure our collective environmental awareness. At the same time, I believe that until we, as individuals and as a society, decide to love the land that supports our landfills, our industrial centers, our inner cities, and our food-producing fields as much as we love the land of our parks, the term "land stewardship" has no real meaning.

1

Conservation Parks

Thousands of acres [elsewhere] have already been rescued from spoliation and subjected to intelligent management, such as will eventually result in the attainment of all the beneficent ends for which public parks exist. In Iowa nothing is done; nothing will be done until somebody or some association of our citizens makes a beginning.   —THOMAS MACBRIDE[1]

The State Park Movement

"Public parks, town parks are *the thing*!" Thomas Macbride exclaimed to the audience assembled in 1922 for the annual meeting of the Iowa Conservation Association.[2] By then, they were so much "the thing" that Iowans had not only embraced the idea of municipal parks, with their emphasis on designed landscapes and city beautification, but they had supported state legislation intended to preserve, in state parks, places of scientific, historic, or scenic value. Although the public at large may not have fully understood the resource conservation motives of those who drafted and lobbied for the state park law, adopted in 1917, the movement's leaders were certain of their objectives. "Now that we have something to conserve," declared Macbride, "we must look to conservation," by which he meant that parks must be responsibly administered to prevent their use for any type of private or commercial gain.[3] Macbride had reason to marvel. When he delivered his 1922 address before the Iowa Conservation Association, the state was busy establishing fifteen state parks. The number would climb steadily throughout the 1920s and 1930s. One contemporary called them "quiet places of natural beauty," although natural scientists always thought of state parks as more than that.[4]

Iowa was not among the first states to establish state parks, but as the movement gained momentum, Iowa took a leading position. Nationwide, the opening phase slowly unfolded after the Civil War, when a few states set aside parks to preserve areas of outstanding natural beauty or with important historical associations, chiefly battlegrounds, forts, and other places of military importance. In 1864, the federal government ceded Yosemite Valley and Mariposa Grove to California for a state park. Twenty years later, in a similar move, the federal government transferred Mackinac Island, a former military reservation, to the state of Michigan for development into a state historical park. In 1883, the New York legislature authorized a reservation of land "to preserve the scenery of the Falls of Niagara." Two years later, New York authorized Adirondack State Forest. Between 1889 and 1895, Minnesota established three state parks to protect scenic and historic areas: Itasca, Birch Coulee, and Camp Release. Itasca preserved the headwaters of the Mississippi River; the other two commemorated Indian-white battlegrounds. New York and New Jersey jointly acquired the nucleus of Palisades Interstate Park in 1895, the first state park created purely for recreational purposes.[5] These and other isolated park creations prefigured more widespread activity in the early twentieth century.

Several interrelated concerns drove state park advocates. In the Pacific Northwest, the initial goal was to protect natural scenery and roadside timber from logging. Similarly, California's movement evolved from the campaign to save the redwoods. Forestry conservation, as opposed to scenic preservation, figured more prominently in the state park movements of New York and Wisconsin. Depredations on the Adirondack State Forest, set aside in 1885, gave rise to the Adirondack Park Association and passage of state legislation authorizing New York's state park system in 1890. The Adirondacks belatedly inspired Wisconsin's state park system. In 1876, the Wisconsin state legislature set aside fifty thousand timbered acres in Lincoln County, which became known as "The State Park." Twenty-one years later, in 1897, the legislature released the park for sale to lumber companies. This action prompted a new effort, which led to land acquisition along the Saint Croix River in what is now Interstate Park. By 1915, state parks, state forests, and fish-and-game laws were jointly administered in Wisconsin by a coordinated State Conservation Commission.[6]

The tie that bound state park advocates, who often were widely separated by geography and sometimes unaware of activities outside their own states, was political progressivism; that is, positive state intervention to improve society. Richard Lieber, an Indianapolis businessman and civic leader, stands out among progressives who called for state parks to provide retreats for urban workers and to preserve relics of Indiana's passing landscape as the state neared its centennial.[7] State park advocates in Iowa were no less progressive in their politics, and they employed all these arguments at one time or another:

forest conservation, scenic preservation, human conservation, and historic preservation. Still, the movement in Iowa was seated in the Iowa Academy of Science. It did not emerge from civic groups or in concert with federal public land actions. Nor was it driven by the threat of losing some particular place of special value. Iowa's first park advocates were natural scientists, and their leadership tended to align the movement much more closely with resource conservation philosophy and goals.

The Dean of Conservation: Thomas Huston Macbride

When Thomas Macbride died in 1934, Bohumil Shimek, his long-time colleague at the University of Iowa, eulogized him as "distinctly the father of conservation in Iowa," who in "his paper on County Parks . . . initiat[ed] the campaign which resulted in our present state-park system."[8] Shimek referred to the speech from which the epigraph to this chapter is drawn. Macbride delivered it before the Iowa Academy of Science in 1895, urging his fellow scientists to launch "an effort to call back into public favor the once familiar public 'common.'" He envisioned a statewide system of county, or rural,

Thomas Huston Macbride, 1848–1934. *Courtesy University of Iowa Archives.*

parks (he used the terms interchangeably). This address and Macbride's call for watershed protection and forest reservations two years later, in 1897, are often cited as the beginnings of the conservation movement in Iowa.[9] But in truth it was a long and complex path from Macbride's call for rural parks in 1895 to the 1917 law authorizing a state park system. Physical scientists, particularly botanists, played leadership roles from the turn of the century into the 1930s.[10] Within this circle of academics, however, Macbride certainly was the "dean" of conservation.

Macbride was of the generation that witnessed the plowing of the prairie, though his own participation in the process was principally that of a youth assigned to farm chores. His father, a Presbyterian minister who farmed a few acres on the side, literally as well as figuratively toiled in the fields for the Lord, a combination of theology and horticulture that instilled in son Thomas a love of the outdoors and a reverence for life. It was an upbringing that surely must have contributed to Macbride's later ambivalence about the fruits of industrial progress. He was born in 1848 to James and Sarah Huston Macbride, who were then residing in Rogersville, Tennessee, where James had a pastorate. A vocal abolitionist sent into proslavery territory, James's tenure as pastor proved to be short. By 1854, it was time to move on. The couple chose to become pioneers in Iowa, settling first in Salem (Henry County), where James, without a congregation, farmed for three years. Eventually, he obtained a charge, and thereafter the family moved from one small town to another as the clergyman's pastorate changed.

When Thomas was sixteen, he entered Lenox College in Hopkinton (Delaware County), where he studied for a year with Samuel Calvin before moving on to Monmouth College in Illinois, obtaining his B.A. in mathematics and languages in 1869. A year later, he returned to Lenox as Professor Macbride and resumed his acquaintance with Calvin. The two of them initiated a habit of field trips through northeast Iowa that led to a lifelong friendship and yielded for Lenox the "natural science cabinet," an assemblage of crustacean and coralline specimens. It was at Lenox College that Macbride also met Harriet Diffenderfer, first his student and then, on December 31, 1875, his wife. During this time he also received an M.A. from Monmouth. Meanwhile, Calvin moved on to the State University at Iowa City in 1873, becoming the university's first professor of natural science. Five years later, he brought his young protégé into the department as assistant professor of natural science, in charge of botany. In effect, Calvin split the "chair of natural science," assigning himself to geology and zoology, and Macbride to botany.[11]

Thus began a long and productive career that combined research, teaching, and applied science.[12] Macbride assisted Calvin in enriching the university's collections and in establishing a school of science that was increasingly specialized, as the natural sciences became more professional in nature and organization. By the time he became a full professor in 1884, the fields of geology, botany, and zoology were firmly established in the curriculum.[13] Throughout his university career, which spanned nearly forty years, Macbride studied and published scientific works on slime molds, fungi, geology, and paleobotany, as well as general botany. In addition, he also was a charter member of the reorganized Iowa Academy of Science and served as its president in 1897–1898. As an advocate of applied science, Macbride served a public role in many capacities, beginning in 1901, when he cofounded the Iowa Park and Forestry Association with Louis Pammel, professor of botany at Iowa State College, and Bohumil Shimek, his former student and, from 1890 on, his colleague in the botany department. He then served as the organization's first president from 1902–1904.

Conservation was only one of Macbride's many "outside" interests, but he was nonetheless its earliest and most eloquent spokesman. He felt an urgency, even a moral obligation, to make science serve the public good. "I am not a very old man," he observed in 1902, when he was 54, "but I have dipped oar in the Des Moines River away down near its mouth and seen the netted sunbeams on its gravel-covered bottom all the way across. Who looking now upon the muddy channel would think such [a] thing possible?"[14] However much he deplored the plundering of nature, he was always careful to spread the blame widely—hoping, one suspects, to arouse equally widespread feelings of personal guilt and responsibility. The homebuilder was just as culpable as the factory owner in Macbride's eyes, and the result was a disgrace. As a people, "we" had moved beyond "destroying natural beauty." Surveying the landscape, Macbride could find "little of it left for our injury or desecration; the prairies are plowed almost to the last acre; the woodlands have been cleared away entirely or converted into pasturelands . . . ; the streams near the town are the dumping place of all uncleanness and in the country are esteemed only as convenient places for watering domestic animals."[15]

Although he could describe a seemingly hopeless state of affairs, Macbride himself was undaunted. Beginning in 1895, he appealed first to his brethren in the Iowa Academy of Science, then began to broaden his audience to include the public at large. He felt a personal and professional charge to save the "little" of unsullied nature that was left, to rescue indigenous remnants of beauty, and to protect them as parks. From the beginning Macbride clearly envisioned parks as the proper vehicle for resource conservation, broadly conceived. He saw a threefold mission for public parks: to promote public health and happiness, to educate, and most important, to preserve "something of the primeval nature" of Iowa.[16] It never occurred to Macbride that there might be inconsistencies in this mission. To him, promoting health and happiness did not mean providing recreational facilities, and serving education did not mean constructing nature centers. In Macbride's view, simply "being" among the trees and birds and clear-running streams was health and happiness defined; to experience the same environment in the company of one who understood nature scientifically was education at its best.

Though certain of his objectives, Macbride was nonetheless flexible in his pursuit. He sought to rouse action at both the local and state levels, and he could see more than one path to solving resource problems. He did insist, however, that his comrades in science be involved. Addressing the Iowa Academy of Science in 1897, he exhorted members to study and promote watershed protection. Excessive drainage for agriculture, he observed, had placed the water supplies of Iowa's communities in jeopardy, an environmental problem that Macbride felt demanded the combined attention of scientists from multiple fields. Original investigation leading to new scientific discoveries was exciting and vital to "progressive research," as he put it, but, he continued, "there are some other considerations to be here noted which seem rather to place us as members of the academy under obligations, especially at the present time, to the accomplishment of work of a practical every-day sort."[17]

It is not clear whether Macbride based his remarks on actual evidence or firsthand knowledge of watershed conditions in Iowa, or whether he was simply reacting to events elsewhere in the country, since he cited the cases of Boston, New York city, and New York state, which had taken steps to acquire vast watersheds in order to protect urban water supplies. In any case, he took this opportunity to point out that the state legislature had ignored the academy's 1896 petition asking for measures to preserve Iowa's lakes. If the state would not protect water resources, Macbride urged his colleagues to press for action on the local level in their respective communities. Taking his appeal to the Iowa Academy of Science seemed the natural thing to do in an era in which the sciences were not highly differentiated and before applied science was turned over to the engineering professions.

As Macbride would learn, however, academic scientists could not by themselves awaken sufficient consciousness to effect change; they would have to find allies outside the academy. Thus, to nonacademicians, Macbride shaped his message to fit the audience. Speaking before the League of Iowa Municipalities in 1906, he appealed to the legalistic bent of civic leaders by asserting what he called the "environmental rights" of citizens: the right to breathe clean air and the right to experience (natural or designed) beauty. Adopting the rhetoric of urban reformers, he charged that science and engineering in the service of industrialism had demonstrated their combined power to transform the world—witness the benefits of railroad and telephone. Yet the advance of civilization had trampled on the environmental rights of citizens. It was up to those who governed to become advocates for the public good. As always, he appealed to moral obligation: "Is it not possible for us as intelligent self-governing people ... to use wealth and opportunity and power in such wise [ways] as to conserve for ourselves and our children all those finer instincts of humanity?"[18]

In 1908, Governor Albert Cummins asked Macbride to chair the Iowa Forestry Commission, established to assist the National Conservation Commission with its investigations.[19] The governor's action came on the heels of the 1906 Forest and Fruit Tree Reservation Act, state legislation long sought first by the State Horticultural Society and then by the Iowa Park and Forestry Association, precursor of the Iowa Conservation Association. Macbride had great hopes that the Forestry Commission would have the power not only to restore Iowa's extirpated forests but would also establish a base within state government for conservation work. On the horizon he saw an "Iowa Commission for the Conservation of Natural Resources," which would "teach the teachable people in the world how to rear and use trees as the profoundest sort of world-economy, the highest sort of earth-culture, and the triumph of enlightenment in this early home which God has given to the sons of men."[20]

Macbride's lofty vision, however, soared further into the future than political minds could follow, and the Forestry Commission faded into obscurity. But there were other fronts on which to pursue conservation goals. After having campaigned for years to get the state legislature interested in water conservation, Macbride was appointed to the Iowa State Drainage, Waterways and Conservation Commission, established in 1909. This was the first state commission with a mandate to study resource conservation issues in Iowa. Buoyed by events on the national front, notably the Ballinger-Pinchot controversy and the hotly debated question of federal versus state control over water resources in the West, the Drainage, Waterways and Conservation Commission pressed to complete its surveys in timely fashion.[21] Macbride oversaw research on the topics of forests and soil conservation, minerals and peat, and lakes and streams, and he authored those sections of the commission's report, which appeared in 1911.[22]

Possibly as a result of his work on the Drainage, Waterways and Conservation Commission, Macbride represented Iowa at the Third National Conservation Congress in 1911. He was also Iowa's delegate to the 1910 and 1911 National Rivers and Harbors Congress, although he attended neither. Despite the organization's name and its motto, "advocates a policy, not a project," this assemblage of handpicked representatives served mainly to promote federally financed waterway development as envisioned by Senator Joseph E. Ransdell of Louisiana, who conveniently doubled as the NRHC president. Ransdell was chiefly interested in flood control in the Mississippi River valley, particularly levee construction along the lower Mississippi River. Thus, in effect, this was an honorary post that mainly put Macbride in charge of statewide fund-raising for the organization, although there is no evidence that he ever did so or that he ever attended any of the NRHC's annual conventions.[23] Single-issue conservation politics held little appeal for Macbride, nor was he inclined, by this time, to divert his energy in support of causes that would be of little benefit to Iowa.

Instead, Macbride turned his attention to fulfilling a long-cherished dream for a field school patterned after Louis Agassiz's summer laboratory on Penikese Island (precursor of Woods Hole). He spearheaded the effort to establish Iowa Lakeside Laboratory on the shores of West Lake Okoboji. Here, in 1909, three hundred miles from the ivy-covered walls of campus, Macbride, Samuel Calvin, Bohumil Shimek, and Robert Wylie began to introduce their students to the natural sciences in situ and, at the same time, to create a scientific laboratory dedicated to the study of practical problems. Determined that the laboratory would succeed despite meager financial support from the university, Macbride not only served as director for four years but dug into his own pockets when necessary in order to pay salaries and operating expenses.

In 1914, Macbride left the classroom to serve as president of the University of Iowa, a post he held for two years before retiring from academic life. He and Harriet then moved to Seattle to be near their children. Though far away and up in years, he somehow managed to spend considerable time in Iowa, and he was constantly in demand as a public speaker. The professor emeritus revered by two generations of students ("he knew every student on the campus by name") returned almost every June to deliver the benediction or the commencement address. In 1919, along with Pammel, Shimek, and others, he helped to organize the American School of Wildlife Protection, a summer field school held annually on the bluffs north of McGregor overlooking the Mississippi River. The field school thrived until its functions were interrupted by World War II.[24]

Macbride never flagged in his long-distance support of conservation efforts at home. When the opportunity for expanding Iowa Lakeside Laboratory came during the late 1920s, Macbride immediately began to assist Bohumil Shimek in writing fund-raising letters, and he eventually contributed a sizable amount toward the purchase price. By then Macbride was in his eighties, and he seems to have realized that this might be the last great contribution he could make. In a 1931 letter to his friend and colleague, he hinted as much. "I see so much in the work to which we have set our hands," he wrote, "—so much that makes for the future well-being of the State of Iowa and the future progress of botanical and zoological science, and conservation generally of the resources of the commonwealth, that I do not intend to spare myself in the least in carrying the thing forward." [25]

"His conception of education was broader than science," wrote a former student. "Under his magic touch botany became . . . a door swinging open to a garden where all the plants from the tall oaks to the mould on a crust of bread carried on their amazing and beautiful existence." [26] To honor his former mentor, and to promote the pending expansion of the field laboratory, Shimek proposed that the name be changed to Macbride Lakeside Laboratory, but Macbride simply would not hear of it, stating instead that he had always intended the laboratory to be a place for all Iowa—an educational facility available to faculty and students from all Iowa colleges and a research station open to all who wanted to study the ecology of Iowa's northern lakes region. Iowa Lakeside Laboratory was the only befitting name, as far as Macbride was concerned, and Iowa Lakeside Laboratory it remained.[27]

Macbride's wide-ranging intellect (he spoke or read German, Spanish, French, Italian, and Swedish and was a lifelong student of literature) and his scholarly research (he was considered one of the foremost authorities on myxomycetes) might have earned him a more luminous position in the scientific world. A leave of absence in 1891 to study botany at the University of Bonn revealed to him that his own research techniques were as rigorous as those of the German masters. This knowledge may have emboldened him, in 1898, to seek the position of chief forester of the United States, a position Bernard Fernow was set to vacate in order to assume an academic post at Cornell University. Macbride wrote to W J McGee, then head of the Bureau of Ethnology at the Smithsonian Institution, inquiring about the position and seeking McGee's assistance in obtaining it. McGee, geologist, hydrologist, ethnologist, Iowa native, and a leading scientist in the nascent conservation movement, was an obvious choice for potential support. Although Macbride did not copy his own correspondence to McGee, the latter's response reveals much about Macbride's stature within the scientific community at the time. Macbride apparently inquired about the salary and, more important, about the "indefiniteness in organization of the forestry work." To the first question, McGee responded that the salary was below Macbride's expectations with little hope of obtaining more. To the second question, McGee observed that it would take "consistent effort" and "sheer force" to overcome the "disorganized condition" of forestry work that had prevailed under Fernow's watch. Given these drawbacks, McGee advised Macbride to remain at the university but promised he would "continue the effort to have the place made available for you" if Macbride so desired. "It is practically certain that, if the place does not go to you," McGee continued, "it will go to the second-best man in the United States, viz: Mr. Gifford Pinchot, who has of late been much in contact with forestry problems."[28]

Indeed, by 1898 Pinchot had himself well positioned to assume Fernow's post, and, in the process, insinuated his way into the center of an emerging policy controversy. In 1896, Pinchot accompanied the newly established federal Forestry Commission on an investigative tour of California, Oregon, and Arizona. A year later, he inflamed John Muir, among others, when he supported the Pettigrew Amendment to the 1897 Sundry Civil Bill, an amendment that began to define forest "reserves" as harvestable sources of timber and watershed protection areas rather than untouchable tracts of the public domain. Shortly thereafter, Pinchot became an unofficial adviser to the Department of Interior for the purpose of reformulating forestry policy.[29] Thus, McGee may have been dispensing flattery in order to ease the way for Pinchot without incurring ill will from his fellow Iowan, but there is little question that he considered Macbride's qualifications for the position to be sound. In a letter to Samuel Calvin, Macbride's superior at the state university, McGee wrote that he was "inclined to think that Macbride can do . . . just what Fernow failed to do," that is, organize the forestry work, then scattered among three government agencies, into a coherent program.[30]

In any case, there is no indication that Macbride pressed his candidacy further. Pinchot won the post, to which he brought a philosophy that Macbride never would fully share. Scattered correspondence between the two men shows that they maintained a polite, professional relationship for many years and that Macbride supported Pinchot's work from time to time.[31] However, Macbride was a natural scientist of the old school by temperament as well as training; his intellect simply could not be narrowed by specialization. The Forestry Bureau's loss was Iowa's gain. After 1898, he devoted his time and energy to education, research, and applied science in his home state. Along the way, he nudged fellow scientists and nature lovers into an active force for conservation. Upon Macbride's death in 1934, Robert Wylie, professor of botany at the University of Iowa, summed up his colleague this way: "He belonged peculiarly to Iowa."[32]

Thomas Macbride gave moral force to the conservation movement in Iowa for at least three decades, and it is his stirring voice of conscience that echoes through the years. In his later years he became the anguished defender of nature unaltered. "We need," he would insist in the mid-1920s, "right where it is, every cedar clinging to the rocks; . . . we need, just where it is, every white oak that does us the honor to show us beauty of bole and leaf; . . . we need all the crabtrees, and the wild plums, and the hawthorn trees, on the flats down at the river's edge." We, meaning the citizenry as a whole for generations into the future, needed places of nature unspoiled in order to check the ill effects of industrialization and urbanization. "The park," Macbride's voice rings even still, "shall set us free." [33]

But his writings and activities were not without certain inconsistencies, a paradox that, on the one hand, reflects the crosscurrents of the conservation movement as a whole, and, on the other, reflects the practical need to find or to create a broad constituency for conservation. During his years of greatest professional activity, he was not always the defender of nature for nature's sake. If he never fully embraced the utilitarian approach to conservation so favored by many applied scientists of his time, he nonetheless was a pragmatist. In 1901, when he first assessed the status of public parks in Iowa, Macbride acknowledged their societal value as recreational lands, although he could not foresee the influence that the automobile would have on public demand for the recreational use of state parks. At the turn of the century, Macbride's view was still shaped by a sense of closely circumscribed communities where inhabitants had to find open space within walking or carriage-ride distance. "How long shall it be said of this proud and wealthy commonwealth that her weary sons and daughters of toil if they would go forth to recreation—recreation, mark you—must betake themselves to the cemetery? It is enough to breed anarchists and suicides, the situation as we see it now." [34] Macbride was among a growing number of people who felt it was inappropriate to use rural cemeteries for secular activities, as had been the case since the mid-1800s. [35]

His notion of benign recreation would change, but other pronouncements more clearly reflect a willingness to adopt pragmatism. In 1908, for instance, after the governor appointed a state forestry commission, Macbride justified the need for forestry in a "prairie state" on the grounds that tree cultivation should be integral to maintaining Iowa's agricultural and industrial sectors. For agricultural purposes, "our people should be encouraged to plant groves in such a way as to bring the highest returns, returns in useful timber and lumber as well as in shade and protection." For industry, "the restoration of the forest will restore the streams; the return of the waters will bring back their constant energy; the possibilities of power will revive the village, and industry will redeem our now wasted river valleys." [36] As a key member of the Iowa State Drainage, Waterways and Conservation Commission, Macbride insisted that practical conservation implied use. "When we advocate the conservation of a lake, . . . we do not mean simply that we would have a body of water occupying so much area on the ground, but we urge that wherever such body of water of convenient size and depth occurs it shall be kept, and *kept in order*, and *used*" [Macbride's emphasis]. Similarly, "our streams are for use. Conservation bids us use them and use them wisely; likewise our forests, these shall not simply stand as in the ages primeval, they must stand and be productive, be used."[37]

In the context of his life's work, however, such statements strike one as the thoughts of a man working out his own land ethic, not as paeans to utilitarianism. Historians, seeking to understand the motives of those who gave shape to the conservation movement, admittedly one of the most amorphous, multifaceted sociopolitical phenomena of the early twentieth century, have unwittingly perpetuated the notion of two camps: utilitarians (conservationists) and aesthetes (preservationists), usually divided and pitted against one another. This distinction is useful for disentangling the many strains of conservationist thought and for dramatizing the historical significance of the infamous Hetch Hetchy controversy, but it simply does not hold up under close scrutiny. The movement was not only multifaceted, but the motives of those involved were inextricably mixed. With hindsight we see the inconsistencies and emphasize the conflicts, but those who lived the moment rarely experienced such divisiveness, at least not in Iowa. In retrospect, Macbride's pragmatism seems more a part of the overall effort to establish broad public interest in and support for resource conservation. The Iowa Park and Forestry Association gave form to that search.

The Iowa Park and Forestry Association

Thomas Macbride became the first president of the Iowa Park and Forestry Association, and in this position he helped to set a course for conservationists that held, albeit sometimes tenuously, for a generation. The organization itself, however, was the brainchild of fellow botanist Louis Pammel. On November 16, 1901, Pammel convened an initial public meeting for the purpose of promoting scientific forest management and municipal parks. He noted that those who attended followed the lead of like-minded citizens in Minnesota, Wisconsin, New York, and Tennessee, where such organizations already existed.[38] Pammel had already selected the organization's name, but the meeting produced a constitution that reflected the broad mission he, Macbride, and others had in mind. While the idea of state parks as a vehicle for achieving integrated conservation goals would come later, the founders nonetheless saw rural parks and forest reserves as part and parcel of wildlife, lake, and stream conservation. They also accepted the aesthetic focus of civic improvement as an equally worthy goal.

Article II listed the association's all-inclusive goals as being (ungrammatically) "to create an interest in, and to encourage the establishment of parks; the beautifying of our cities, the better care of cemeteries, the planting of trees in country homes for aesthetic purposes as well as for the supply of timber for commerce; the proper utilization of our remaining timber, and to assist in the inauguration of rational methods of forest management and thus help in the protection of our wild game and song birds; the creation of one or more state parks in the vicinity of our lakes and streams; [and] to encourage state and national legislation for rational forest management, and the creation of more forest reserves."[39] In essence, the Iowa Park and Forestry Association emerged as the political arm of Iowa's conservationists. During the twenty-five years of its existence, the association experienced several ups and downs as it sought a stable constituency and, in the process, underwent two name changes, emerging in 1917 as the Iowa Conservation Association.

Macbride was elected the first president of the IPFA; Louis Pammel as secretary. Pammel then took over as IPFA's second president in 1904. These two men were to become the most influential personalities in the Iowa conservation movement. In fact, academic and applied scientists affiliated chiefly with the State University of Iowa, Iowa State College, and the Iowa Geological Survey dominated the leadership of the Iowa Park and Forestry Association from its inception in 1901 through the mid-1920s.[40] The IPFA was thus instigated and led by scientists who realized they must attract broad public support in order to achieve political goals. Neither the Iowa Academy of Science nor the State Horticultural Society provided an appropriate institutional framework for political action, but the IPFA drew fully half of its charter members from both of these organizations.

To achieve the desired balance between academic scientists and civic leaders, Macbride and Pammel sought members who would "avoid as much as possible everything that looks like commercialism in our association" and to "look out lest men use us for interests of a selfish nature."[41] The State Horticultural Society, organized in 1866 to collect and disseminate information on fruits, flowers, and trees suited for Iowa's soils and climate, was a logical ally. During the 1870s and 1880s, when Charles Bessey was struggling to create a botany program at Iowa State College that was both scientifically rigorous and yet responsive to the needs of a developing agriculture state, the State Horticultural Society repeatedly petitioned the legislature to fund experimental research.[42] Various members of the society also assisted with the work carried out through the agricultural experiment station attached to the land-grant college. From the early 1870s on, the society likewise called for the creation of a forestry commission, appointment of a state forester, and state appropriations for a forestry program. In these efforts, they were unsuccessful.[43] Nurserymen, orchardists, and some farmers thus understood the economic benefits of applied science, especially for agriculture, and they were veteran lobbyists, but their interests were narrowly focused.

The Iowa Academy of Science, on the other hand, offered breadth of scientific knowledge. It was always the more distant associate, however. Fellows and associates of the IAS may have had similar educational backgrounds, but their research interests were diverse. Moreover, their livelihoods rarely depended upon purifying municipal water supplies, controlling soil erosion, inhibiting plant diseases, or stemming the loss of fisheries. Founded in 1875 by Bessey and eleven other professors and physicians, the IAS met semiannually during the next decade. An anemic organization—papers were rarely of high scholarship—the group foundered when Bessey left for the University of Nebraska in 1884. Reorganized in 1887 by Macbride, Pammel, and others, the IAS then became an annual forum for scientists, and those interested in scientific work, to share their research and observations.[44]

IAS members were not entirely drawn from colleges and universities, but primarily so. A few were affiliated with the Iowa Geological Survey, the State Weather Service, and the Dairy Commission, as well as the State Horticultural Society, and a good many were unattached to any institution. Macbride's call for "practical" research in 1897 prompted only a modest shift toward applied science research among his colleagues, if one is to judge from the published proceedings from 1900 to 1920. Meetings always included reports of a "technical" nature, such as "The Sanitary Analyses of Some Iowa Deep Well Waters," "Municipal Hygiene," "The Sioux City Water Supply," "Some Forestry Problems on the Prairies of the Middle West," "An Ecological Study of a Prairie Province in Central Iowa," or "The Pollution of Underground Waters with Sewage Through Fissures in Rocks."[45] Members of the academy reportedly took a keen interest in natural resources, but there is little evidence that such academic interest translated into a "great deal of influence" as has been claimed.[46] The IAS, for instance, never sponsored directed research for the purpose of addressing environmental problems, never organized symposia to focus attention on conservation issues, and did not even establish a conservation committee until 1915. And although the IAS passed resolutions urging state and federal legislators to take action, the organization never participated in drafting conservation legislation during this period.

Thus, the purpose of the Iowa Park and Forestry Association was to bridge the gap between the limited focus of the State Horticultural Society and the disparate, as well as self-distanced, nature of the Iowa Academy of Science. Macbride envisioned the IPFA as a forum for vigorous, high-minded interdisciplinary discussion. He proposed, for instance, that the 1903 annual meeting include "our most distinguished lawyers setting forth the Legal Rights possessed by our citizens who undertake public improvement; another which might set forth the Parks as a Sanitary Necessity to be offered by a distinguished member of the Medical profession; another on Parks and Morals by some distinguished clergyman; . . . and a series of papers illustrating the absolute necessity of forests and forest preservation for the maintenance of the highest type of agriculture."[47] At the same time, he felt that the IPFA should keep the spotlight on "Iowa conditions" and not get sidetracked debating situations elsewhere. IPFA's publications had to be "Hawkeye from cover to cover" in order to "win the support of Iowa people."[48]

Macbride's vision notwithstanding, the IPFA appears initially to have cultivated eclecticism rather than a sense of common purpose. The mixture of papers presented during the first few annual meetings surely reflects diversity, but not quite the intellectual forum Macbride had in mind. Titles such as "Dendro-Chemistry" by J. B. Weems, "The Phenology of Our Trees" by Charlotte M. King, and "Forestry and Its Effect on Western Climate" by H. C. Price, all of Iowa State College, reflected IPFA's scientific element. City beautification, however, received a great deal of attention—everything from "Street Trees and Parkings" to "Outdoor Art and Morals"—and the 1907 meeting included an entire session devoted to papers on civic improvement in Madison, Wisconsin, and Iowa City. The important topic of forestry in agriculture typically came under the guise of farmstead improvement; at each meeting someone usually addressed the topic of farm woodlots, and J. W. Kime, M.D., once spoke on the subject of "Farm Sanitation." In the minds of many Iowans, forest conservation still encompassed small-lot tree-planting for ornamental as well as commercial purposes, as evidenced by a plethora of offerings along these lines.[49] In short, judging from the titles of papers presented at its annual meetings, one has difficulty discerning what direction the IPFA was headed; it proved to be a sundry group, too. Another attribute of the organization was a preponderance of older members, as acknowledged by all-too-frequent memorials to those departed.

A sustaining membership proved elusive. Macbride buoyantly predicted 300–400 members by the end of 1903. However, despite diligent canvassing by himself, Louis Pammel, and Bohumil Shimek, halfway through the first year Macbride was forced to acknowledge that there was "not near the support from the State that the cause merits." By early 1903, even charter members were beginning to drop their support (table 1). "So far[,] that list of charter members is silent as the grave," Macbride wrote to Pammel. "Perhaps they are tired of us. I was hoping not; for as you and I realize, there is no discharge in

Table 1.

Iowa Park and Forestry Association Charter Members, 1901–1902[50]

Name	Location	Other Affiliation(s)	Occupation, If Known
Blumer, J. C.	Luverne		
Bomberger, W. M.	Harlan	SHS	
Brumagin, M. K.	Ames		
Budd, J. L.	Ames	ISC, SHS	professor of horticulture
Burnap, W. A.	Clear Lake	SHS (director)	
Donelsen, W. C.	Ogden		
Duebendorfer, A.	Ames		
Erwin, A. T.	Ames	ISC, IAS, SHS	professor of horticulture
Fulmer, J. T. D.	Des Moines		
Greene, Wesley	Davenport	SHS (secretary), IAS	
Hoffman, A.	Des Moines		
Little, E. E.	Ames	ISC, IAS, SHS	experiment station
Lummis, G. M.	Ames		
Macbride, Thomas H.	Iowa City	SUI, IAS	professor of botany
Mosier, Cyrus A.	Des Moines	US General Land Ofc.	GLO special agent
Pammel, Louis H.	Ames	ISC, IAS, SHS	professor of botany
Parker, G. F.	Hartford, CT		
Price, H. C.	Ames	ISC	
Reeves, Elmer	Waverly	SHS	
Ritzman, E. G.	Maquoketa		
Roberts, T. G.	Marathon		
Secor, Eugene	Forest City	SHS (director)	state legislator
Sexton, J.	Ames	SHS	
Sheldon, De La	Ames		
Shimek, Bohumil	Iowa City	SUI, IAS	professor of botany
Stevens, S. W.	St. Louis, MO		
Thompson, W. F.	Cambridge		
Van Houten, Geo. H.	Lenox	SHS	
Watrous, C. L.	Des Moines	SHS	nurseryman
Wilson, Silas	Atlantic	SHS	state legislator

SHS = State Horticultural Society, IAS = Iowa Academy of Science, SUI = State University of Iowa, ISC = Iowa State College.

this war."[51] There are no surviving official records, other than published proceedings, to chart the ups and downs of membership and finances, but the correspondence among Macbride, Pammel, and Shimek suggests that they struggled to maintain enough members, about 100, to cover most operating costs. Dues, set at $1.00 per year, were never sufficient. At times, the organization resorted to borrowing money to pay bills, and one generous soul or another continually subsidized the treasury.

The IPFA survived, in part, because it enjoyed a particularly close relationship with the State Horticultural Society. In fact, the proceedings of IPFA's annual meetings for the years 1908–1910 and 1916–1917 were published as part of the horticultural society's annual reports; no proceedings at all were published for the years 1911–1913. Because of its close alliance with horticultural interests, the IPFA appears to have given priority, during the early years, to securing forestry legislation; but it also appealed to the legislature to create a state park around the capitol in Des Moines, to authorize townships and counties to purchase land for parks, and to transform Camp McClellan at Davenport into a state park.[52] These initial efforts failed to arouse much political interest or public support, causing Shimek to admonish the members that "vigorous action" was required "if the activity of this Association is to consist of something more than the mere annual parading of theories and ideals which are of little value if not carried out in practice."[53] Redoubling its efforts, the IPFA drafted legislation that would grant favorable tax advantages to private citizens who planted forest and fruit trees, the so-called "Secor bill" because it was introduced by Assemblyman Eugene Secor, also a member of the association. The bill passed in 1906.[54] A year later, the association again petitioned the governor and members of the assembly to secure "more ample grounds, and better approaches" to the state capitol "in the interest of both economy and landscape art."[55]

IPFA's recommendations concerning the capitol grounds probably had less influence than Charles Mulford Robinson's 1909 planning report recommending an entire system of parks linked by boulevards.[56] In any case, Governor Warren Garst urged the General Assembly to establish a commission with authority to purchase property adjacent to the capitol in order to "provide a beautiful boulevard of approach and surroundings." The General Assembly responded by establishing such a commission, but its mandate could hardly have satisfied the IPFA. Its charge was simply to erect a monument to the late Senator William B. Allison somewhere on the capitol grounds "or any extension thereof." The commission, as it turned out, paid more attention to the "extension thereof" clause and proceeded to draft an ambitious plan for expanding and developing the capitol grounds.[57] The General Assembly eventually fell into line and authorized improvement of the grounds in 1913. To plan the landscaping, Governor William Harding subsequently appointed a Plant Life Commission, which consisted of E. R. Harlan, curator of the State Historical Department, Thomas Macbride, and Louis Pammel. Harlan contended, and his fellow commissioners agreed, that "the planting plan should illustrate typical Iowa." To a certain degree, this approach was carried out by landscape architect L. E. Fogelson, who created a parklike setting around the state capitol.[58]

The Plant Life Commission helped Macbride and Pammel cultivate political relationships in Des Moines, although the appearance of the capitol grounds was the IPFA's least concern. Creating a stable and active membership took more time and energy than Macbride, Pammel, and Shimek could spare from their academic and research responsibilities. By 1910, the Iowa Park and Forestry Association was moribund. There was little business to transact, membership had dwindled to the party faithful, and the entire proceedings were easily conducted in a couple of half-day sessions. Moreover, it became increasingly difficult to distinguish the interests of the IPFA from those of the State Horticultural Society.[59] Still, its progenitors had not lost faith in the cause, even if the organization had devolved to a special interest group within the SHS. In November of 1913, Louis Pammel wrote to Macbride asking, "Can we not revive the Iowa Park & Forestry Association?" To do so, he proposed publishing a tribute to the conservation work of former U.S. Congressman John F. Lacey, who had died two months earlier. Macbride thought it a fine idea if they could manage the funding, and they agreed to have it published "like the old Proceedings."[60] Lacey had cut an important legislative path for the conservation movement on the national level, although there is no evidence that he participated in the movement as it played out in his home state. Nevertheless, his fellow Iowans determined to claim his legacy.

John Fletcher Lacey was not a native Iowan but moved to the state in his boyhood. Born in New Martinsville, West Virginia (then Virginia), in 1841, he moved with his family to Oskaloosa in 1855. Thereafter, Oskaloosa was home. During the Civil War, he compiled an impressive battle record in the Union army, mustering out with the rank of Brevet-Major in 1865, a rank that dressed his name for the remainder of his life. That same year he was admitted to the Iowa bar and married Martha Newell. During the next twenty-three years, he practiced law in Oskaloosa and launched his political career, beginning with a term in the Iowa legislature (1870–1872). As a lawyer, state politician, and conservative Republican, Lacey established a reputation for defending railroad interests and high protective tariffs. As a legal scholar, he compiled a widely used two-volume legal history of railroad decisions in U.S. courts, published under the title of *Lacey's Railway Digest*.[61]

His early career thus gave no hint of the political interests he would advance after he entered the U.S. House of Representatives in 1888. While serving in that body, he turned his intellect to the conservation of natural, cultural, and human resources. Addressing concerns for miners' working conditions, he was the first to introduce legislation to create a bureau of mines within the Department of Interior. As a long-time member of the House Indian Affairs Committee, he opposed attempts to cut appropriations for Indian schools, favored local schools over boarding schools, and supported citizenship for Indians, positions that were considerably more enlightened than prevailing attitudes of the day. Lacey wielded his most enduring influence within the House Public Lands Committee, however, which he chaired for twelve years. From this committee came the Forest Reserve Act of 1891 (which gave rise to the U.S. Forest Service) and the 1902 Alaska Game Law. It also produced the Lacey Act of 1900, which extended the commerce clause of the U.S. Constitution to permit federal entry into wildlife conservation legislation. Summing up his career near the end of his life, Lacey would say that the law bearing his name was "one of the most useful of all my Congressional acts."[62] Lacey also sponsored legislation to establish game preserves in specific national parks and forests, as well as bison breeding grounds in Yellowstone National Park and Oklahoma's Wichita Forest Reserve. After visiting pueblos and archaeological ruins in New Mexico, he introduced legislation to protect historic and prehistoric sites on public lands. Under the Antiquities Act, passed in 1906, more than two hundred national monuments were established during his lifetime.[63]

Although Lacey eschewed the term "progressive" (which contributed to his defeat in 1906), he nonetheless supported the utilitarian philosophy of his friend Gifford Pinchot and worked to transfer the Bureau of Forestry from the Department of Interior to the Department of Agriculture in 1905. After leaving office in 1907, he continued to lobby Congress as a representative of the League of American Sportsmen's Committee on Conservation. In this capacity, he urged legislation to establish game preserves in national forests and to protect migratory birds. Lacey lived just long enough to see the fruits of his labor realized in 1913 with passage of the Weeks-McLean Migratory Bird Game Act. He died at home, in Oskaloosa, on September 29, 1913. William T. Hornaday, director of the New York Zoological Park and one of Lacey's chief allies in the fight for wildlife conservation laws, eulogized him as "the first American congressman to become an avowed champion of wild life."[64]

The *Major John F. Lacey Memorial Volume*, which the IPFA published in

1915, contained a collection of Lacey's major addresses on conservation along with several tributes to him, including an assessment, by Pammel, of his legislative accomplishments.[65] A list of IPFA members inserted into the pages also revealed that the organization still fell considerably short of one hundred. Publication of the Lacey volume provided not only an opportunity to revive the IPFA but, once again, to distinguish the organization's work from that of the State Horticultural Society. Shortly before the book appeared, Pammel announced that the IPFA would change its name. "The name of the Association is to be changed to that of Forestry and Conservation," Pammel wrote E. R. Harlan, curator of the Iowa State Historical Department, in order "to bring in every interest. We thought best to cut loose from the Horticultural Society because they have given us very little support."[66] The name change, however, proved to be both confusing and short-lived. During the next three years, members used both names interchangeably, then finally resolved their indecision by adopting a streamlined moniker, the Iowa Conservation Association, in 1917.

It is unclear whether the Lacey volume attracted the attention that Pammel, Macbride, and Shimek hoped it would, or whether grassroots interest in parks and conservation was coming into its own. In any case, at about the same time the Lacey volume appeared, support for parks began to emerge from other organizations. The most important of these was the Iowa Federation of Women's Clubs, which became a major ally of the IPFA in 1916. During the early years of the Iowa Park and Forestry Association, few women participated in the meetings. A conspicuous exception was Charlotte M. King, a student of Pammel's who presented scholarly papers even during the very early years. By 1907, women were presenting more papers, a change that surely signified that the association attracted mainly those interested in the "softer sciences," along with nature lovers, city beautifiers, and tree planters.[67] When the IPFA reorganized in 1913–1914, women began to play a greater role by participating in committee work. They did not, however, occupy positions of leadership in the IPFA, which means that they were not involved in setting policies and goals, at least not openly. This would change with the Iowa Federation of Women's Clubs.

Iowa clubwomen were not unique in their efforts on behalf of conservation. In 1896, the General Federation of Women's Clubs resolved to study forest conditions as a result of efforts by New York and New Jersey clubwomen participating in the campaign to save the Palisades of the Hudson River. Two years later, the GFWC recommended that state federations establish committees to gather information on the "beauties and resources" of their respective states. From then on, clubwomen throughout the country gradually became

active at both the national and state levels, lobbying for legislation to extend
forest reserves and establish national parks, to reclaim arid lands through irrigation, and to protect water resources in the national domain. Their interest
in conservation was recognized in 1908, when the GFWC was invited to send a
delegate to the first National Conservation Conference.[68] Several organized
women's groups participated in these national conferences, and the Iowa Federation of Women's Clubs sent its own delegate to the Third National Conservation Congress, held in Kansas City in 1911.[69]

In 1908, the IFWC established a standing committee on the conservation of
natural resources, with five subcommittees on forestry, soils, minerals, waterways, and birds. Iowa claimed to be "the first state to have a Conservation
Committee, preceding a similar committee in the General Federation by a
year."[70] The claim might be substantiated on technical grounds, but women
elsewhere certainly shared their concerns and were just as active. Nonetheless,
Iowa clubwomen were early participants in the conservation movement. In
1908, the Forestry Subcommittee promoted tree planting and related public
education activities. The next year, they joined many other conservationist
groups, including the Iowa Park and Forestry Association, in lobbying for legislation to create a national park in the Appalachians.[71] This was the first nationwide campaign to establish a national park outside the West and to confer
such status on a landscape that was not marked by rugged, scenic grandeur.[72]
It is of more than passing interest to note that Iowans were moved to promote
a national park in the East because in a few short years the IFWC would be
among the Iowa organizations campaigning for a national park in the Midwest. The IFWC played a prominent role in securing passage of 1924 federal
legislation creating the Upper Mississippi Valley Wildlife and Fish Refuge, part
of a much larger and longer (and unsuccessful) effort to establish a national
park in the Upper Mississippi River Valley. The riverine refuge, one of the
largest in the United States, and Effigy Mounds National Monument, which
protects a string of American Indian mounds along the river bluffs, were the
tangible results of this long-running political engagement.[73] Parks, however,
were not the only conservation interest the Iowa women supported. In 1915,
for instance, the IFWC Conservation Committee drafted a bill designed to protect the "purity" of drinking water in Iowa and then lobbied (unsuccessfully)
for its passage.[74]

Iowa clubwomen were politically astute and active in many arenas during
the early twentieth century—an important force in the whole reform impulse.
But Cora Call (Mrs. Francis E.) Whitley of Webster City took a special interest
in the work of the IPFA during her tenure as president of the Iowa Federation
of Women's Clubs from 1915 to 1917. Eventually, her involvement would engage

Cora Call Whitley, 1862–1937. *Courtesy*
Iowa Women's Archives, University of Iowa.

a succession of women in conservation work, and together they would help shape the state park system. In her personal papers there is a well-used copy of Mrs. Fred Tucker's *Handbook of Conservation*, published by the Massachusetts State Federation of Women's Clubs in 1911, attesting to her strong interest in conservation. It was one of two areas in which she concentrated her public service work, the other being child welfare.[75] As president of the Iowa Federation, she promoted both.

At the urging of Mrs. John Sherman, chair of the General Federation's Department of Conservation, Whitley established a new subcommittee on the conservation of natural scenery.[76] That was merely the beginning, though. After reading a newspaper article by Louis Pammel, she wrote him inquiring whether his organization would want the IFWC's cooperation. Pammel answered immediately in the affirmative and asked if she would give a short presentation at the upcoming meeting of the IPFA in February of 1916. Mrs. Whitley agreed, despite other speaking engagements crowding the same week, and then let drop that she had the ear of "17,000 and more well organized club women."[77]

At the February meeting, women were a greater presence than ever before,

May H. McNider. From a 1918 article in
Iowa Conservation profiling the ICA's new
vice president. *Courtesy State Historical Society
of Iowa, Iowa City.*

reinvigorating the organization and establishing themselves as determined
comrades in the cause. So much so, that during the business meeting, Pammel
moved to have the incoming president, Fred Lazell, appoint an executive committee that included members of associated organizations.[78] Whitley subsequently served as vice-president and Rose Schuster Taylor of Sioux City as
treasurer. Taylor was a long-time friend of Pammel's. They had studied botany
together at the University of Wisconsin during the 1880s. May (Mrs. C. H.)
McNider of Mason City, who succeeded Whitley as IPFA vice president and
then served on the executive committee for several years, was equally active in
the Iowa Federation of Women's Clubs. Key individuals, such as Whitley, Taylor, and McNider, formed a bridge between the two organizations. McNider
chaired the IFWC conservation committee established in 1915, which also included Roxanna Lazell of Cedar Rapids, who participated alongside her husband, Fred. Jane Parrott of Waterloo, who with her husband, William F., was
active in the IPFA, likewise carried her support to sister organizations. She not
only was active in the IFWC, but, in addition, served as secretary of the Iowa
Audubon Society.

The Iowa Federation thus brought prominent women into the conservation constituency at a time when women's clubs in general provided an outlet for middle-class women who wanted to be involved in social and political reform.[79] During Cora Whitley's two-year tenure, for instance, the Iowa Federation lobbied the state legislature to establish the Child Welfare Research Station, petitioned the same body to restore funding for the women's reformatory at Rockwell City and the Travelling Library Commission, sponsored conferences on public health education and well-baby campaigns, initiated a regional Upper Mississippi Valley Conference of State Federations, continued the long-running campaign for women's suffrage, and, when war erupted in Europe, participated in all the "war-time activities in which the women of the country were called upon by the government to engage."[80]

That Whitley placed conservation uppermost among so many pressing concerns is testimony to the strength of her commitment. The title of the committee she established, though, Conservation of Natural Scenery, indicates the particular emphasis she promoted. "Is there need of a campaign to stimulate interest in the preservation of natural beauty?" she asked. To find the answer, one had only to "look at the communities where the only natural grove near enough to be available for . . . neighborhood festivities and the greatest ornament to the countryside has been chopped down to make a few cords of fire-wood." If that common sight did not stir one's sensibilities, "look at the places where the banks of the river winding about or flowing through the town have been made the common dumping ground of everything that is ugly and unsightly."[81] Whitley appealed to the aesthetic senses, which suited many women drawn to clubwork. The campaign for parks also provided a tangible, comprehensible goal. Toward this end, the members of the Iowa Federation surveyed their local areas "to help determine what tracts of beauty or historic interest should be preserved for parks."[82]

Women, in particular, also urged an approach that linked natural resource conservation with social reform. President Theodore Roosevelt first linked them in 1908, when he advocated a federal public health program as necessary for the "conservation of human health." Contemporary writers thereafter began to speak of "human conservation," and the National Conservation Congress even made "the conservation of human life" its theme in 1913.[83] "Human conservation" struck a responsive chord in women. As Rose Schuster Taylor phrased it for Iowans, "The waste of our fields, our soils, our forests, our birds, our flowers, has led us to emphasize conservation. The waste of human life needs also the cry of conservation for its protection and preservation."[84] Women, it appears, responded to the campaign for state parks because they perceived it as a means to address multiple concerns. By promoting the preservation of scenic and historic places, women could feel they were promoting better public health and protecting women and children as well.

Through Jane Parrott and T. C. Stephens of Morningside College, ornithologists and birders also began to ally themselves with the conservation group. Actually, Parrott, long-time secretary of the Iowa Audubon Society and one of the first women to become active in the IPFA, took the additional step of asking the IPFA simply to absorb the society and "appoint a special committee to look after the Audubon work and interests." She noted that Audubon members were scattered and their financial resources meager. In addition, she reckoned that the IPFA was "well equipped to carry on the conservation work," and, perhaps most important, that she was weary of shouldering the administrative burden for an organization whose work she perceived as becoming increasingly duplicative. This did not happen, although the Audubon Society's work thereafter would be regularly noticed in the IPFA's new quarterly, *Iowa Conservation*, published from 1917 to 1923.[85]

T. C. Stephens, a professor of biology, also became more active in the IPFA. He earned a reputation as a seasoned conservationist when he went head-to-head with the state game warden in a legislative committee hearing as part of a determined bid to secure passage of the 1917 Turner Quail Bill. This piece of legislation, introduced by Representative Fred G. Turner of North English, was supported by the IPFA, as well as many farmers. The state fish and game warden, E. C. Hinshaw, opposed it, as did one faction of organized sportsmen. Its passage is considered a turning point in wildlife conservation in Iowa because the Turner Quail Bill marked the first time the state legislature endorsed a policy of game management by authorizing a five-year closed season in order to give a threatened species time to repopulate. For this achievement Stephens was awarded a gold medal by the Permanent Wild Life Protective Fund.[86]

Others urged parks specifically for their recreational value. Chief among them was A. H. Carhart of Mapleton, who later became a "recreation engineer" for the U.S. Forest Service. Noting that Scotland, Germany, and other industrialized European nations had state-sponsored recreation systems, he castigated America's preoccupation with industry and commerce, which had, in his words, "strangled any desire for play and pleasure."[87] He therefore advocated a complete system of national, state, and local rural parks "for outdoor sport and rest." For Carhart, resource conservation was an added benefit of parks; aesthetics, the central reason for their being. The national park movement, which placed emphasis on preserving unique environments and scenic grandeur, encouraged this response, and Carhart was among those who campaigned for a national park encompassing the Upper Mississippi River

Valley. "These parks," wrote Carhart, meaning the fifteen national parks established by 1916, "not only preserve the best scenic beauties of the country but they also act as educational storehouses of nature's wonders and as a sanctuary for the fast disappearing wild animal, bird and flower life. In this manner they serve in as great a measure as in their aesthetic uses."[88]

The 1917 State Park Law

When the IPFA reorganized as the Iowa Conservation Association in 1917, it had a stronger, more vocal constituency. It also had more focused goals, chief among them promoting a complete system of parks—local, state, and national—throughout Iowa. Importantly, the affiliations had changed. The State Horticultural Society no longer was its prominent ally, although certain individuals still held memberships in both organizations. Women who sought entry into the political process began to take the place of horticulturists, who were increasingly less interested in advocacy. The Iowa Conservation Association's leadership, however, remained principally in the hands of academic scientists, as evidenced by the roster of officers and advisory board members (table 2).

The 1916–1917 campaign for state parks meshed most closely with a long-standing effort to preserve and enhance Iowa's lakes. As early as 1896, the Iowa Academy of Science had appealed to the state legislature to protect the lakes "in order to maintain some of the original conditions of the state" and, additionally, to develop them into "pleasure resorts" for the citizens of Iowa.[89] Similarly, in 1901 the IPFA adopted as one of its goals the creation of state parks near lakes.[90] Later, Macbride advanced the idea of lake preservation and state parks through the Iowa State Drainage, Waterways and Conservation Commission. The commission's 1911 report urged that Iowa place its lakes under the jurisdiction of a custodial agency empowered to arrest the loss of these public waters.[91] In 1915, the legislature responded to the commission's report by directing the Highway Commission to coordinate a study of approximately seventy lakes to which the state still held sovereign title.[92]

Originally, the state owned no fewer than 109 lakes covering approximately 61,000 acres, plus an assortment of ponds and lagoons along the Mississippi and Missouri rivers. In 1850, some of these lakes were reclassified as swampland and subsequently sold. Early in the twentieth century the state drained several smaller lakes and sold the land, and in 1913 it authorized the sale of abandoned river channels, which affected a few more bodies of water. Because the remaining lakes typically were surrounded by privately owned land, either in whole or in part, they had been subjected to numerous abuses over the years, chiefly by farmers. Farmers, conversely, were of the opinion that lakes

Table 2.

Iowa Conservation Association Officers and Advisory Board, 1917

Name	Location	Other Affiliation(s)	Occupation, If Known
Bennett, George	Iowa City		former clergyman
Conard, Henry S.	Grinnell	Grinnell College, IAS, SHS	professor of botany
Lazell, Fred J.	Cedar Rapids	CR Dept. of Parks (head)	newspaper editor
Macbride, Thomas	Seattle	SUI, IAS, SHS	professor emeritus, botany
MacDonald, G. B.	Ames	ISC, IAS	professor of forestry
NcNider, May H.	Mason City	IFWC	
Pammel, Louis H.	Ames	ISC, IAS, SHS	professor of botany
Parrott, Wm. F.	Waterloo		newspaper editor
Pearson, R. A.	Ames	ISC	president ISC
Sanders, Euclid	Iowa City		banker
Shimek, Bohumil	Iowa City	SUI, IAS, SHS	professor of botany
Spurrell, John A.	Wall Lake		farmer
Stephens, T. C.	Sioux City	Morningside College	professor of biology

SUI = State University of Iowa, ISC = Iowa State College, IAS = Iowa Academy of Science, SHS = State Horticultural Society, IFWC = Iowa Federation of Women's Clubs.

indirectly led to crop losses by providing blackbird nesting sites; they, therefore, wanted more lakes drained and sold for cropland. The lake issue placed farmers on one side, conservationists and sportsmen from the cities on the other.[93]

The purpose of the lake study, therefore, was to assay the state's holdings and determine which lakes should be maintained and which should be sold. Over a two-year period the chief highway engineer, Thomas H. MacDonald (also a professor of engineering and Iowa State College), coordinated a variety of surveys and studies. Louis Pammel, G. B. MacDonald, and H. D. Hughes, the latter affiliated with the Agricultural Experiment Station at Iowa State College, assisted in these investigations, which yielded detailed topographic maps covering approximately 90,000 acres. In addition, they produced individual lake studies and separate reports on crop surveys taken near various lakes

Left to right: archaeologist Charles R. Keyes, botanist Bohumil Shimek, botanist Louis H. Pammel, geologist George F. Kay, and ornithologist T. C. Stephens, photographed together at the July 1919 meeting of the Iowa Conservation Association, held at McGregor. *Courtesy State Historical Society of Iowa, Iowa City.*

(Hughes), the vegetation of Iowa lakes (Pammel), and the potential for improving lake shores through forestation (G. B. MacDonald). The Fish and Game Department also assisted with survey work. When the Highway Commission reported its findings in 1917, it recommended, with few exceptions, that the state retain its lakes. More specifically, the commission recommended that concrete dams should be constructed at all lake outlets to preserve water levels, that the state should acquire acreage along lakeshores for public parks, that roads should be constructed to connect parks with existing highways, and that trees should be planted in parks wherever the natural timber had been cut down. Crop surveys, vegetation studies, and even an analysis of the stomach contents of about 1100 blackbirds found no substantive link between lakes and crop loss patterns.[94]

Conservationists seized on the lake study to clarify their arguments and rally support for park legislation. G. B. MacDonald advocated public control over lakes and woodlands not only to preserve "the most beautiful spots in the country," but to provide refuges for wildlife, and ample fishing grounds as well.[95] Pammel, who staked out a position somewhere in the middle, searching for a "happy medium" to the lake issue, nonetheless was forced to admit, on the basis of preliminary studies, that many aquatic plants once abundant in

Iowa, such as wild rice and the white water lily, were now rare. Other plants, such as rushes, wild celery, and freshwater eel grass, served as food or breeding places for waterfowl and fish, and all were equally in need of protection from a scientific standpoint.[96] State parks, in essence, were seen as the means for accomplishing several interrelated resource conservation goals, and lake parks became the ideal.

The lake study thus provided whatever official justification was needed for park legislation, although no one in the Iowa Conservation Association had any fixed idea about what should be in the bill. MacDonald thought perhaps control should rest with local boards, advised by a panel of "experts" from the state's educational institutions, with land acquisition funded by automobile taxes and hunter's license fees.[97] C. F. Curtiss, dean of agriculture at Iowa State College and director of the Iowa Agricultural Experiment Station, felt that it was "logical and proper" to use an existing fund of about $150,000, accumulated from the sale of hunting and fishing licenses, to "develop places that would conserve fish and game and forestry resources" and see that they were "properly guarded by state policy."[98] Pammel, thinking in smaller terms, thought $6,000 to $8,000 would be sufficient to purchase a few areas to be set aside as state parks under the control of "educational institutions for the use of all educational interests."[99]

When the state legislature convened in 1917, the Iowa Conservation Association executive committee met in Ames to plan a political strategy. Several men were selected to form a special legislative committee that included Pammel, Shimek, and G. B. MacDonald, together with E. R. Harlan, C. F. Curtiss, Dr. James H. Lees, assistant state geologist, and Thomas MacDonald. Macbride, who was finishing up his tenure as president of the University of Iowa and preparing for retirement, was unavailable to assist in this effort, though he certainly followed its progress. Their strategy was to meet with the Senate's Fish and Game Committee and with a similar committee of the House in order to cultivate an interest "in preserving historic places in the state," as Pammel later recalled, although what the committee really had in mind was legislation authorizing a state park system.[100] When Pammel talked of "historic places" he conceived of natural history and cultural history as two sides of the same coin.[101] Passage of the National Park Service Act in 1916 most certainly gave them confidence, or at least hope, that action finally was possible on the state level.

Coincidentally, in 1917 Perry Holdoegel of Rockwell City was elected to the Iowa Senate and, immediately upon taking office, was appointed chair of the Fish and Game Committee, the same committee that heard debate on the Turner Quail Bill that session. Holdoegel, as luck would have it, expressed greater interest than the ICA committee expected. He agreed to call a meeting at his quarters in the Savery Hotel, which was attended by the ICA group, Senator Byron W. Newberry of Strawberry Point, Representative Orville Lee of Sac County, and Representative B. J. Horchem of Dubuque. The meeting turned into a working session that produced a draft legislative proposal. According to Pammel, Holdoegel first proposed a bill focused on acquiring land for the fish and game warden and for improving lakes. To this initial proposal was added a provision to appoint a commission (unpaid) with authority to utilize state agencies for the purpose of identifying and investigating "beauty spots and other places of historic interest."[102]

Holdoegel subsequently introduced the bill, which easily passed the Senate (without dissent) and the House (63–25). Governor William L. Harding signed the Holdoegel Act into law on April 12, 1917. It appropriated an annual sum from the state's fish and game protection fund and authorized the state fish and game warden, an existing state officer, to establish public parks wherever there was reason—by virtue of scientific interest, historic association, or natural scenic beauty—to locate them. Importantly, the legislation also authorized the Executive Council to establish a Board of Conservation, whose duty it would be "to investigate places in Iowa, valuable as objects of natural history, forest reserves, as archaeology and geology, and investigate the means of promoting forestry and maintaining and preserving animal and bird life in this state and furnish such information to the executive council for the conservation of the natural resources of the state."[103] Except that primary authority for parks was vested in the state fish and game warden, and not the Board of Conservation, the state park law contained nearly everything that conservationists had hoped for.[104]

It is uncertain whether Perry Holdoegel saw the state park idea as a convenient issue on which to build his political career, or whether he was genuinely concerned about resource issues. Certainly, his political inclinations and ambitions were largely unknown to members of the ICA. There is no evidence that he participated in any activities of the conservation movement in Iowa before entering the legislature in 1917, and nothing in his background suggested him as an ally. During his early career he taught school, then entered school administration, but he is chiefly remembered as a prominent pioneer in the independent telephone company business in Iowa.[105]

However, once he took hold of the state parks bill and secured its passage, he proved to be a constant friend to park promoters, as well as sportsmen, throughout his tenure in office, which lasted until 1925. Among other things, he initiated the effort to establish a state park, Twin Lakes, in his home district.

As a measure of his integrity, Holdoegel then agreed to act as honorary custodian after the park was formally acquired in 1923. In that capacity he worked to see that the park was landscaped, fenced, and improved with camping and picnicking facilities. He also helped to acquire additional park acreage in 1924. When Louis Pammel delivered the park dedication speech in 1926, he paid sincere tribute to Senator Holdoegel not only for his role in creating Twin Lakes State Park, but for the entire system that was then beginning to take shape.[106]

Conservation Parks

Just as he had been the first to call for "rural" parks, so too was Thomas Macbride the first to try his hand at defining the concept for common understanding in a 1922 article entitled "Parks and Parks." It was a difficult task. Macbride appeared to be on solid ground in distinguishing what he now called "conservation parks" from baseball parks, athletic parks, and auto parks, but when it came to describing the precise attributes of a "conservation park," that is, a "real" park, he resorted to vague language and retreated from the pragmatism of earlier years. To Macbride, a park must have three essential qualities: it must have "beauty of location," it must be "roomy," and it must be "quiet." A conservation park was to Macbride the metaphorical equivalent of a "great cathedral" in which congregations might commune together. "A great solemnity seems to attend them both; the grove rather better, I incline to think, because it has a higher ceiling, is cleaner and has far finer ventilation."[107]

This divinely inspired concept was hardly sufficient to grapple with the very real questions and issues that arose when conducting the sometimes complex negotiations required to transfer privately held land into the public trust. Equally important was the very real problem of trying to instill in park users a sense of stewardship. Here Macbride clearly sat on the proverbial horns of a dilemma. On the one horn, he had always maintained that parks should be for all the people, not one class of society. On the other, he realized that people in the aggregate—i.e., "civilization"—were responsible for nearly obliterating the landscape that had provided complete sustenance for Native Americans and that had greeted Euro-American settlers. Where to draw the line? Once a state park system seemed assured, Macbride fretted more intensely over this dilemma without reaching a satisfactory solution. Indeed, he began to advocate open and closed seasons for parks in order to protect "the original proprietors," that is, the "hawthorn thickets, orchards of crab, clusters of red cedar"; the "vireos, warblers, thrushes, . . . orioles, tanagers, wood-peckers, cuckoos," and other forms of wildlife that flourished in places unoccupied by humans. He also advocated an abundance of small parks, "one in every township," in order to avoid crowding and overuse. "Only abundant parks, parks little used by man, can for our children, our grand-children, our successors all, save the wonderfully beautiful, interesting birds of Iowa." [108] But by the time the state park system was within grasp, Macbride had removed himself from the circle of action. It was Louis Pammel who would work within the State Board of Conservation to define state parks on a much more prosaic plane.

2

Pammel's Way

The rural park of Iowa . . . should present Iowa as it was at one time. There have been presented, to the State Board of Conservation, many propositions that seem to the board to partake . . . of recreational work for the city.

—LOUIS H. PAMMEL[1]

Louis H. Pammel

To a large degree, the state park system that took shape in the 1920s reflected decisions influenced by Louis Pammel's unique combination of vision and practicality. To understand the way in which the system emerged, therefore, it is important to understand the man most responsible for making it happen. "In his lifetime," writes Marjorie Conley Pohl, "Louis H. Pammel was the conservation movement in Iowa." She overstates his role in the long view, but during the last two decades of his life, Pammel *did* make his name nearly synonymous with state parks in Iowa. His domineering personality was legendary; his energy, seemingly boundless; and his work, prolific. Pohl describes him as "a person of volcanic, almost furious activity, always moving on to another activity while finishing a first and contemplating the third. He proceeded through his life almost as if he were afraid he would run out of time before he finished everything on his personal agenda." But the pace of his life was balanced with, and perhaps driven by, uncommon foresight. His collected papers reveal a person of well-developed ego, but also one who understood quite clearly that he was participating in—indeed sometimes directing—the course of history insofar as conservation and parks in Iowa were concerned. Pohl attributes his "large vision" to extensive reading, not only in his own field of botany, but in history and biography as well. "From his continuous reading," she surmises, "he enabled himself to understand the significance of his own work and to see the areas which would develop from it in the future."[2]

Whatever the source of his vision and his energy, Pammel the conservationist was a man with a mission. It was he who shaped the state park system during the late 1910s and 1920s. It is tempting to say that Pammel did so almost single-handedly, but that would be denying his immense interest in people, including the younger generations (he worked with youth groups and took the time to answer letters from children), and his tremendous ability to work with others. No, the original park system was not created solely by Pammel, or even solely as he envisioned it. But he was *the* person who held state parks on a steady course while he chaired the State Board of Conservation (1918–1927), and his guiding hand enabled Iowa to assemble an impressive array of parks during this period. Anyone who worked with him in this endeavor would have agreed that the park system, as it stood in 1931 when he died, had been built by following "Pammel's way."

Louis Hermann Pammel's childhood, like Macbride's, shaped his professional career in ways subtle yet deep. He was born in 1862, the second of five children born to Louis Carl and Sophie Freise Pammel. His parents were Prussian immigrants who settled in La Crosse, Wisconsin, and became prosperous farming several hundred acres in Shelby Township near the Mississippi River. As a young boy, Louis H. reportedly was an instinctive student of his surroundings, freely roaming his family's large farm, examining the flora, gathering American Indian artifacts from the fields, exploring the varied topography along the Mississippi River, and observing the abundant wildlife that inhabited the area. As the eldest son, he was also expected to follow in his father's footsteps, so after completing the fifth grade in 1872, he spent the next six or seven years as his father's apprentice, a stretch of time that removed him from a formal educational setting but hardly stilled his inquisitive mind.

Louis Carl was considered a "progressive" farmer and a community builder, so his son's apprenticeship was far more than vocational training. The father experimented with new crops, new farm implements, and purebred cattle. The son developed a keen interest in bees and honey and was permitted to establish and maintain his own apiary. The father subscribed to German-language newspapers, magazines, and agricultural publications. The son read widely from the library his father assembled. Louis Carl insisted that his children be fluent in both English and German. He also served on the township board and as a school director. In the latter capacity, he stocked the library bookshelves with works of literature and science, including works by Alexander von Humboldt, the great nineteenth-century pre-Darwinian naturalist who urged an integrative and comparative approach to the study of natural phenomena.[3]

Despite inadequate formal preparation, Louis H. determined that he would go to college. In a move that may have been calculated to impress upon his parents his sincerity, he published, at age seventeen, a "Letter of inquiry about bergamot" (a honey plant) in the *American Bee Journal*. If so, the ploy worked, for he apparently had no difficulty persuading them that he had the makings of a scholar. He was permitted to leave farming behind, although at the time he thought he might eventually return, and so for the next several years he spent his summers working with his father on the farm. Considering the difficulty of the task he was about to undertake, his uncertainty was justified. It took two intense years of private tutoring in addition to coursework at the La Crosse Business College before he was ready to take the entrance examinations for the University of Wisconsin.

Pammel entered the university in 1881, the same year that William Trelease, trained at Cornell, Harvard, and Johns Hopkins, arrived to teach botany. It was a fortuitous coincidence. Trelease served as an inspiring mentor, encouraging Pammel's interest in parasitic fungi and seeds, and exciting in him an interest in bacteriology. Later, Trelease used his connections to help Pammel establish his own academic career. At Madison, Pammel studied from the newly published text by Charles E. Bessey, *Botany for High Schools and Colleges*, little knowing that he would one day follow in Bessey's footsteps at Iowa State College. He also met Rose Schuster, a classmate who would remain a lifelong friend and eventually become a colleague in Iowa's conservation movement. During his undergraduate years, Pammel also began to develop the avenues of social and intellectual interchange that would so characterize his entire professional career. Along with Trelease, he organized the Natural History Society, spent much of his spare time on field trips, debated with the Hesperian Society, and was active in the Unitarian Society.

During his senior year, he met Augusta Marie Emmel at a dance in Chicago that he attended with college friends. The two were equally smitten with one another, so much so that, after he graduated in 1885, he took a job with a Chicago seed company in order to be near her. The scholar in him, however, was not satisfied, so he also matriculated at the Hahnemann Medical College. Classes had barely started in the fall of 1885 when William G. Farlow of Harvard University contacted him, at William Trelease's suggestion, about coming east to work as his assistant. Flattered at the offer and not wishing to disappoint his undergraduate mentor, Pammel decided to postpone matters of the heart. He spent a year at Harvard, where he indexed fungi for Farlow, audited classes, and met the eminent Dr. Asa Gray, who flattered him even more by acknowledging that he had read Pammel's article, "On the Structure of the Testa of Several Leguminous Seeds," his undergraduate honor's thesis published in the *Bulletin of the Torrey Botanical Club* in 1886. Any lingering thoughts about returning to the farm seemed to evaporate at Harvard.

Pammel might have taken up graduate study at Harvard except for Trelease, who by now had moved on to become director of the Missouri Botanical Garden, affiliated with Washington University in St. Louis. Trelease needed an assistant, and he tapped Pammel for the job. Pammel accepted, and for the next two and a half years he worked with Trelease. He also began graduate work at Washington University, where eventually he earned his doctorate, in 1899. In the summer of 1887, he returned home to marry Augusta. They started a family immediately, and the pattern of their life together began to emerge: she taking charge of the children (they would have six by 1897) and he charging about establishing his scholarly career. The summer of 1888 prefigured much of what was to come. Within three weeks of their first daughter's birth, on June 1, Louis took Augusta and baby Edna in tow on a collecting trip to Cahokia, Illinois. It was only a short trip across the Mississippi River, but on the way back a fierce rainstorm nearly capsized the ferry on which they were riding. No sooner had they recovered from that adventure than all three boarded a train bound for College Station, Texas, where Pammel had accepted a summer job with the Texas Agricultural Experiment Station to study the cause of root rot in cotton plants. Pammel must have had precious little time for his wife and infant daughter, for he spent the summer traveling throughout Texas examining cotton plants (and along the way collecting for his own herbarium). By summer's end, though, he had succeeded in isolating and culturing the parasitic fungus responsible for threatening much of Texas's cotton crop. As a result, the Texas Agricultural Experiment Station invited him back the next summer to continue his research, which he later turned into his thesis for a master's degree from the University of Wisconsin, awarded in 1889.

Pammel's career opportunity came in late 1888 when a position teaching botany at Iowa State College became available. Trelease once again assisted his young protégé, as did William Farlow. With their strong letters of recommendation, as well as the favorable impression Louis and Augusta made on the Board of Trustees, Pammel won the position. He assumed his post as professor of botany in February 1889, more than a decade after Thomas Macbride had assumed a similar position at the State University in Iowa City. This brought to Iowa State a person predisposed, by nature and training, to follow the lead established by Charles Bessey.

It had been Bessey's lot to forge an applied science curriculum that would fulfill the mission of Iowa's newly founded land-grant college, when the only courses of instruction were "agriculture" and "engineering," with botany part of the former. This would prove to be a difficult task because there were few models to follow. Higher education in the sciences was in the midst of maturing, and applied science was in its infancy. There was, in addition, considerable disagreement about how the science of botany fit into the scheme of practical education. Organized agricultural groups, particularly the Grange, and more than a few legislators were vocal in their insistence that the college focus on "how to" courses in crop management, stock breeding, soil preparation, pomology, and the like. Bessey, however, refused to concoct what he considered a "cheap" vocational program. In his view, "the agricultural college should concentrate on the basic sciences, award the student a college degree, and then provide an apprenticeship on a successful farm."[4] However, he also understood the need to mold instruction in the basic sciences toward their practical application. Accordingly, for fifteen years he persisted in building a science curriculum that stressed laboratory work (Iowa State and Harvard University were the first institutions to establish botanical laboratories), that provided for serious instruction in and research on lower forms of plant life (especially fungi, which caused considerable crop damage), and that concentrated on surveying and classifying regional vegetation (thus laying the groundwork for region-specific research). In addition, he became an active member of those "practitioner" organizations, such as the Iowa State Horticultural Society, that welcomed dialogue with the new college faculty. Although he was never able to establish a program that would satisfy the college's critics (a factor in his decision to leave Iowa State for the University of Nebraska in 1884) Bessey nonetheless established a rigorous program in applied science that Pammel was well-suited and well-prepared to continue.

At Ames, Pammel threw himself into teaching, into research and writing, into professional organizations, into community service, into church work, into public speaking, and into the fray of politics. He was a forceful, enthusiastic educator, both inside and outside the academy, and there was virtually no arena in which he did not participate. As a teacher and researcher, he carried on his expansive interests in economic botany, plant pathology, bacteriology, mycology, horticulture, forestry, bees and pollination, seeds and germination, flowers, grasses, climate, ecology, and conservation. Summers often were spent in the employ of the U.S. Department of Agriculture conducting research. These summer sojourns not only supplemented the income needed to raise and educate six children, but they also enabled him to build the collections of the college herbarium and to gather material for his own research and writing. During his lifetime, four plants were named in his honor. Between 1896 and his death in 1931, he authored or coauthored six scholarly books (a seventh was published posthumously), and two reminiscences. In addition, he edited the *Major John F. Lacey Memorial Volume* for the Iowa Park and Forestry Association. A prolific scholar and writer, he wrote nearly 700 articles, research notes, reports, educational circulars, and addresses. Many of his publications were coauthored with one or more of several research collaborators, including Charlotte M. King, botanical artist for the Agricultural Experiment Station, and Ada Hayden, one of his finest students who, after receiving her doctorate, joined the botany department as a faculty member.

As a public speaker, Pammel was in great demand. Whenever asked to do so, which was often, he spoke before chambers of commerce, men's groups, women's clubs, and campus organizations, at high school and college graduations, and at church. Over the years, he developed a repertory of topics that covered the gamut of his scientific and personal interests. His addresses, dedications, memorials, and sermons, it is said, were always delivered with authority, with engaging disposition, and without notes—although typically he prepared a text. Somehow, he also found time to write an astounding volume of correspondence that would, in time, amply document every aspect of his multifaceted career.

Pammel seems to have found the extension work of land-grant colleges easy to adopt, perhaps because his father had worked with state officials in Wisconsin to conduct agricultural experiments. He willingly made his services, and those of his students, available to municipalities and to the State of Iowa, generally without fee. He put his expertise in bacteriology to work analyzing public water supplies and sewage disposal systems. For the state legislature, he helped write bills addressing agricultural and horticultural needs. At the state's request, he prepared for the annual Iowa State Fair exhibits and educational pamphlets on weeds and wildflowers, and later established a plant laboratory on the fairgrounds. He also directed the preparation of exhibits on crop diseases as part of Iowa's displays at the 1893 World's Columbian Exposition in Chicago and the 1904 Louisiana Purchase Exposition in St. Louis. He initiated annual plant disease surveys for the state. Eventually, his research and public service in plant pathology brought national and international recognition, and in 1919 he was called upon to serve as one of four distinguished scientists on the American Plant Pest Committee, a joint U.S.-Canada initiative.

Pammel gave as much time to public and professional service as he did to teaching and research. He served as president of the Iowa Academy of Sciences (1892–1893; 1923) and as president of the Iowa Park and Forestry Association (1904–1906). He also served on a variety of state commissions: the State Forestry Commission (1908–1929), the State Geological Board (1918–1929), the Plant Life Commission (1917), and the State Board of Conservation (1918–1927). As chair of the latter, he made his most enduring contributions. He gave so much of his time to conservation work, in fact, that when he was reelected chairman of the board in 1922, he announced his intention to relinquish his position as head of the botany department in favor of "doing something of real value for the citizens of the State."[5] If he followed through with a formal request to be relieved of college administrative duties, it was denied, as were similar requests after his health began to decline in 1924. Nonetheless, despite a heavy teaching and administrative load, Pammel continued to give unsparingly to board work, even after suffering a heart attack in 1925.

The State Board of Conservation

Governor Harding, who signed the Holdoegel bill into law in April of 1917, and who would later boast that the State Park Act passed during his watch, took no immediate steps to appoint the Board of Conservation so authorized by its provisions. The reasons for this appear to be mixed. First, the governor himself became the target of considerable controversy and ridicule when, in May 1918, he issued a wartime ban on the use of foreign languages in churches, on the radio, and over the telephone. Before this controversy had died down, Harding pardoned a convicted rapist before the man even began serving his sentence, a move that provoked the House to initiate impeachment proceedings. The governor survived this scandal with nothing more than an official censure; however, it placed him in a tough campaign for reelection in 1918. As a Republican incumbent, his bid should have been effortless, but it was only with diligent campaigning that Harding managed to win a second term.[6]

Given the imbroglio of 1918, one might have expected the governor to avoid additional conflict. This is precisely what he did when E. C. Hinshaw, the fish and game warden, openly opposed the State Park Act. Hinshaw based his opposition on its funding mechanism, which required the fish and game warden to divert $50,000 from his budget, derived solely from fishing and hunting licenses, in order to fund parkland acquisition. Claiming that this would "not leave sufficient funds to carry on the work of the [fish and game] department," Hinshaw recommended that the funding provision be "repealed."[7]

To a degree, Hinshaw's opposition was understandable. Drafted under the auspices of the Senate Fish and Game Committee, the State Park Act initially vested authority for establishing state parks with the state fish and game warden and the Executive Council.[8] Funding for state parks was to come from the Fish and Game Protection Fund, an existing source of money. One section of the act, however, provided for a new Board of Conservation as an investigative and recommending body that would report to the Executive Council, and not necessarily in consultation with the fish and game warden. By avoiding the appearance of creating a new government agency requiring its own appropriation, the framers of the law, a group that did not include Hinshaw, were able to skirt any number of challenges that might have led to defeat.

There is little doubt as to why Hinshaw was excluded from the committee that drafted the state park bill. Most certainly it was because he opposed another controversial piece of legislation, the Turner Quail Bill, which also passed in 1917. Early in 1917, at the urging of a few individual sportsmen, the Iowa Conservation Association began backing legislative bills designed to establish five-year closed seasons for prairie chickens and quail. Previous attempts to pass stronger laws to protect these badly decimated species had failed. However, in 1917 the ICA, with financial support from the Permanent Wild Life Protective Fund, got behind the effort and pushed hard. G. B. MacDonald of Iowa State College and T. C. Stephens of Morningside College spent much of the legislative session in the State Capitol keeping support strong. Their chief opponent was E. C. Hinshaw, who took the position that it was unfair to force hunters to purchase a license and then deny them the right to use it. After acrimonious debates in committee hearings and on the floor, the Quail Bill finally passed both houses.[9] By the time the legislature adjourned, it had passed laws that established a five-year closed season not just for quail and prairie chickens, but for Hungarian partridges, Mongolian ringneck pheasants, and English pheasants as well.[10] Conservationists were ecstatic over this victory, even more so when the Permanent Wild Life Protective Fund awarded a gold medal to T. C. Stephens for his role in directing the campaign.[11] Hinshaw was left to contend with some disgruntled hunters.

The Quail Bill controversy reflected more than a contest between the Iowa Conservation Association and the fish and game warden; it also exposed a deep rift among sporthunters that left legislators divided in their loyalties. Between 1907 and 1909, the Iowa Fish and Game Protective Association, an organization largely composed of sportsmen, had successfully campaigned for tougher fish and game laws. During that period, the association reportedly was the main force behind new laws that required Iowa residents to purchase hunting licenses, to prohibit the year-round sale of game, to establish bag limits on ducks, and to place deputy wardens on a salary compensation basis (replacing a fee-system basis that was rife with abuse). The Fish and Game Protective Association, however, did not represent the views of all sporthunters, and the fish and game warden, that is, E. C. Hinshaw, was widely perceived as pandering to the side that opposed increasingly restrictive game laws.[12]

While Hinshaw opposed protective game laws for indigenous game birds, he also worked with private landowners to introduce new species, notably the Hungarian partridge and the ringneck pheasant. This position was consistent with those who saw no long-term harm in the liberal harvest of native species; they could simply be replaced or augmented by introducing new species. In a different vein, Hinshaw supported restoring Iowa's natural lakes. He even advocated establishing public parks near lakes. To pay for them, he recommended increasing the Fish and Game Protective Fund by requiring fishing licenses in addition to hunting licenses. However, he also advocated that those who purchased licenses receive "the entire benefit" of such parks.[13] Hinshaw's views not only clashed with those of the Fish and Game Protective Association, they also clashed with those advanced by the Iowa Conservation Association, the organization whose views stood to gain greater currency with passage of the State Park Act. As it turned out, though, Hinshaw's tenure as fish and game warden was cut short for other reasons. During his six years as warden, he also acquired a reputation for spending more than his department earned in receipts. Charges of financial mismanagement ultimately led to his departure in 1919.[14] It is therefore possible that Hinshaw also opposed implementing the State Park Act because it would have drained a fund that was fiscally unsound.

These jurisdictional conflicts and financial problems had to be resolved before the law could be implemented. For a time, it looked as though the governor might not appoint anyone to sit on the Board of Conservation. Although frustrated by Harding's lack of action, conservationists first avoided doing anything that would further embarrass the embattled governor or lose the support of those sportsmen who had been behind the Quail Bill. When a year passed, though, and the governor still had not acted, the Iowa Academy of Science moved. In mid-1918, the IAS adopted a resolution, drafted by Bohumil Shimek, calling upon the governor to appoint the Board of Conservation and to "take such steps as are necessary to insure the scientific treatment of all conservation problems, such as the preservation of game, fish and birds, forests, streams and lakes, natural scenery, etc." The IAS expected the Iowa Conservation Association to join in the resolution, but Pammel counseled otherwise.[15] Whatever strategy the ICA pursued, it appears to have been carried out quietly, most likely by working through "friends" in the state legislature. The ICA never acted on the academy's proposed resolution, and its news organ, *Iowa Conservation*, never hinted that political problems were the cause of delay.

Governor Harding finally got around to appointing the first Board of Conservation late in 1918, at which time he asked Louis Pammel to serve on it, although Pammel himself supported other scientists for board positions.[16]

Actually, Pammel dearly wanted a board composed entirely of scientists, but Harding appointed as the other two members John Ford, mayor of Fort Dodge, and Joseph Kelso, a banker from Bellevue. These were men who were widely seen to represent the interests of sportsmen, although it is not clear whether either of them had supported Hinshaw or the policies of the Fish and Game Department.[17] E. R. Harlan, curator of the State Historical Department, served as ex officio member and secretary of the board. Pammel's new colleagues selected him as chairman.

During the 1920s, the Board of Conservation matured from an advisory body with vague authority into the state's leading resource conservation agency. As articulated in the 1917 State Park Act, political expediency dictated that initial authority over the future park system rest with the fish and game warden. Once organized, however, the board took action to change this. In 1919, the board easily convinced the state legislature to amend the State Park Act, transferring the authority to establish parks from the fish and game warden to the board with the consent of the Executive Council. Additionally, the state legislature amended the funding provision, establishing a direct appropriation to the Board of Conservation, an amount initially set at $100,000 per year.[18]

The 1919 amendments clarified the Board of Conservation's role and funding, but it left hanging another important issue: lakes. Although public and political interest in lake preservation had helped to pass the state park law, lakes were not specifically mentioned in the 1917 statute. This omission presumably stemmed from the ambiguous provisions that placed the fish and game warden and the not-yet-created Board of Conservation in competition for authority and financial resources. For decades, the fish and game warden had command over the handling of fish in lakes and rivers, and as early as 1914 or 1915 Hinshaw had anticipated the authority to restore lakes and to establish lake parks under the auspices of his office. Because the amendments did not address the lake issue, the board initiated a study to determine its legal and fiscal status with respect to lake improvements. It also initiated a policy of cooperating with the Fish and Game Department, through the new warden, W. E. Albert, in "matters relating to dams, water levels, riparian rights, dredging, [and] reclamation." Within the next year, the board began making recommendations for land acquisitions along the shores of more than twenty lakes and for beginning the process of lake restoration through dredging or by constructing dams to raise water levels. One of these proposed acquisitions, a large area adjacent to Rice Lake in Winnebago County, was envisioned as a wildlife sanctuary.[19] These activities led to additional amendments in 1921 that legally resolved the lake issue. The board was given clear authority over "all meandered lakes and streams . . . and state lands bordering thereon." This authority included the right to establish state parks on "bordering" lands.[20] In essence, the 1921 amendments eliminated the remaining legal obstacle to implementing lake improvements that had been recommended several years earlier in the 1915–1917 Highway Commission study.

Further amendments in 1923 increased the board from three to five members with staggered three-year terms, an important structural change that gave the board a measure of continuity from year to year. In addition, the board was authorized to call upon the State Highway Commission for general engineering assistance.[21] The latter provision gave the board greater control over road improvements in parks. M. L. Hutton, the highway engineer assigned to work with the board, proved to be a decent choice. Among other things, he understood the aesthetic considerations of building roads and other improvements in state parks. He also provided technical knowledge and expertise that board members lacked, expertise that was invaluable when proposals for power dams began to occupy so much of the board's time. By the late 1920s, the board considered Hutton indispensable. In 1929, he left the highway commission when the board offered him a job as the first superintendent of parks, a position he held until his death in 1941.[22]

As of 1923, the state legislature fully recognized that state parks had become a permanent public institution and that authority over them should be vested in one body, the Board of Conservation. By law, however, all board resolutions, expenditures, and land acquisition recommendations still had to be approved by the Executive Council. Legislative changes from 1919 through 1923 also pointed toward the eventual emergence of a comprehensive resource conservation agency, which is what many conservationists had hoped for since the 1911 report of the Drainage, Waterways and Conservation Commission. But it was not to be in the 1920s. After a rocky start, the Board of Conservation and the Fish and Game Department cooperated in a number of ways: to establish fish hatcheries in state parks, to improve the quality of lakes, and to establish wildlife refuges. Throughout the 1920s, their functions and activities increasingly overlapped, and in 1930 Albert was given authority to deputize park custodians in order to enforce fish and game laws more efficiently.[23] Nonetheless, the two agencies remained separate. In the area of forestry, Pammel and G. B. MacDonald sought a legislative mandate to establish a forestry program under the Board of Conservation, but by the end of the decade that goal, too, remained largely unrealized. Likewise, even though the legislature had given the board jurisdiction over lakes and rivers, it vested authority over water quality more firmly in the Department of Health. By 1930, the Board of Conservation certainly was the leading resource agency, but jurisdiction over resource policy was far from centralized. Chapter 3 will treat the politics of environmental policy during the 1920s in greater detail.

What the Board of Conservation lacked in centralized authority, Pammel attempted to strengthen through personal commitment. His position as chair gave him some, though not unlimited, influence over future board appointments. Scattered correspondence indicates that he worked to secure board members who held conservation values similar to his own and who also had the means to devote the long hours of unpaid work that board responsibilities demanded. He was not always successful, as when he sought to include more scientists like himself on the board, and when he tried to persuade Governor Kendall that a board made up of representatives from the University of Iowa, Iowa State College, the State Horticulture Department, and the Executive Council would "bring the various organizations into harmony."[24] Pammel also attempted to coordinate conservation initiatives by bringing the fish and game warden on to the board. While there was widespread sentiment to merge the two entities, W. E. Albert did not favor a seat on the board, even in an ex officio capacity. Possibly, he saw potential conflict of interest, that is, participating in the formation of policy that he would have to carry out as a state employee. More likely, he was sensitive to divided points of view within his own constituency. In any event, he dissuaded Pammel from pursuing this board arrangement.[25] Albert, however, routinely attended Board of Conservation meetings until the Fish and Game Commission was established in 1931.

Pammel did succeed in importing to the board conservation allies from the Iowa Federation of Women's Clubs. May McNider, chair of the IFWC's conservation committee and former vice president of the Iowa Conservation Association, was appointed to the board in 1922. After the board was legally expanded to five members in 1923, Governor Kendall appointed Mary C. Armstrong of Fort Dodge, also active in the IFWC, to fill one of the new seats. When May McNider decided to give up her seat in 1926, the IFWC backed Margo Frankel of Des Moines as her replacement. Four other women sat on the board from 1929 until it was reorganized as the State Conservation Commission in 1935: Mabel Volland (1929–1931), Mrs. C. C. Laffer of Sigourney (1931–1933), Grace Gilbert King (1932–1934), and Helen Taylor of Bloomfield (part of 1935). None of them, however, assumed a leadership position that equaled Frankel's.

Others who served during the 1920s included Euclid Saunders, a banker from Emmetsburg and former president of the Iowa Conservation Association; Clifford Niles of Anamosa, who was instrumental in securing convict labor to make certain park improvements; State Senator Willis G. Haskell, a well-known businessman from Cedar Rapids, who served as a member of the standing committees on appropriations, highways, mines and mining, public utilities, and railroads; George Wyth of Cedar Falls, president of Viking Pump Company and the driving force behind Cedar Falls's city park system; and former State Senator Byron Newberry of Strawberry Point, who had supported the state parks bill in 1917 and was instrumental in establishing the first state park, Backbone, in Delaware County. Thus, while Pammel did not achieve the scholarly scientific weight on the board that he sought, he successfully engineered a membership that was, for the most part, sympathetic to his way of thinking. Besides, his imposing presence more than made up for whatever the board lacked in academic credentials.

The Parks Mission

As long as Pammel was at the helm, the Board of Conservation defined parks broadly as preserves for rare, unusual, or threatened natural resources and for cultural antiquities. Reflecting on the intent of the state park law some years later, he wrote: "The persons who framed the law had in mind the preservation of animals, rare plants, unique trees, some unique geological formations, the preservation of the Indian mounds, rare old buildings where Iowa history was made. . . . The framers of this law wished to show generations yet unborn what Iowa had in the way of prairie, valley, lake, and river." [26] Noticeably missing from this statement was any mention of soil conservation and forestry, an omission worth noting since Pammel directed much of his research toward science and agriculture, and because he, along with many others, was concerned about soil erosion, which resulted not only from farming practices but from the immense loss of forest and prairie cover and the drainage of sloughs, marshes, and bogs for agricultural purposes.

Horticulturists, some of whom farmed, supported certain aspects of resource conservation, notably forestry. The Iowa Park and Forestry Association, however, had never succeeded in attracting ordinary farmers to the conservation movement. There were, correspondingly, decided limits on what the Board of Conservation could hope to accomplish. Pammel, who had reasonably good ties to those who promoted scientific agricultural practices, would have been the person best situated and best suited to attempt some liaison with agricultural leaders. However, even he seems to have had difficulty seeing any way to integrate agricultural concerns into the conservation movement, or vice versa. At most, he thought the state should "work out a constructive program [of] bringing all the divergent interests together," but he offered no specific plan for doing so. [27] Whatever chance there might have been for yoking scientific agriculture to conservation, admittedly small at best, Pammel would not attempt to bring interests quite so "divergent" together on the Board of Conservation. Pammel's way of dealing with the competing goals of agriculture and conservation was to avoid stepping on farmers' toes as much as possible. One way to do this was to avoid creating parks on land that had agricultural potential. For the most part this was an unwritten rule. Only once did Pammel openly admit that his tacit strategy was to establish parks on land that had no agricultural value. "We want only waste land that can present interesting facts in geology, botany, and history out of respect of the pioneers and the race that preceded us" he wrote in a 1921 report to the board.[28]

Also missing from Pammel's statement was any mention of parks for recreational purposes, although the Board of Conservation faced this issue from the very beginning. By mid-1919 Pammel was advising that "the Board plans to have three systems of parks; state parks, in regions of special recreational, historic and scientific interest to preserve for future generations some of the things which made Iowa great in its history and science; . . . highway parks, . . . to be located in every county as funds are made available for this purpose; . . . [and] lake parks, . . . wooded tracts on the shore[s] of the larger lakes." [29] Although there is no indication in the board minutes that any formal discussion of a three-tiered park system ever took place, Pammel restated this plan in essentially the same words a year later at the 1920 dedication ceremony for Backbone State Park.[30] Already he was beginning to distinguish between what Macbride would, in 1922, call "conservation parks" and those other places where the public could engage in social and recreational activities.

Pammel elaborated on his ideas for highway parks, the second category, at another early meeting of the board. In his opening remarks to a group that included several guests who were there to present petitions on March 8, 1919, Pammel announced that "we," meaning the board, "propose to have at least one county park in every county in the state, and in some cases perhaps two or three, and hope that the areas will not be less than fifteen acres in each case. It is the hope of the Board," he continued, "that the counties or communities will, themselves, take care of the erection of buildings. The management of parks, etc., will be under state supervision." [31] In the past, the term "county park" had been used simply to designate places outside municipal boundaries that were desirable for public access. The growing popularity of the automobile, however, could not be ignored. As Pammel used the term, "county parks" often meant "highway parks."

Few states, Iowa included, were prepared for the tremendous burst of tourism that began in the wake of World War I and increased rapidly during the 1920s. Several factors combined to create the new middle-class tourist boom, including changing social values that placed greater emphasis on individual

freedom, rising incomes for a good portion of the American public, the availability of automobiles at successively lower prices, and the corresponding
movement to build good roads on which to drive them. Neither the Board of
Conservation nor its major supporting organizations actively enlisted the aid
of "good roads" advocates in Iowa, as happened elsewhere.[32] In Washington
state, for instance, the Natural Parks Association teamed up with the Auto
Club and other organized groups to bolster support for stronger state park
legislation, which came in 1919. The Natural Parks Association also received
support from Stephen Mather, chief of the new National Park Service, who
was, at that very time, promoting a federal highway to link all the national
parks in the West. To build support for his effort, Mather spent considerable
time in the Pacific Northwest, "encouraging scenic preservation efforts at the
state and local levels."[33] By 1920, automobile travel and parks were inextricably
linked in the public's imagination, and Iowa, like many other states, was in the
midst of a highway frenzy.[34]

One certainly cannot deny that inexpensive and easily available outdoor
recreation became the chief attraction of state parks during the 1920s. Pammel,
who never learned to drive an automobile himself, nonetheless accepted the
obvious. Early in 1919, he advocated placing a state highway commissioner on
the Board of Conservation "because we must have good roads to have successful parks."[35] He may have seen small parks in rural areas, yet adjacent to
highways, as a way to accommodate tourists and, at the same time, keep the
larger, implicitly more important, parks relatively uncluttered with campgrounds and other tourist facilities. In any case, he clearly envisioned "highway parks," or "county parks," as distinct from "state parks," in terms of their
size, frequency, location, method of acquisition, and use. The State Highway
Commission, in turn, urged the Board of Conservation to do something about
lake access. Early in 1919, highway commissioner R. W. Clyde appeared before
the board asking that it take the first step to acquire land along lake shores for
public parks; the Highway Commission, he said, would provide access roads
from public highways.[36] Automobiles, improved highways, and lake access
were the handmaids of outdoor recreation and tourism. Thus, within limits
the board would work to initiate scientific forestry, to preserve native flora, to
restore fish and wildlife habitats, and to protect lakes and waterways.

Pammel's early statements about the park system's three-tiered organization also conveyed a greater sense of coherence to the board's work than actually prevailed. At the initial meeting on December 27, 1918, all agreed that
the first order of business was to begin acquiring land, but of course the board
had no real land acquisition policy, only the mandate contained in the 1917
law. Although the minutes record that "matters of general policy" were "dis-

cussed at some length," there was greater appetite for acquiring the Devil's Backbone region in Delaware County as soon as practicable. Even as the political problems of getting the Board of Conservation launched were being worked out during much of 1917 and 1918, informal negotiations were underway to secure this area—long coveted by natural scientists—as a state park.

Thomas Macbride first visited the Backbone in 1864, so named for a long, narrow limestone ridge along the Maquoketa River, and in subsequent years he returned as often as he could. He also urged neighboring farmers to keep their cattle out of the area.[37] The unglaciated cliffs and river valley contained species of flora quite unlike those of the surrounding prairie. In particular, the area held one of the few remaining stands of native white pines in the state. Geologist WJ McGee investigated the region between 1876 and 1881 and described its features in his *Pleistocene History of Northeastern Iowa*. Botanist Samuel Calvin first called public attention to its potential park value in 1896, noting that "the beauty, the seclusion, that attractiveness of the place, are certain to be appreciated more and more as the years go by, provided short-sighted, unaesthetic avarice does not transform its forest lands into pastures, or does not attempt to 'improve' it for the sake of converting it into a profitable summer resort. If it can only be let alone, it will remain a source of purest pleasure."[38]

Backbone State Park dedication, May 28, 1920. A "great concourse of people" assembled near the park superintendent's cottage, according to a lengthy article in *Iowa Conservation*. *Courtesy State Historical Society of Iowa, Iowa City.*

By 1918, more than two decades had passed since Calvin had hinted that Backbone might be threatened by development. Time, therefore, was of the essence. What transpired at the December 1918 meeting indicated that policy, land acquisition or otherwise, would evolve from rather than guide the board's work. In its desire to take action, the board resolution to acquire the Devil's Backbone was represented as "the general policy which [the board] regarded as essential for the proper development of the state's resources."[39] The resolution, however, contained little that could be construed as general policy. By resolving to purchase "not less than twelve hundred (1,200) acres," the board implied that state parks would be relatively large tracts of land. By noting that the area had been "thoroughly examined by members of the Board," the resolution implied that personal examination of proposed parks would be required before the board would recommend land acquisition. These, however, were little more than vague guidelines. The clearest statement of policy embedded in the resolution said nothing about what types of land would or would not be acquired, or what procedures would be followed. Rather, it recommended that all land acquisitions be dedicated as "Iowa Memorial Parks" out of deference to the men and women who served their country in World War I. Such "policy" was little more than a high-minded gesture of the moment that would promptly be forgotten.

Board members took a more serious stab at policy a few days later. At E. R. Harlan's suggestion, the board agreed to publish a circular stating that it would "search for" and recommend acquisition of lands suited "for public gathering places such as reunions, celebrations and picnics or to commemorate any worthy person or historic event"; areas where the natural history of the state could be studied; areas "for camping, hunting, fishing, bathing," and other such recreational pursuits; and areas suitable for "the preservation or propagation of species of wild, native animals and plants otherwise rare or in danger of extermination." Recreation clearly ranked higher in Harlan's scheme of the park world than it did in Pammel's. Perhaps sensing which direction the board was drifting, Pammel immediately took the additional, and critical, step of moving that the board prepare "a list of places now thought suitable" for acquisition.[40]

"The list," then, would serve as a substitute for any articulated statement of land acquisition policy. Working from research notes and reports issued over the years by Bohumil Shimek, Thomas Macbride, James Lees, Rose Schuster Taylor, May McNider, Eugene Secor, and others, the board compiled a target list of ninety-eight areas, reasonably distributed across the state, that were desirable for acquisition.[41] When the report was published, there was no mention of recreation in the title: *Iowa Parks: Conservation of Iowa Historic, Scenic and Scientific Areas*, which carried an unwieldy subtitle, "Suggested by Responsible Citizens as Suitable for Public Park Purposes and So Regarded by the Board of Conservation, From Which Selections Will Be Made, Not Yet Acquired for the Want of Appropriate Conditions Found or Created." Nor was recreation mentioned as the prime reason for placing any of the ninety-eight areas on the list. Nor were there many historic places on the list, although there had long been sentiment to preserve Fort Atkinson and Camp McClellan. In any case, the places that were officially published as desirable for state parks were lakeshores, wooded creeks, ice caves, limestone ledges, Indian mounds, fossil beds, bluffs, ravines, knobs and kettles—in short, the places that best revealed Iowa's geology and natural history.[42]

Wartime exigencies delayed public issue of the report until 1920, by which time the state had acquired, or was in the process of acquiring approximately 3,500 acres of land in nine different locations. The list of ninety-eight proposed acquisitions, therefore, did not include these areas. Considering subsequent actions, it is unlikely that some of them would ever have been considered except that several individuals and organizations immediately expressed interest in donating land, generally in small parcels, such as a few acres of native timber in Hardin County and the former chautauqua grounds in Oakland. These early acquisitions, or proposed acquisitions, nonetheless revealed high public interest in the state parks program. They also foretold some of the difficulties the board would face trying to work with specific members of that vast, great public that state parks were to benefit, and to figure out just what a state park should be. The "public-spirited" donor of fifteen acres along the Shell Rock River in Floyd County, for instance, also donated "reasonable rules recognizing and differentiating Sunday as the one day on which pastimes and performances of all sorts shall be in harmony with the mental attitude of devout people," a condition the board appears to have accepted without question at the time. Private gain, not public morality, surfaced as an issue in Pottawattamie County, where the owners of land adjacent to the Oakland Chautauqua Association grounds were holding out for "an exorbitant price." Since the chautauqua grounds had been donated on the condition that the state acquire additional acreage in order to create a roadside park, the future of this proposed park appeared dim, although the board would not admit so in its 1919 report.[43]

Of the nine initial properties entrusted to the Board of Conservation, six are still part of the state system, although the state does not manage all of them. Backbone State Park in Delaware County, Lacey-Keosauqua State Park near Keosauqua, Wildcat Den State Park in Muscatine County, and the preserve known as Woodman Hollow near Dolliver Memorial State Park are owned and managed by the state. Lepley State Park in Hardin County and Oakland Mills State Park in Henry County are managed by county conservation boards. The other three either never were established as state parks or have been transferred out of the system. Indian Lake County Park is owned by the City of Farmington, the City of Oakland owns Chautauqua Park, and the acreage along the Shell Rock River in Floyd County reverted to private ownership. Various circumstances contributed to the weeding-out process, but size and location, as well as the nature of local support, often determined what would or would not remain in the state system.

On balance, though, the 1919 report laid the foundation for much of the board's work over the next decade. A tone of urgency pervades its pages. Having spent two decades getting into a position where action was possible, conservationists would not delay. The haste to preserve whatever remained of forests, waters, prairies, natural scenery, and wildlife before farmers, tourists, and land developers made the job impossible, accounts, in part, for the lack of attention given to defining the ground rules. In so doing, park advocates also tended to gloss over the inherent conflict between using public parks for both preservation and recreation. Instead, they placed great faith in nature's spontaneous regenerative processes.[44] The Board of Conservation, for instance, predicted that in state parks "nature will reassert herself and recreate or reproduce some of the stately forests that formerly fringed our river banks and lake shores and dotted over our hills." When that happened, which it was assumed might take twenty-five years, "bird, animal and plant life there will return much of the early wild life that has gradually disappeared from the wooded hills and vales of Iowa." However, in the next paragraph the board could assert with apparent equanimity that "our wooded river banks and lake shores are ideal for recreational parks as well as valuable for study of natural history, forest research, geology and propagation of wild life, and furnishing splendid fields for the students of plant life also."[45]

It was the beginning of a new decade, a new era, and the years ahead seemed ripe with possibilities. Pammel refused to be fettered by fears of where it all might end. The political challenges, the financial hurdles, the conflicts of interest, the philosophical differences would be faced as they arose. Above all else, Pammel took a practical approach to the task of creating a state park system. Preservation of scientific, historic, and scenic resources would be the guiding principle, but he would bend that principle a little here, a little there, to keep the goal intact.

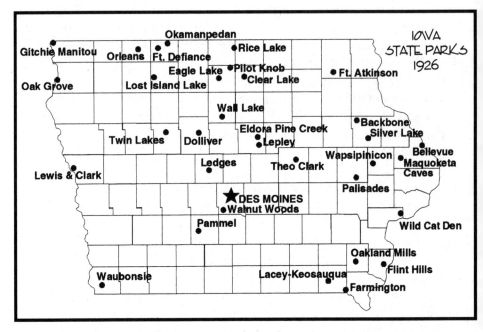

Iowa state parks as of 1926. Based on information in *Bulletin, Iowa State Parks*, January–February 1926. *Courtesy Iowa Department of Natural Resources.*

Compromising the Ideal: Lacey-Keosauqua State Park

During the 1920s, Thomas Macbride's lofty vision of "conservation parks" underwent considerable transformation. To some of his academic colleagues, at least, Macbride's vision must have seemed unrealistic even at the time. Nonetheless, many conservationists, particularly the academic scientists, held great hope that state parks would be the remedy for a host of nature resource problems. Ideally, state parks would preserve indigenous plant species, provide sanctuaries for wildlife, showcase important historical places, enable the board to initiate a reforestation program, and give the state an incentive to restore the water quality of lakes and rivers. The evolution of Lacey-Keosauqua State Park exemplifies the degree to which public demand for recreation forced the Board of Conservation to compromise this ideal.

Lacey-Keosauqua was one of the first parks to be acquired, in part, with local assistance. Board secretary E. R. Harlan, a native of Van Buren County, initially worked with a small group of local men who were successful in raising $6,400 toward land acquisition, a sum that paid for 160 of the approximately 1,200 acres in the original park, acquired in several parcels throughout 1919 and 1920.[46] Named in honor of Major John F. Lacey, the park comprised sev-

eral prior farmsteads located southwest of Keosauqua in what was known as
the "great horseshoe bend country" along the Des Moines River. Its hilly terrain supported a variety and abundance of trees, wildflowers, shrubs, and
wildlife. In addition, a string of American Indian mounds and a prehistoric
village site lay near the river. A spot known as Ely's Ford marked the place
where pioneers and early settlers crossed the river.

Lacey-Keosauqua also was one of two state parks dedicated in 1920; the
other was Backbone State Park in northeast Iowa. The latter was purchased
entirely with state funds. However, because it worked with a limited budget,
the board encouraged cities, counties, and private citizens to assist with the
purchase and, to a certain degree, the development of state parks. Lacey-
Keosauqua provided a different model for developing parks under a state-
local partnership. These two parks shared some important qualities. They
were the largest state parks in the system. In addition, both contained areas
highly desirable for their botanical variety and their scenic qualities. The
method of funding land acquisition nonetheless set them apart from the very
beginning. Backbone came with no strings attached. Conversely, the board
would spend the better part of the 1920s detaching the strings that came with
the $6,400 in local subscriptions for Lacey-Keosauqua. Most important for
Lacey-Keosauqua's future, certain members of the Executive Council in office
during 1919–1920 apparently agreed, though not in writing, "that suitable
tracts within the park be reserved for a public golf course and summer home
sites."[47]

Lacey-Keosauqua's dedication, a mixture of conviviality and formality, provided additional hints of the compromises that were in store as the realities of
the partnership approach were revealed during the next several years. The
two-day affair began on the afternoon of October 26, 1920, with a "discussion"
on the question of whether pheasants and foxes could "thrive together" in the
park. The ultimate answer to this (entirely rhetorical) question was provided
in the form of a fox hunt, the purpose of which was to clear such "predators"
from the new park. With prohibition newly in force, members of the "fox
hunting fraternity" were reminded to refrain from imbibing intoxicating bev-
erages at this public event, but the opening was nonetheless a lively event. In
contrast, the Board of Conservation planned a sedate inspirational and educational program for the following day. A guided tour through the park occupied the morning hours, with various professors speaking about the plant
and animal life, geology, archaeology, and scenery. ("Absolute quiet will be
enjoined upon assemblage during the addresses of these eminent leaders" the
program read.) After a picnic lunch, the crowd gathered for an "inspirational
hour" with Thomas Macbride presiding. Governor William L. Harding then

formally accepted the park, and the Reverend Charles S. Medbury of Des Moines closed the day with a sermon.[48]

These were the two sides of the public dedication: local organizers focusing on recreation, the board on natural resources. Moreover, while professors were holding forth on the natural attributes of the area, the Executive Council and a local group of park "trustees" were beginning to conduct park administrative matters informally. On the day of the dedication, E. H. Hoyt, state treasurer and a member of the Executive Council, gave the local group permission to purchase a team of horses and equipment for maintaining a road through the park. As events transpired, the Executive Council never did consult the Board of Conservation about establishing a local oversight committee, nor did the council bother to make formal appointments. For the most part, both the local committee and the Executive Council simply ignored the Board of Conservation and conducted park business with verbal agreements. The $500 cost of the team and road equipment was paid from state park funds without a bill ever passing by the Board of Conservation. To care for the team, the "trustees" then authorized the park custodian, Herb Rees, also a local man, to plant a portion of the park with hay and grain crops.[49]

Park administration continued in this manner for about a year. When a group of Masons examined the park as a spot for their annual picnic, they turned it down, initially, because "there was no place cleaned off where they could get to the river, no swings, nor furnaces." Not wishing to disappoint or offend the Masons, H. E. Blackledge, a realtor and head of the local "trustees," organized a work crew, which included park custodian Herb Rees. They cleared a picnic area, built swings and teeter boards, and installed a firebox for outdoor cooking. The cost of the materials, about $200, was paid for locally with private funds. Later, State Treasurer Hoyt authorized more land cleared for picnic areas, the construction of additional fireboxes as well as picnic tables and benches. Under ambiguous authority, a well was drilled, fences were removed, and a new park road was constructed. Some of this work was paid for privately, some from the state park fund.[50] None of these "improvements" were cleared in advance by the Board of Conservation, and the board learned about them only after the fact. Legally, of course, the board was only an advisory body to the Executive Council, so Hoyt generally was acting within the letter of the law. Practically speaking, though, the informality and the lack of courtesy extended to the board led to considerable confusion.

More important, the lack of coordination precluded any possibility of undertaking park improvements based on scientific studies and approved development plans. When a new administration entered office in 1921, the Board of

Conservation seized on the opportunity to clarify the lines of authority with the new Executive Council. After several conferences, both bodies agreed that park administration would rest with the Board of Conservation, subject to approval by the council. Then fell to the board the task of determining where it stood with respect to the administration of individual parks. E. R. Harlan, the board member assigned to oversee Lacey-Keosauqua, initiated a series of interviews that enlightened the board not only about park administration but about local views concerning the value and use of state parks.[51]

Harlan interviewed Blackledge and the other two "trustees," druggist J. Henry Strickling and banker S. W. Manning. The transcripts read like a comedy of manners, neither side wishing to render personal offense but both intent on making their positions clear. At certain points, though, courtesy gave way to plain speaking. "The local committee and I should say nine out of ten people who contributed the $6400.00," J. Henry Strickling finally offered bluntly, "having started this park knew absolutely nothing of the scientific feature[s] of this park and know comparatively little of [its] condition; they care but little for the historical, but they are ardently interested in the recreational phases." With regard to recreational improvements in the park, Strickling informed Harlan that the local committee had been "deluged with complaints and suggestions." Strickling made it clear that the local committee considered the picnic grounds, play equipment, and fireboxes as both "necessary" and "well earned" by those who had donated money for park acquisition. If Harlan was taken aback by these revelations, he nonetheless managed to remain polite. In reply, he assured Strickling that he "appreciate[d] the justice" of his words and "the truth of the relative possibilities of the recreational as compared with the scenic, scientific and historical elements of a state park." Harlan even went so far as to predict that board members John Ford and Joseph Kelso would agree with Strickling on the need to develop parks for recreational use, "while Dr. Pammel will concede the point subject to the appropriate care of objects and areas of scientific interest."[52]

If the Board of Conservation ever was in doubt about the disparity of vision concerning the future of state parks, the Lacey-Keosauqua interviews provided unequivocal clarification. With the question of administrative authority settled, the board chose, however, not to stir already troubled waters by requesting the removal of improvements already in place. Things remained quiet for the next year, until the board learned that a local American Legion post had cleared yet another area for picnic and recreational use. This time, Pammel investigated.[53] In the fall of 1922 and spring of 1923, he traveled to Keosauqua more than once in order to determine the state of affairs and to formulate recommendations pertaining to future development. Three reports Pammel filed on Lacey-Keosauqua not only codified his thinking about the degree and type of development that should take place there but contributed more generally to the process of formulating board policies concerning recreational development use in all state parks.

Pammel was particularly irked by the verbal agreement to allow private cottages in the park, an issue that was not confined to Lacey-Keosauqua. Local residents near Eldora Pine Creek State Park were then in the process of subdividing private land adjacent to the park and making plans to develop a landscaped, citylike park between the proposed summer residential tract and an artificial lake inside the state park. On private land adjacent to Backbone State Park, several people had built "shacks" in order to take advantage of a well, a picnic area, and a campground located just inside the west entrance to the park. Even though such encroachment threatened the resources of any park, the board had no legal authority to control land use outside park boundaries. Development within parks was another matter. Not only was the board's administrative authority clear, Pammel asserted that the board had an *obligation* to ensure that all people had equal rights within state parks. He therefore recommended "emphatically" that no summer homes be permitted in Lacey-Keosauqua, regardless of promises made extralegally in the past.[54] Although the board took no immediate action on this recommendation, in mid-1926 it adopted a blanket policy of not permitting private cottages in any state park.[55]

The development of group camps and lodges posed different considerations, since the line between public and private use was not always so clear. Pammel, like many middle-class reformers of his day, subscribed to a belief in the efficacy of organized youth groups, particularly the Boy Scouts and the Camp Fire Girls, to guard against the decline of American civilization. "Indeed," he wrote, "there is no better method of offsetting this calamity [the degeneration of civilization] than by the establishment of various group camps, where the boys and girls may learn and benefit through the true value of outdoor recreation." Without considering that group camps inherently gave park-use preference to certain members of the public, Pammel identified specific areas inside the park he thought suitable for organized youth camps.[56] The board would vacillate for several years on this issue. On the one hand, park advocates genuinely felt it was important to instill a sense of stewardship in young, impressionable minds. What better way to do this than through organized youth camps and outdoor nature education in state parks. On the other hand, the obligation to provide equal public access was quite real. The board could hardly grant Boy Scouts and the Camp Fire Girls special-use privileges one day and deny permission to an American Legion post for a dedicated picnic ground the next. For the next five years the board debated this issue back and forth, finally adopting a policy, in 1928, that disallowed further organized group facilities in all state parks.[57]

The golf course at Lacey-Keosauqua State Park Pammel conceded, since he had no other choice. He did, however, insist that it be "strictly for public use," that the "confines of the course area . . . be specifically designated," and that "no additional land be granted in the future for this project."[58] From Pammel's perspective, the golf course set an unfortunate precedent; and, indeed, by the end of the decade, public golf courses were constructed in three more parks: Bellevue State Park in Jackson County, Flint Hills State Park north of Burlington, and Wapsipinicon State Park near Anamosa. Private golf courses also were located adjacent to Eldora Pine Creek State Park in Hardin County and Rice Lake State Park in Winnebago County. Pammel took the golf course in stride as a necessary political trade and set out to see that its placement and design would at least be part of a general development plan.

Before any park planning studies could be completed, the unofficial local "trustees" initiated, in 1924, another effort to "improve" Lacey-Keosauqua. Claiming that they had been "promised some game a long while ago," H. E. Blackledge apparently arranged through political connections to request "a pair of deer, some pheasants, and some pheasant eggs" from Game Warden W. E. Albert. The Board of Conservation learned about the request only after Albert pointed out that the board would have to approve it first. Blackledge then contacted Pammel, but he did not request the board to introduce game into the park; rather, he requested that the board supply enough materials to fence a ten-acre deer park.[59] Blackledge clearly assumed that there could be no objection to placing deer and pheasants in the state park; he therefore saw this as a simple request that involved only the use of a small number of acres, some materials, and the ongoing provision of food for deer.

Opinion among board members about fencing wild animals in state parks proved to be divided and passionate. Wm. E. G. Saunders of Emmetsburg, a former state representative who joined the Board of Conservation in 1921, presented the reality-of-politics argument. "We must get some sympathy from somewhere for the Conservation Board," he cautioned Pammel, an oblique reference to political maneuvers that were behind a 1923 cut in the board's appropriation. "You cannot get votes in the House of Representatives for these parks unless you are willing to grant the people some recreation in connection with them. . . . If the Conservation Board takes the position of opposing recreational spots, they might just as well make up their minds that the appropriation will be stopped, and to be perfectly frank, I do not think it is unreasonable for the people of the State of Iowa to expect ten acres out of fourteen hundred to be devoted for recreational purposes for the children." [60]

E. R. Harlan, who retired from the board in 1923, but whom Pammel contacted out of courtesy, presented the "parks-as-sanctuaries" argument. He stated "again" that he was "first and last opposed to introducing animal and plant life that is not and never was part of the natural history of that region." Noting that southeast Iowa was the "quail center" of the state, Harlan wondered what would happen if pheasants proved to be "obnoxious to the Ruffed Grouse or to the Bob White." With respect to deer, Harlan noted that both deer and elk once had inhabited the region but reintroduction would require restraint. Small petting zoos he found abhorrent; the animals either escaped, were turned loose after the novelty had worn off, or became "pitiably unrepresentative of their species and habitats." The alternative was to fence the entire park so the deer could range freely. "But that would be to invite such as I to hunt them at will, to risk my safety in their mating season, and annoy them in yeaning time. It would mean foot-prints . . . where isolation sacred to your plants and other Divine things should always [be]." [61]

Pammel's solution to irreconcilable points of view was to engage his son, Harold, newly graduated from Iowa State College with a degree in landscape architecture, to study the plants, animals, and history of the area and to design a park development plan. To avoid any possible political controversy, Pammel sidestepped the Executive Council by seeking outside sources to pay for Harold's services. To the park's local supporters, Pammel stressed that the plan would be designed "from a recreational standpoint." To others, he stressed that it would be "worked up from the standpoint of conservation." To everyone, he stressed repeatedly that Harold's plan would provide a "model" for all state parks in Iowa. [62] Indeed, the board was then in the process of working out arrangements with Iowa State College for landscape architecture services on an ongoing basis, and in July 1924 the board adopted a formal policy calling for all landscape work in state parks to be based on topographic surveys and studies of indigenous species. [63]

Lacey-Keosauqua's "trustees," who could always find money for the improvements local citizens wanted, respectfully declined to help pay for the park development plan. Pammel ended up paying his son a modest fee of $100 for the plan; E. R. Harlan later picked up some of the expenses. [64] Whatever the source, it was money well spent. Harold proved to be the balm that soothed irritable relations. Moreover, his plan represented the first attempt to balance local demands for a golf course and deer park with the board's intent to protect the Indian mounds and to establish plant preserves, wildlife sanctuaries, forest reserves, an experimental orchard, and youth camps. More important for the park itself, however, was the custodian's unanticipated leave of absence. For three months in late 1924 and early 1925, Harold filled in as temporary park custodian while Herb Rees was off touring national parks. During this time, he was able to supervise some of the initial work and to restore good will. When he left at the end of February 1925, over fifty people signed a letter of "keen appreciation" and petitioned the board for his continued service. By that time, however, Harold was preparing to continue his studies at Syracuse University. Rees returned to his post with a new perspective on parks. His tour of the West and South exposed him to the need for forest conservation and reforestation programs, an experience that gave him greater appreciation for the goals of the Board of Conservation.[65]

Harold Pammel's interlude in the park marked a turning point in relations between local citizens and the Board of Conservation. Thereafter, a spirit of cooperation took hold, although there were still many problems to be addressed. The golf course was laid out by May 1925, and plans were underway to establish a tent camp for Boy Scouts and to construct a shelter house.[66] Pammel arranged for the Extension Service to take over some old orchards for experimental work in horticulture.[67] He also began working with Rees to

Lacey-Keosauqua State Park, golf course, lodge (*right, background*), and old farm building, c. 1927. *Courtesy Ralph Arnold.*

establish an arboretum in the park, and he arranged to have wild turkeys rein-troduced.[68] To solidify the new spirit of goodwill, the Board of Conservation staged a special conservation program in the park on September 10, 1925. The one-day event featured a morning field trip and an afternoon of guest speakers that included native son E. R. Harlan, Raymond Torrey from the National Conference of State Parks, and professors from several colleges.[69]

Plans to construct park buildings and improve park roads took longer to realize because the Executive Council repeatedly failed to approve board-recommended expenditures. In lieu of constructing new buildings, the board endeavored to have several existing farmhouses relocated and remodeled for use as a clubhouse and other visitor facilities. Even this proved difficult to achieve with limited funds, so the board negotiated for convict labor to carry out road improvement and building projects.[70] Most of this work was done in 1926 and 1927. During the summer of 1926, the board dedicated a huge boulder containing a bronze plaque commemorating the conservation work of the park's namesake, John F. Lacey.[71] Later that year, shortly before he retired from the board, Pammel worked with G. B. MacDonald to begin reforestation work in the park.[72]

By the end of the decade, Lacey-Keosauqua was one of the board's most highly prized state parks. Here, through much effort, all interests had been accommodated, even if not satisfied. Lacey-Keosauqua had historic sites, areas for scientific study, spectacular vistas along the Des Moines River, and abun-dant recreational facilities. The 1931 report of the Board of Conservation de-scribed Lacey-Keosauqua as "one of the finest real conservation areas in the state," noting that "the greater portion of the park lies in a solid block of flower and shrub filled woods, never entered by the casual visitor."[73] There were miles of hiking trails over tree-covered hills, along river bluffs, by the Indian mounds, all connecting campgrounds and picnic areas. There was a lodge available for public use. And, of course, there was the nine-hole golf course with its club house. There were, however, no private summer homes in Lacey-Keosauqua or in any other state park. Nor was there a deer park.[74] Lacey-Keosauqua might not have exemplified Thomas Macbride's ideal of a "conservation park," but conservation interests had been served.

The Park Mission as Realized

Despite political pressure to bend toward more and greater recreational use, the Board of Conservation moved boldly under Pammel's guidance. It is doubtful whether any other person could have accomplished so much. Pam-mel's name, if not the man himself, was well known among academic scien-tists, horticulturists, sporthunters, farmers, businessmen, newspaper editors, clubwomen, and especially politicians. He cultivated friendships and contacts in every quarter. Respect for his opinions ran deep, and no other conservation leader in Iowa held as much power to influence legislative policy in the post–World War I era. He was, in many ways, a rarity: an academic scientist who also understood the business of politics, that is, how to accumulate and trade political "capital." Certainly he made mistakes, and he lost some key political skirmishes. Equally important, ardent conservationists sometimes felt that he sometimes traded away his principles in the political process. In his search for practical solutions, however, Pammel never lost sight of the goal.

In the late 1920s, when he took stock of what had been accomplished since 1895, Pammel recalled that the framers of the State Park Act "felt that a part of this heritage left to us was not only for the present generation, but that [Iowa's] citizens of the future had a just claim on this heritage."[75] It was this vision that sustained not only him but a host of park advocates. Like most visions, though, it was a bit vague, a condition that allowed everyone to keep moving forward despite disagreements and disappointments. When the Devil's Backbone area of Madison County was renamed and dedicated as Pammel State Park in 1930, Margo Frankel orchestrated a program that honored the man as much as the park. William Trelease, Pammel's mentor of many years past, traveled from Illinois to surprise the guest of honor; and Jay N. Darling, who would carry the torch for conservation in the 1930s, gave the keynote address.[76] The event was a fitting and timely capstone to Pammel's career. Less than a year later, he died. In a lengthy resolution commemorating his service, the Board of Conservation cited Pammel's work "for the cause of conservation" as "the most valuable single influence in this movement" in the State of Iowa.[77]

Although the board was still far from being a comprehensive resource conservation agency in 1931, it nonetheless had made remarkable progress in establishing a state park system. When the National Conference on State Parks made its first sweep through the country to assess the status of state park development in 1925, Iowa ranked fourth in the nation in terms of numbers. Only New York, Michigan, and Texas had established more state parks.[78] At that time, forty-three states were engaged in acquiring and developing state parks; by the late 1920s, forty-five states had some form of park or recreation system, and Iowa was still helping to set the pace.[79] In 1931, the Board of Conservation issued its second major report. In slightly more than a decade, forty properties had come under its jurisdiction, and they were divided into two categories: parks and wildlife preserves, although there were only three of the latter. The total included six of the original nine areas acquired or under consideration in 1919. Eleven other parks could be called "lakeshore parks," representing as they did the first steps toward acquiring public access to lakes and rivers. By including the lakes and streams that were placed under the board's authority in 1921, the total state park "system" comprised 7,500 acres of park area, nearly 41,000 acres of lake waters, 800 miles of rivers, and 4,200 acres of drained lake beds.[80]

The degree to which the park system reflected what the board set out to create can be measured in general terms by comparing the "wish list" of 1919 with the 1931 report of accomplishments (table 3). A rough analysis reveals that about forty of the parks or preserves, as of 1931, had been listed in the 1919 report. Another measure of achievement is to consider those parks and preserves that were acquired for their scientific value; that is, they possessed geological features, rare plants, stands of native timber, or other natural features that scientists wished to preserve. While it is true that the board often found something of scientific interest to justify each land acquisition, only fifteen of the forty parks and preserves appear to have been acquired because they held significant value for natural scientists. Eleven others, however, were acquired chiefly to provide lake or river access, another important goal the Board of Conservation sought. By this measure, the board's success ratio was slightly better.

Although no formal categories had been established to distinguish the types of areas that had been accepted into the park system, by 1931 the Board of Conservation was calling them "areas of historic, scientific, scenic, or *recreational* value" [author's emphasis]. At the time, recreation generally meant hunting, fishing, camping, and swimming, although public demand forced an expansion of the definition to include golf. Eventually the golf courses would go, but they surely signaled a growing acceptance of outdoor recreation in parks. In addition to golf courses, lakeshore and river access parks also opened the door to recreational use. Even so, under Pammel's direction, the Board of Conservation would limit recreational potential by limiting the construction of facilities.

State parks as of 1931 also were spread very unevenly across the state. Arrayed on a geologic map of Iowa, they revealed a bias for land acquisition in certain landform regions: the Des Moines Lobe of the Wisconsin Drift, the Paleozoic Plateau (i.e., the driftless area in northeast Iowa), and the eastern reaches of the Southern Iowa Drift Plain. From the standpoint of scientific value, these landform regions contained areas of greater interest to students of natural science, but the resulting pattern of land acquisition meant that vast stretches of the public were nowhere near a state park.[81] This imbalance became an issue as visitation increased. During the 1923 season, an estimated 232,000 people visited Iowa's state parks, representing practically every county,

Table 3.

Status of State Park System, 1931 [82]

Park/Preserve as of 1931	1919 List (Y/N) or "Original Nine"	Primary Reason(s) for Acquisition	Present Status o=owned
Ambrose A. Call	N	scenic	state park
Backbone	original nine	scientific, scenic	state park
Barkley Memorial Preserve	N (donated)	scientific	state forest
Bellevue	Y	scenic	state park
Bixby	Y	scientific, scenic	state preserve
Theodore F. Clark	N	scenic	o-Tama Co.
Clear Lake SP	Y	lake access	state park
Dolliver Memorial	N	scientific, scenic	state park
Eagle Lake	Y	lake access	state park
Elbert Tract (Walnut Woods)	N	river access	state park
Eldora Pine Creek	N	river access, recreation	state park
Flint Hills	N	recreation, scenic	o-City of Burlington
Ft. Atkinson	N	historic	state preserve
Ft. Defiance	N	historic, scenic	state park
Gitchie Manitou	Y	scientific	state preserve
King (Springbrook)	N	scenic	state park
Lacey-Keosauqua	original nine	scientific, scenic, rec.	state park
Ledges	Y	scientific, scenic	state park
Lepley	original nine (donated)	scenic	state park
Lewis and Clark	Y	lake access, recreation	state park
Lost Island	N	lake access	o-Palo Alto Co.
Maquoketa Caves	Y	scientific, scenic	state park
Oak Grove	N	scenic	state park
Oakland Mills	original nine	scenic	state park
Okamanpedan	Y	scientific, scenic	state park
Palisades-Kepler	Y	scientific, scenic	state park
Pammel	Y	scientific, scenic	state park
Pillsbury Point	Y	lake access	state park
Pilot Knob	Y	scientific	state preserve
Rice Lake	Y	lake access, recreation	state park

(continued)

Table 3. (Continued)

Park/Preserve as of 1931	1919 List (Y/N) or "Original Nine"	Primary Reason(s) for Acquisition	Present Status o=owned
Silver Lake	N	recreation	o-Delaware County
Storm Lake	Y	lake access	o-City of Storm Lake
Twin Lakes	Y	lake access	state park
Wall Lake	Y	lake access	wildlife area
Wapsipinicon	N	scenic, recreation	state park
Waubonsie	N	scenic	state park
Wildcat Den SP	original nine	scientific, historic	state park
Woodman Hollow Preserve	original nine	scientific	state preserve
Woodthrush Preserve	N (donated)	scientific	preserve, o-City of Fairfield

over thirty states, and several foreign countries. Thereafter, the number of visitors steadily climbed. Visitor tabulations also became more reliable as park staff began to establish common procedures. Rough calculations showed that the number of visitors escalated from about 1.5 million in 1928 to 1.8 million in 1930.[83] During the 1930s, user demand for outdoor recreation would become the driving force in park development and expansion, both on the national and state levels. The availability of federal funds through various New Deal programs enabled state park agencies to meet that demand. In Iowa, the vision of 1917 would be challenged anew.

3

Reshaping Park and Conservation Goals in the 1920s

> In the creation of state parks, recreation is important but we should not lose sight of the fact that conservation, too, is fundamental. Our modern life has brought so many changed conditions that conservation must be urged more strongly than it has been in the past. — LOUIS PAMMEL[1]

Centralized versus Decentralized Authority

Establishing state parks was only part of the story in the 1920s. This was a decade of struggle to expand the park mission into a set of state resource policies. The Board of Conservation carried a broad mandate that reached far beyond creating public parks. It had the authority to seek out places valuable for their archaeological or geological features, to establish forest reserves and promote forestry, and to preserve native animal and plant life as well. Several competing forces determined the degree to which the board could meet this mandate. Public demand for outdoor recreation was of course a major force, resulting in a trend toward emphasizing the aesthetic aspects of parks vis-à-vis their resource preservation and conservation values. However, this was not the only force that caused Pammel and others to bend their ideas about the ideal park system. Fiscal constraints also played a large role in the decision-making process; more often than not, available funding determined whether the board could acquire desired lands or embark on programs designed to preserve or restore resources. In addition, the complex history of the conservation and parks movement in Iowa during the 1920s reveals much about the way political ideology reshaped environmental values. Throughout the decade there was constant tension between forces pushing the state toward consolidating authority over resource policy and opposing forces inclined to hold the status quo.

The degree to which party politics shaped environmental policy in Iowa is less clear because Iowa was so thoroughly dominated by the Republican party in that era. On the national level, it has been argued, the Republican "ascendancy" of the 1920s, with its focus on voluntary cooperation rather than government regulation, changed the character of conservation policy. Many Progressive Era conservation organizations claiming a nationwide constituency fell by the wayside during the late 1910s, and uncompromising positions staked out by key national figures made it difficult to reestablish and maintain broad-based organized support for conservation measures. Without a cohesive, outside force to apply political pressure during the 1920s, agency heads often determined whether the federal bureaucracy advanced the conservation agenda or placed obstacles in its path.[2] One thing is clear: the making and implementation of conservation policy at the national level was anything but centralized. A host of agencies located in three separate departments of government—Interior, Commerce, and Agriculture—carried various charges with respect to protecting or managing particular resources. Moreover, working at cross-purposes, more often than not, was the Army Corps of Engineers, located in the War Department. During the 1920s, certain agencies of the federal bureaucracy, notably the National Park Service, the Forest Service, the Bureau of Reclamation, the Bureau of Biological Survey, and the Bureau of Fisheries, carried out their duties independently. With resource conservation agencies split among departments of government, there was no real mechanism for coordinating policies.

In contrast, Iowa's Board of Conservation answered directly to the governor and the Executive Council. Its place in the organization of governance made the board vulnerable to gubernatorial preferences and politics, but it also created a real possibility of centralizing authority over resource protection and management. The tension between two opposing forces, centralized versus decentralized authority, is a theme that runs throughout the history of Iowa's state park system in the 1920s. It would be wrong to assume, though, that key personalities consistently articulated either one pole or the other. This was not the case. No one person campaigned vigorously for a centralized resource agency, although Pammel and others worked behind the scenes trying to consolidate and expand the power of the Board of Conservation. Rather, the tension can be seen as part of the political process itself. In this respect, the environmental politics and policies woven through the board's history provide evidence that Iowa's progressive political tradition remained reasonably strong, though not unchallenged or unchanged, throughout the 1920s. The subtleties were played out in three areas of special concern: park aesthetics, including historic sites; forestry; and water resources.

The Conservation Constituency

Importantly, the Board of Conservation always saw itself as upholding values that were above special, or "selfish," interests. Increasingly throughout the 1920s, the "public" interests championed by the Board of Conservation were those expressed through certain organizations. Three organizations in particular defined the conservation constituency to which the board responded: the Iowa Conservation Association, the Iowa Academy of Science, and the Iowa Federation of Women's Clubs. Other conservation organizations gained importance during the 1920s, notably the Iowa Division of the Izaak Walton League, organized in 1923 as the first statewide affiliate.[3] Numerous bird clubs and sportsmen's organizations maintained local interest in wildlife, though not necessarily in wildlife conservation. None of these organizations, however, had the same level of influence with the Board of Conservation as did the ICA, the IAS, and the IFWC. The activities and positions of these three groups, therefore, are important in order to understand the shifting of resource conservation goals that took place in and around the Board of Conservation.

The degree to which national park issues and debates influenced thinking on the state level is hard to determine, since discussions rarely were framed in such a context. Certainly, though, many conservationists in Iowa followed national events. Likewise, the 1919 state park report, outlining as it did such an ambitious agenda, focused nationwide attention on Iowa, prompting the Department of the Interior to select Iowa as the location for the first national conference on state parks. E. R. Harlan worked with Secretary of the Interior John Barton Payne and National Park Service Director Stephen T. Mather to organize the conference, which convened on January 10, 1921, at Des Moines. Approximately 200 delegates from twenty-four states and the District of Columbia attended the three-day event to discuss issues of common importance in the growing state park movement. Out of the Des Moines meeting came a formal organization, named after the convening body, the National Conference on State Parks.[4] Casting the spotlight on Iowa, even for this brief period, no doubt raised Iowans' awareness of events and issues beyond their own state's borders, but sustained awareness depended upon strong statewide organizations.

Iowa Conservation Association

For several years after passage of the State Park Act, the Iowa Conservation Association maintained a flurry of activity. Annual meetings were well attended, and they provided a forum for the usual presentations meant to inform as well as uplift. More important, meetings were a nexus for coordinat-

ing action on various issues since many ICA members also were active in the
IFWC and the IAS. Recent legislative actions were reviewed, missives to the state
legislature were composed, resolutions were passed, and legislative proposals
were drafted. State senators and assemblyman often attended, as did the heads
of various state agencies, especially the Fish and Game Department and the
State Highway Commission.

Academic scientists continued to provide much of the leadership within
the ICA. Scientists dominated the Advisory Council and rotated through the
ranks of officers. The same names appeared year after year: Pammel, G. B.
MacDonald, and R. A. Pearson from Iowa State; Bohumil Shimek, George
Kay, Thomas Macbride, and Robert Wylie from the University of Iowa; Henry
Conard from Grinnell; T. C. Stephens from Morningside; Charles Keyes from
Cornell; and James Lees from the Iowa Geological Survey. The bond between
clubwomen and the ICA also remained strong, chiefly through the efforts of
May McNider, who served as an officer for several years, and through Cora
Whitley, who often spoke at annual meetings. In 1917, the ICA initiated a new
quarterly, *Iowa Conservation*, which kept members abreast of pending legisla-
tion, carried news about local conservation clubs as well as the activities of the
IFWC, profiled the organizations' leaders, and reported on recent field studies.

Until the mid-1920s, the ICA was Iowa's chief conservation organization.
The ICA organized support for a proposed Mississippi Valley National Park
and, when that idea faded, campaigned to help pass the 1924 federal law establishing the Upper Mississippi River Wildlife and Fish Refuge. Working
through the ICA, G. B. MacDonald helped to mobilize support for the 1924
Clarke-McNary Act, which carried some small benefit for forestry in Iowa. On
the state level, the ICA promoted closer ties between sportsmen and conservationists through its annual meetings, through the pages of its newsletter, and
through its support of the American School of Wild Life Protection, a field
school held each summer at McGregor beginning in 1919. Rev. George Bennett, founder of the field school, also edited *Iowa Conservation*. The ICA also
backed any number of initiatives to advance the work of the Board of Conservation. In 1921, for instance, it urged the state legislature to adopt a forestry
policy and implement a reforestation program through state parks. The ICA
also urged the Board of Conservation to acquire abandoned schoolhouse
grounds and vacated roadways for wayside parks, and to acquire historic sites,
particularly Indian mounds.[5]

Despite its strong comeback between 1915 and 1917, however, the ICA never
did find a permanent, sustaining membership. Part of the reason may have
been that the leadership was too narrowly held to attract that wide cross-
section of members everyone so desired. It simply didn't happen. Early in 1921

the secretary, G. B. MacDonald, reported that paid memberships stood at 425, down from 804 a year earlier.[6] Declining membership placed funding of the quarterly newsletter on an increasingly precarious basis, especially considering that the organization already was in chronic debt for publication of past issues. The reason for this, in turn, seems to have been Rev. Bennett, the ICA's field representative in charge of membership building and newsletter publishing. As founder of the American School of Wild Life Protection, Bennett enjoyed a spot among the inner circle of conservationists in Iowa, but he was not a young man when he took on the responsibilities of ICA field representative. His age may or may not have had any bearing on his performance, but in any case, he was much more interested in putting out a good-looking newsletter than he was in scouting the field for new members. As things turned out, an attractive newsletter was not enough. ICA secretary G. B. MacDonald tried to keep the magazine financially solvent, but two more years of membership decline made that impossible. The executive officers finally voted to suspend publication and retire the debt.[7]

Iowa Conservation officially ceased publication with the last issue of 1923. Its passing marked the loss of an important link with members, although the gap was filled by the new *Bulletin: Iowa State Parks*, which commenced publication that same year. Since Pammel edited the *Bulletin*, one can assume that he probably did not mourn the passing of *Iowa Conservation*. Suspending its publication not only relieved the ICA of a longstanding financial burden, but it gave Pammel an opportunity to focus the messages that went out to the conservation constituency. Nonetheless, *Iowa Conservation* was the one tangible connection that ICA members had with their organization. Without it, the organization foundered. The ICA remained active for a few more years, holding its regular annual meeting in 1924 and sponsoring a short-course in cooperation with Iowa State College in 1925.[8] Sporadic references to the organization after 1925 indicate that the ICA remained a recognizable identity in conservation affairs until about 1930, but the organization had no public visibility.

The *Bulletin*, published by the Board of Conservation between 1923 and 1927, contained information about board activities, reports on studies of proposed park areas and state park dedications, descriptions of the flora and outstanding geological features to be seen in state parks, news of related conservation activities such as the continuing effort to establish a national park in the Upper Mississippi River Valley, and articles on a variety of conservation-related topics. As editor, Pammel emphasized the importance of conservation vis-à-vis recreation mainly by failing to mention the latter. This was in keeping with his normal diplomatic approach to the work of the board, which often produced carefully composed statements and a tendency to circumnavigate controversy until consensus could be reached.

During these four years, the *Bulletin* served as the principal means of keeping the public informed about the board's progress in establishing state parks. The Iowa Federation of Women's Clubs, for instance, urged its members to use the *Bulletin* in developing local programs. Publication, however, demanded more time and money than it was feasible to give, especially after Pammel suffered his first heart attack in 1925. Nonetheless, he carried on as best he could until his retirement from the board in 1927, when the *Bulletin* ceased. Its demise severed the last direct link between the board and its constituency.[9] In 1928, Bennett tried to resurrect *Iowa Conservation* under the name of *Wildways*. The first issue opened with a message "To Members of the Iowa Conservation Association," indicating that it still functioned in some capacity.[10] Bennett's own death in 1928, however, effectively put an end to *Wildways* after just two issues.

Iowa Academy of Science

Passage of the 1917 State Park Act renewed academic interest in the conservation movement, and members of the Iowa Academy of Science stayed abreast of conservation issues throughout the 1920s. Membership in the IAS also grew substantially, a marked contrast to the Iowa Conservation Association. In 1919, the IAS, with 350 members, was the largest statewide professional organization of scientists; a year later the IAS was among the first state-level organizations in the country to affiliate formally with the American Association for the Advancement of Science. By 1928, IAS membership had grown to more than 600.[11] Considering its size, the IAS might have been a powerful force for conservation in the 1920s. As an organization, however, its support was of a different order, detached from the realities of politics or economics. In addition, resource conservation was but one interest among many. As the sciences became more specialized, the IAS became more sophisticated in its operation, its membership more diversified. Members met annually to share information about and discuss topics in botany, chemistry, physics, geology, zoology, mathematics, psychology, bacteriology, and archaeology.[12]

Those academic scientists who addressed conservation topics tended to focus their attention on water resource issues: the quality of drinking water, overdrainage of low-lying lands for agricultural use, and stream pollution. A notable exception was geologist James H. Lees, who in 1917 urged preserving "in a series of . . . parks" several geologic "phenomena" from the northeast counties comprising the "Switzerland of Iowa," as Samuel Calvin called it, to the loess bluffs along the Missouri River.[13] More typical was Samuel W. Beyer's

1919 presidential address, in which he called attention to studies indicating that groundwater levels in Iowa had dropped more than twelve feet between 1860 and 1910. In order to assure the long-term viability of agriculture in the state, he stressed the need for a hydrometric survey, which he believed "would demonstrate the wisdom of preserving large tracts of land in their natural state." [14] Five years later, Bohumil Shimek tried once again to rouse interest in a statewide investigation of groundwater resources, believing it would expose the fallacies "of the present reckless system which operates on the groundless assumption that *all* drainage is beneficial." [15] In the wake of a disastrous fish kill in the Shell Rock River caused by industrial pollution, George Bennett called upon the state to clean up lakes and streams in order to protect wildlife. [16] After A. H. Wieters assumed the post of state sanitary engineer in 1926, he made a courtesy speech before the academy to assure members that the department was making headway in dealing with widespread stream pollution insofar as treatment was "economically feasible." His qualifying remark probably evoked a disheartened response, especially when Wieters went on to note that at least 200 small communities as well as the state's largest sixteen cities were still dumping untreated sewage into Iowa's waterways. [17]

While the IAS provided an important, if relatively closed, forum for debating resource problems and issues, a cadre of members—including Pammel, G. B. MacDonald, T. C. Stephens, Bohumil Shimek, and Henry Conard, among others—also used the organization as an umbrella to advance specific policy recommendations. In 1920, for instance, the IAS recommended that natural lakes and streams should be preserved, that plant and game preserves should be established, that hunting laws should be strengthened, and that the state should adopt legislation to preserve Indian mounds and burial grounds. [18] In 1924, the IAS asked the state to enact legislation giving the Board of Conservation power to review and approve or deny all proposals to drain lakes and swamps or to change the course of rivers. [19] That same year, the Biological Survey Committee called upon the state legislature to fund a natural history survey of the state for the purpose of gathering data pertinent to the conservation of fish and game resources as well as native plant species. Working through Louis Pammel, the IAS succeeded in getting an appropriations bill to fund such a survey. [20] The legislature failed to act on the request the first time, in 1925; four years later, however, it appropriated $6,000 for the Iowa Geological Survey to undertake biological survey work. [21] The IAS also encouraged its members to speak on behalf of conservation in the public schools, a cooperative effort involving the State Fish and Game Department, the Board of Conservation, the State Teachers' Association, and the Iowa Federation of Women's Clubs. [22]

While Louis Pammel was at the helm of the Board of Conservation, the IAS enjoyed direct access to policy-making channels. As a measure of the respect IAS members accorded Pammel and his efforts on behalf of conservation, they elected him president for a second time in 1923; he was the only member to hold that office twice. Nonetheless, the Board of Conservation often fell short of meeting the expectations of the scientific community. Scientists, more than any other group, favored centralized jurisdiction over resource matters. In addition, scientists looked askance at the increasing recreational use of state parks. After Pammel retired from the Board of Conservation, the IAS made sure the board continued to hear its concerns. In 1927, the IAS Committee on Conservation urged the Board of Conservation to set aside areas to "serve as sanctuaries for the remnants of our native plant and animal life for scientific and general conservation purposes, as was originally intended." In order to acquire more preserves, the IAS further urged the board to discontinue "making so-called improvements, and particularly the building of expensive highways, the sole purpose of which is to attract crowds." [23] Two years later, the IAS and the Iowa Conservation Association issued a joint statement to the Board of Conservation asking it to take stronger action "in regard to preserving the natural plant and animal life in the state parks." [24]

It is difficult to know whether such pressure influenced board policy to any appreciable degree from 1927 on, since Pammel had tried, with some success, to pack the board with people who had a genuine commitment to maintaining the spirit and intent of state park law. The late 1920s saw the board opposing power dams in rivers just as strongly as it had under Pammel. The board also took the lead in pressing for water pollution control measures. Certainly, though, missives from the IAS let the Board of Conservation know that its actions were being monitored. It is therefore possible that the IAS played some part in board decisions to allow no further Boy Scout buildings in state parks; to set aside Woodman Hollow as the first preserve; to prohibit private concessions near state park entrances; to begin working with the Fish and Game Department to establish wildlife refuges at Wall Lake in Sac County (now Black Hawk Lake), Palisades State Park, and Ledges State Park; to investigate the "desirability" of wildlife refuges in all state parks; to impose fees and time limits on the use of campgrounds and lodges; and to impose rules and regulations for the operation of power boats on lakes and streams. [25]

Iowa Federation of Women's Clubs

Support from the Iowa Federation of Women's Clubs proved to be equally durable. May McNider, Mary C. Armstrong, Margo Frankel, and Louise Parker gave the Iowa Federation of Women's Clubs a continuous voice on the Board of Conservation from 1922 to the late 1940s. When Louis Pammel acknowledged the contributions of women to the conservation movement in Iowa, which he did often, it was not empty praise. The letters that Pammel wrote to May McNider and Mary Armstrong during the 1920s reflect a good measure of respect for their opinions, their knowledge, and their abilities. Certainly, he also valued the influence that clubwomen wielded in local communities, or in political circles, for all of them were married to prosperous business and professional men. Despite such connections, however, the women of the board did not engage in the face-to-face political negotiating that became a necessary part of board operations.

Actually, Pammel handled most of the statehouse lobbying himself, and he corresponded regularly with May McNider, as he did with other board members, to keep her apprised of legislative matters, She did not hesitate to make suggestions concerning proposed bills, and he kept her up to date on the progress of important legislation. The 1923 bill revising the code to expand the board reveals much about the policy role that clubwomen played through the Board of Conservation. Introduced at the behest of the board, Pammel desired the change not only because it would give that body more continuity but because it also would "give the Governor a chance to appoint several women on the board. This it seems to me is important," he continued. "Of all the people in the state to have taken [an] interest in the park movement the women have been the most energetic and interested." [26] Indeed, after the bill passed, the governor appointed a second woman, Mary C. Armstrong, to the board. When Armstrong's first term ended, Pammel went out of his way to persuade her to remain on the board for a second term.

In terms of political involvement, women had earned a place on the board. In addition, both McNider and Armstrong earned Pammel's respect because they shared his values; that is to say, they followed "Pammel's way" easily. When Margo Frankel and J. G. Wyth joined the board early in 1927, just after Pammel retired, Mary Armstrong offered Pammel an early assessment of the change: "Mrs. Frankel takes hold of the work with great enthusiasm and Mr. Wyth seems much interested but the flower bed idea looms high in his mind as a means of *Beautifying* the parks. I think he has been in consultation with Mr. Fitzsimmons [consulting landscape architect] and also with Mr. Hutton [consulting engineer] and I am sure they will both help him to a better point of view." [27] Iowa may not have been the only state in which women were included in making conservation policy, but there is no question that, in Iowa, their influence extended into the policy realm. [28] In contrast to the pattern that has been observed on the national level, Iowa women were given a place within the power structure that debated, created, and implemented state conservation policy.[29]

Cora Whitley had made conservation a top priority issue during her tenure as president of the Iowa Federation, 1915–1917; and, although she was not among those women who served on the Board of Conservation, she remained active in conservation circles until her death in 1937. Her lasting influence was of a different nature, but it was equally important. During the 1920s, she took her own voice to the national level. From 1920 to 1924 she chaired the General Federation's Conservation of Natural Resources Committee. She succeeded Mrs. John Dickinson Sherman, who, through her well-orchestrated campaign to support passage of the 1916 National Parks bill, had made the Conservation Committee one of the strongest GFWC committees. It remained a strong, active committee under Whitley's direction. After 1924, Whitley chaired the Forestry and Wildlife Refuge Committee. In these capacities, she acted as a spokeswoman for the General Federation at appropriate national conferences. In 1924, for instance, she represented the federation at the Izaak Walton League's national convention, where, in the course of events, she was elected to serve on its board of directors (Bohumil Shimek also was elected to the board that year). In 1926, she spoke at the annual meeting of the American Forestry Association.[30]

As part of her work with the General Federation, she oversaw the preparation of programs and study outlines for dissemination to local clubs. This was typical of the way the General Federation promoted social activism through its complex organizational structure. Various departments and committees developed continuing education materials for use at the local level. Education was not the end goal, though. Women were encouraged to use this knowledge in their local communities, often to take direct action aimed at influencing public opinion at the state and local levels. Whitley took the local education responsibility quite seriously. She worked with the U.S. Forest Service and with state forestry associations in order to provide clubwomen with up-to-date information, thereby increasing the chance for effective grassroots work. Women were encouraged to study the issues (bibliographies were provided), write letters supporting or opposing proposed legislation, sponsor tree planting projects, organize field trips, or write newspaper editorials—activities that could be carried out effectively at the community level.

Whitley not only helped to prepare educational materials for clubwomen, she also demonstrated how to use them to best advantage. She appeared before congressional committees. In February 1923, for instance, she traveled to Washington, D.C., to testify in support of establishing a federal wildlife refuge in the Upper Mississippi River Valley, an effort supported by Iowa clubwomen as well as the Izaak Walton League and the Iowa Conservation Association.[31] She spoke at the annual meetings of the Iowa Conservation Association.[32] She wrote articles for the IFWC's quarterly newsletter, for *Iowa Conservation*, and for the *Bulletin*. In 1927, she used the occasion of Forest Week to deliver a radio address to citizens of her own state on the value of woodlands for soil, water, and wildlife conservation.[33]

When she took over as chair of the forestry committee, she made sure that foresters understood the value of having the GFWC as an ally. She reminded them that, since the 1890s, clubwomen had undertaken many reforestation efforts, supported and opposed legislation, and launched countless educational initiatives. Women from Connecticut to Kentucky, from Mississippi to Minnesota, from Colorado to California had worked on behalf of forestry for three decades. Disputing the notion that clubwomen were only sentimental nature lovers, she asserted that "women have not shown their interest, as has sometimes been imagined, simply by expressing their love for trees or by planting memorial avenues; they have tried to study the question from the utilitarian, the economic, as well as the esthetic stand-point."[34]

All of these activities were important to Whitley, but nothing else commanded more of her attention or her enthusiasm than the Outdoor Good Manners campaign she conceived in 1925 and promoted through the GFWC Forestry and Wildlife Refuge Committee. "Each year's output of new automobiles means new throngs of tourists or campers," she wrote to clubwomen across the country. "The devastation of natural beauty which they often leave to mark their presence is disheartening."[35] Whitley's Outdoor Good Manners campaign brought domestic hospitality and housekeeping fundamentals to Mother Nature. It was conservation with a decidedly feminine twist. The object was to teach children, especially, "that to abuse the hospitality of parks and forests is just as really a breach of good manners as would be such conduct in a home where one has been treated kindly."[36]

With the backing of the U.S. Forest Service, the American Forestry Association, the American Nature Association, the Boy Scouts, the Camp Fire Girls, the Izaak Walton League, the Parent Teacher Association, and superintendents of public instruction across the country, Whitley and the GFWC launched a massive effort to get the Outdoor Good Manners message into schoolrooms, libraries, and Sunday schools. Three million clubwomen were urged to give classroom talks, develop library programs, sponsor poster and essay contests, show films, and promote Forest Week. Whitley personally carried the message to clubwomen in fifteen different states. In her home state, she offered a prize to the fifth, sixth, or seventh grader who designed the best poster illustrating

If We Treated Our Homes As We Do Our Woods

What A Lucky Thing Folks Never Took To Holding Picnics
In Other Folk's Houses

Moral: Take Your Indoor Manners With You When You Go Outdoors

Jay N. "Ding" Darling cartoon for Outdoor Good Manners Campaign. *Courtesy J. N. "Ding" Darling Foundation.*

Outdoor Good Manners, or the lack thereof.[37] To aid women in their local efforts, the GFWC printed posters and palm cards bearing the Outdoor Good Manners message urging young and old alike "to leave the woods and parks as beautiful as you find them."[38]

Editorial cartoonist Jay N. Darling lent assistance with a two-frame drawing urging not just children but the public at large to "Take Your Indoor Manners With You When You Go Out Doors." Harrison Cody of Life Publishing Company, "Kettner" of the Western Newspaper Union, and Dornan H. Smith of the National Education Association joined Darling to expose ersatz "nature lovers" as thoughtless "nature vandals." The *Philadelphia Inquirer* and *Saturday Evening Post* ran cartoons.[39] Secretary of Agriculture Wm. M. Jardine commended her "interest in the forest problems that confront us in the United States."[40] Whitley's press releases were picked up by major newspapers. The *Christian Science Monitor,* for instance, ran her picture alongside an article entitled "Etiquette of the Outdoors Invoked Throughout Nation." Her homestate newspaper, the *Des Moines Register,* also backed the campaign, as did WHO radio station.[41]

The General Federation of Women's Clubs and its statewide affiliates supported many causes in the pre–World War II era, providing an outlet for educated, middle-class women who wanted to be involved in everything from education reform, public health, and social welfare to civic affairs and international relations to fine arts and gardening. Whitley was among those who carried the GFWC's banner for conservation. The national program she helped to develop was infused with her personal ideas and values, but it also was shaped by broader trends in the conservation movement as a whole. Most notably, the Outdoor Good Manner campaign reflected important changes in the U.S. Forest Service, which was one of the GFWC's principal allies in the educational campaign. Under the direction of William B. Greeley, who became U.S. Forester in 1920, the Forest Service promoted fire prevention rather than strict regulation of timber cutting as a means to conserve trees. Fire prevention was Greeley's main policy goal, but he also sought a policy of federal-state cooperation in order to expand forest reserves, research, and forestry education.[42] The chief means of promoting volunteerism was through education and persuasion, and the General Federation of Women's Clubs was perfectly organized to deliver public service programs. It was, therefore, no coincidence that the Outdoor Good Manners message also incorporated the Forest Ranger's Rule: "Always leave a clean camp and a dead fire."

The degree to which the Outdoor Good Manners educational campaign influenced tourists, campers, and hikers to douse their campfires, carry out their litter, and refrain from trampling wildflowers is, of course, impossible to measure. That battle has proved to be never ending. More to the point, though, the Outdoor Good Manners campaign reflected the aesthetic dimension of the conservation movement turning sour. The "back to nature" movement blossomed fully in the wake of World War I, encouraged by the National Park Service's "See America First" campaign and aided by increasingly affordable automobiles and "good roads" on which to drive them. But too many "nature lovers" who toured parks were urbanites who liked the *idea* of nature more than the reality of wilderness. "Nature," to many tourists, meant spectacular scenery and little else. They had little understanding of or respect for the actual flora and fauna, little skill in outdoor living, and little interest in camping etiquette. By 1920, the campgrounds of Yellowstone, Yosemite, and Mt. Rainier were littered; the streams, polluted.[43]

What was true of national parks was also true of state parks. "Motoring, from the standpoint of the conservationist," she wrote, "has been like the ancient goddess [*sic*] Siva, both the destroyer and life-giver; the automobile carries the crowds to the distant solitudes where formerly wild life, birds, flowers, blossoming shrubs were comparatively safe." Yet Whitley believed that a campaign of education, not visitor restrictions, could turn things around. "[T]he great increase of tourists on business or pleasure bent, has given a new impetus to an interest in natural scenery, in the beauty, or the barren monotony of the country side as well as in the quality of the roads."[44] If people could practice good manners at home, they could learn to do the same in parks.

Although the women of the Board of Conservation received far less publicity than Cora Whitley for their efforts on behalf of conservation at the state level, they were no less dedicated. Their views, like Whitley's, often reflected the trend toward emphasizing park aesthetics. For instance, women took a greater interest in park building design and pressed for landscape architects to be involved in park development. Women also were among the first to join Pammel in opposing commercial enterprises in parks.

Policy by Design and by Default

In varying ways, the men and women of the Iowa Conservation Association, the Iowa Academy of Science, and the Iowa Federation of Women's Clubs sought to shape resource policies through the state park system. They served to articulate positions, advance ideas, and critique failure, which is to say that they participated in reshaping conservation goals during the 1920s. In the process, certain philosophical differences were sharpened and political realities emerged. Park aesthetics and amenities, forestry, and water resources were issues that commanded much attention. By the end of the decade, the debates over these issues would reveal that state parks provided only limited means for addressing multiple resource concerns.

Park Aesthetics, Amenities, and Historic Places

"It was our early policy to devote our efforts to the acquisition of park areas in the state," Pammel wrote to E. R. Harlan at the close of the board's first year.[45] Park improvements, however, could not be long ignored. For one thing, park visitors began demanding roads, picnic areas, and camping grounds. For another, there were requests to introduce all manner of exotic plants and animals into park areas. And there was a constant stream of requests for privately owned lodges, cabins, and concessions. Those requests that came from private individuals were easy to turn down, but the board had a harder time saying no to the Boy Scouts, church groups, and war veterans. In the early years, the board also had to worry about losing control over park development to the Executive Council, which proceeded to appoint local park trustees and approve building projects without consulting the board. Local communities and individuals who had donated money or land to establish state parks often sought a *quid pro quo*, and the Executive Council was more responsive than the board to political pressure.

Pammel recognized that park improvements would have to come sooner or later, but he tried to delay the construction of roads and buildings until the board could frame some general guidelines. "I agree fully that the State Board of Conservation should, in the future, provide proper buildings in the larger parks in the State," he confided to Harlan. But roads, he felt, should not be constructed without benefit of topographical surveys and buildings should not be erected willy-nilly. While this was a prudent approach, Pammel nonetheless had a limited notion of what improvements would satisfy public demand. The public buildings he envisioned in parks were something akin to visitor centers that would "contain 'in a nut shell' the scientific aspects of the region, botanical, plants, rocks and soils."[46] Visitors and local patrons, however, would demand much more than this.

Park landscaping came up for discussion shortly after the board got down to work. At its 1920 annual meeting, the ICA urged the Board of Conservation adopt a "policy of natural parks as opposed to landscaped parks," noting "that the primary purpose of State Parks is the preservation of certain areas in unmodified condition."[47] This formal recommendation coincided with informal discussions then taking place concerning the need for a consulting landscape architect. After the board received a number of unsolicited inquiries from professional firms offering their services, Pammel quietly let it be known that the board would not squander public funds on expensive professional fees.[48] Writing to A. T. Erwin, a colleague at Iowa State, Pammel stated that the Board of Conservation did not want the "ordinary kind of landscaping" but wanted all state parks to "remain as nature left them with simple driveways through the premises."[49]

Despite these early exchanges, the board took no immediate steps to adopt a policy on park landscaping. By 1923, however, there was need to take action. In March, the board invited several park custodians to attend a meeting for the purpose of discovering from those in the field what services and amenities would make "their" parks more useful to the public. Roads, shelters, latrines, fences, and garbage disposal ranked high on the list of priorities from the custodians' perspective.[50] Realizing that park improvements must proceed in an orderly fashion, May McNider, who joined the board in 1922, approached Francis A. Robinson of the firm Pearse-Robinson, with offices in Des Moines, Chicago, and St. Louis, about providing landscape architectural services. She also urged the board to consider his offer.[51] Engaging an outside expert was not an unsatisfactory arrangement from a professional standpoint, but the board, conscious of budgetary priorities, felt it would be more advantageous to have the landscape architecture department at Iowa State College provide these services.[52] College administrators and faculty, however, were not as anxious to be so burdened. They expressed sympathy for the board's position, but refused to take on added responsibilities without some formal directive, although E. A. Piester did provide limited assistance for road design. In 1923, he reviewed plans for a road through Backbone State Park and oversaw construction of a road through Pilot Knob State Park.[53]

While the board worked to overcome whatever objections the landscape architecture department had to providing professional services, Pammel enlisted his own children, Lois and Harold, to lay some groundwork. In 1923, Lois, working with J. J. Beard, designed a "General Plan" for Theodore F. Clark State Park, near Traer. Harold, who had just graduated from Iowa State with a degree in landscape architecture, was first assigned to prepare a report on the "recreational resources" of Wapsipinicon State Park.[54] After that, he reviewed plans for a trout hatchery in Backbone State Park.[55] A year later, in June 1924, Pammel engaged his son to design the plan for developing Lacey-Keosauqua that was to be a model for subsequent park plans.[56]

Finally, in early 1924 the state legislature authorized the Board of Conservation to call upon Iowa State for landscape architectural services, which were to be rendered gratuitously, with the board paying expenses only. Again, some members of the faculty fussed about the directive, but by midyear the department had agreed to assign faculty members on a rotating basis to work with the board.[57] With landscape architectural services assured, Pammel drafted a formal policy on "Landscape Work in State Parks" which the board adopted in July 1924. It called for landscape plans, as well as land acquisitions, to be based on topographical surveys. Landscape architects were to plot overgrazed, eroded, or otherwise denuded spots "and then with the assistance of [a] botanist (ecologist) . . . replant these areas with plants suitable for such areas." Suitable plants meant "native plants." In parks where there was evidence of native "boreal or northern plants," these were to be restored. In parks where there were "prairies" and "dry ridges with sparse growth," these plant communities were to be maintained. The placement and design of roads, campsites, buildings, markers, and the like would proceed from preliminary surveys and the development of landscape plans.[58] The language reflected not only Pammel's ideas about what improvements were appropriate, but the ICA's recommendations of four years earlier urging resource preservation be given higher priority than aesthetic values. The new policy statement also reflected recommendations contained in four reports filed by landscape architect Henry F. Kenney. After studying the vegetation of Lost Island Lake, Eagle Lake, Theodore F. Clark, and Eldora Pine Creek state parks, Kenney recommended that certain areas be set aside as plant and animal reserves and that camping be allowed only in designated areas.[59]

The landscape architecture policy fit with the board's go-slow approach to park development. Its adoption also marked the beginning of new construction in state parks. In 1925, for instance, the board finally approved construction of several buildings in Backbone State Park: the trout hatchery, remodeling of a stone barn for use as a custodian's residence, a new barn, entrance pillars, and a shelter house, all to be built by Anamosa prison inmates. Other projects approved in 1925 included a shelter house at Pilot Knob State Park.[60] During the next several years, the Iowa State landscape architecture department turned out a variety of studies and building plans for state parks. Among those assigned to the board, John R. Fitzsimmons appears to have developed the closest working relationship. Many of the buildings constructed between 1925 and 1931 were of his design.[61] Fitzsimmons, in turn, used state parks as a laboratory for training future landscape architects. Under his direction, students designed lodges and shelter houses as well as standard plans for picnic tables, fireboxes, fences, gates, park signs, trail markers, trail steps, drains for trails, and erosion control barriers.[62]

Considering that Iowa was in the vanguard of state park development, it is not surprising that landscape architects at Iowa State were involved in promulgating a new aesthetic in park design. What became known as the "rustic style" grew out of the English landscape gardening tradition as interpreted in

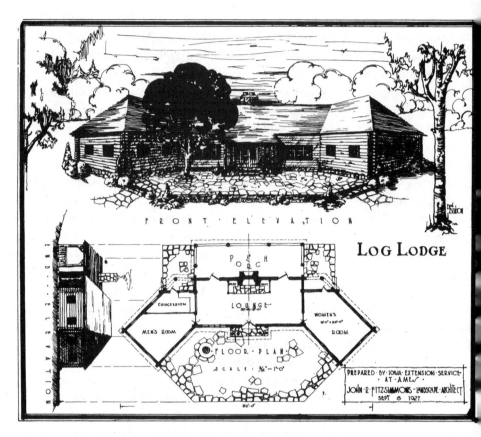

Dolliver Lodge, 1927 conceptual plan, John Fitzsimmons. *Courtesy Iowa Department of Natural Resources.*

the United States principally by Andrew Jackson Downing and Frederick Law Olmsted, Sr. Based on the use of native materials to blend buildings into their natural surroundings, rustic architecture was perfectly suited to America's new national parks. In 1918 the National Park Service formally adopted a policy that called for harmonizing all improvements—roads and trails as well as buildings—with the natural landscape. At the time, Iowa State, Harvard, the University of California at Berkeley, Cornell, and the University of Illinois were the leading landscape architecture schools. P. H. Elwood, chair of the Iowa State landscape architecture department, was considered one of the most capable educators embracing the rustic architectural style for park design.[63]

Although the Board of Conservation spent comparatively little on park development during the 1920s, by 1931 the trend to facilitate public access and recreational use was evident. Six parks contained lodges, or enclosed shelters, for public gatherings. There were food concession buildings in Ledges and Maquoketa Caves. One dedicated Boy Scout cabin remained in Wapsipinicon. A large stone auditorium in Backbone was destined to become the only building of a planned nature education complex ever to be constructed. In Palisades-Kepler, several cottages and an old tavern recalled the days when a summer resort flourished on the banks of the Cedar River. There were "pavilions," or open-walled picnic shelters, in Theodore F. Clark, Clear Lake, Eagle Lake, Farmington, Lost Island, Oak Grove, and Pilot Knob state parks. At Okamanpedan, picnickers could tuck themselves inside a picturesque stone cottage, a remnant of former times. Fireboxes, picnic tables, and latrines were generally available in all parks. Thirteen parks permitted camping, but only in designated areas, in keeping with a 1929 decision to impose camping fees and limit campers' stays in parks.[64]

These facilities would pale in comparison to the extensive park improvement program carried out in the 1930s under the auspices of various New Deal agencies, but the early buildings constructed in Iowa's state parks reflected an adherence to the principles of rustic landscape and architectural design. All new buildings were of stone and log construction. Likewise, roads never were paved. When, in 1927, Bohumil Shimek made published remarks accusing the board of favoring paved roads in parks, Pammel, newly retired from the board, was quick to point out Shimek's error. Hinting that Shimek "seem[ed] to have slipped over," Pammel defended his former colleagues: "I know such a thing never entered the minds of any one. It is absurd."[65] There was no escaping the importance of park roads to automobile tourists. In reality, however, only forty-six miles of unimproved or graveled roads, total, were open to auto travelers in state parks as of 1931. Some parks had no roads at all.[66]

Though few in number and reflecting the highest standards of design, the buildings and other conveniences constructed between 1925 and 1931 nonetheless constituted an acknowledgment that the public would not be satisfied merely with access to parks for the "quiet" pursuits of picnicking, hiking, fishing, and the like. If one considers the number of golf courses within or adjacent to state parks, typically accommodated with a lodge, it is clear that more than a few people were beginning to see state parks as something akin to country estates for the middle class.

The board's initial mandate also included acquiring places of historical interest, which, practically speaking, would come to mean archaeological sites, chiefly Indian mounds, and old military forts; but this became clear only after a period of time. As a case in point, the remains of Fort Atkinson and the site of Fort Defiance were acquired as parks during the 1920s, although neither had been included on the 1919 list. The 1919 report did include, however, three

Mill at Wildcat Den State Park, c. 1931. *Courtesy Iowa Department of Natural Resources.*

American Indian mound sites: Toolesboro Mounds near the mouth of the Iowa River in Louisa County, Fish Farm Mounds near Lansing in Allamakee County, and unnamed mounds near Eddyville in Wapello County. Presumably, Effigy Mounds along the Mississippi River was excluded from the state park target list because conservation and park advocates had every hope that this area would be included in the proposed Mississippi River National Park. Other historic places came into the system coincidentally, such as the mill and dam located on land donated by Emma and Clara Brandt to establish Wildcat Den State Park near Muscatine. The recognized antiquity of the mill and dam, and their association with early Euro-American settlement, were sufficient to confer historic value by standards prevailing in the 1920s. Other places that might be considered valuable historic resources by today's standards—farmhouses, barns, and other rural structures—were simply treated as "old" or "charming," even though they might have been of considerable age.

The term "places of historic interest" must, therefore, be understood within the context of the time. Indian mounds represented cultures past, so of course these were historic. Beyond that, there was general agreement that

Iowa, as a state, was finally old enough to have some history of its own—Euro-American history, that is. However, the definitions of "historic place" tended to be personal and fused with a desire to build state patriotism on something tangible. In a sense, the search for Iowa history, as opposed to local history, began in the post–World War I era. It started from a logical point: saving those places that symbolized "the struggle of pioneer days."[67] Never mind that the same pioneers whose "struggle" was now revered were responsible, in the aggregate, for the tremendous loss of timber, prairie cover, wild game, and wildflowers that conservationists decried. This irony was not entirely lost on certain individuals, but the urge to establish a heritage for future generations was greater. If coming generations were to have pride in their state, the legacy must encompass a tangible, cultural history as well as a natural history.

But what was Iowa's Euro-American cultural history? Now that the pioneer generation itself had largely passed on, the search began for places that loomed large in memory. Of course, the memory that loomed large in one locale was not quite the same memory that passed from one generation to the next in another locale. For one individual, Iowa's distinctive history was present in "a group of immense boulders . . . on the crest of a hill" that was a legendary "camping place for pioneers and settlers crossing the flat plains over the untried road of fortune and home finding." For another, it was "the old trail which crossed the prairie to Boone" and "that beautiful old state capitol upon the grounds" of the University of Iowa.[68] In truth, history was in hundreds of places that held symbolic or personal meaning to hundreds and thousands of people. How could the Board of Conservation become the caretaker of them all? Obviously it could not, but it could pursue certain places that loomed large in the collective memory.

Fort Atkinson in Winneshiek County was the historic place most in need of rescue in the 1920s. Moreover, everyone could agree that it was an important symbol of the "Indian troubles" that had plagued Iowa settlers.[69] Established in 1840 and constructed of native stone, Fort Atkinson was intended to protect white settlers, especially those connected with a missionary settlement near the mouth of the Yellow River. Soldiers stationed there also tried to keep the peace among the Winnebago and their enemies until the Winnebagos were removed from the area in 1848. Fort Atkinson's omission from the 1919 report is especially puzzling because there had been calls to preserve the site ever since 1853, when the federal government ignored a request to turn over the fort to the state for an agricultural school. Instead, the War Department sold the buildings and land at public auction. By 1918, "Old Fort Atkinson had become a pigsty." There were pigeons in the blockhouse, chickens in the powder house. The roofs were full of holes, the stone walls crumbling, and the

Fort Atkinson, farm house and block house, c. 1931. *Courtesy Iowa Department of Natural Resources.*

whole was surrounded by a "forest of weeds."[70] Whatever the reason for its omission, the oversight was soon corrected. In 1920, the Board of Conservation acted to acquire the fort.[71]

The intent was to acquire about fifteen acres of land, enough to take in the old fort and its environs, but the board settled for about five acres, which contained four deteriorated buildings of the fort complex and the mission school buildings. Plans for reconstructing the blockhouse and the barracks were requested in 1922.[72] The board's policy of deferring improvements in favor of land acquisition, however, left little in the budget for reconstructing and rehabilitating the buildings. As time and the fort wore on, it became clear that the task of restoring the fort and mission school, estimated at $40,000, was beyond the board's financial capabilities. Recognizing its limitations, the board struck upon a plan that, if successful, would enable it to care for this historic park and thus meet its responsibilities. In 1928, the board petitioned Congress and the War Department "to repair [the fort] and return [it] to a fair state of preservation." If the federal government would fund restoration, the Board of Conservation would thereafter maintain the historic fort. The federal government declined involvement, however, on the basis that it no longer owned the property. This was simply a dodge, as quickly became clear when federal agencies also declined the board's offer to transfer the property back into federal ownership on the condition that the fort be restored and maintained as a national monument.[73] The fort would continue to crumble away until the availability of New Deal funds rekindled the restoration effort.

When the State Board of Conservation took stock of its holdings in 1931, only Fort Atkinson and Fort Defiance had been established as state parks solely for their historical value. The latter, however, located in Emmet County near Estherville, was a historic site only; there were no visible remnants of the fort. If the park system was lacking in actual historic places—not one of the three mound sites listed in the 1919 report had been acquired—several parks nonetheless were deemed to have historical associations. Lewis and Clark State Park, on the shores of Blue Lake, an oxbow of the Missouri River in Monona County, commemorated the explorers who had camped in that region in 1804. Lepley State Park in Hardin County reputedly contained a tract of virgin timber preserved by pioneer settler Manuel Lepley. Some parks had historical association in name only, such as Dolliver Memorial State Park in Webster County, so named for Jonathan P. Dolliver, U.S. Congressman from 1889 to 1900 and U.S. Senator from then until his death in 1910. Ambrose A. Call State Park preserved in perpetuity the name of the first white settler to claim land and build a log cabin in Kossuth County. History at all levels was neatly wrapped in Lacey-Keosauqua State Park. Named in honor of John F. Lacey, the park also held Ely's Ford, a shallow spot in the Des Moines River that had served as a river crossing before roads were built, a prehistoric American Indian village site, and several Indian mounds.[74]

In this fashion Iowa's state park system came to incorporate a smattering of history, a trend that was noticeable in other states as well. Actually, the creation of historical parks predated the setting aside of public land for scenic, scientific, and recreational purposes. New York had the distinction of establishing the first state historical park in 1849, when it purchased the site in Newburgh where George Washington had headquartered during the Revolutionary War. After the Civil War, states began to acquire battlefield sites; sites associated with the Indian Wars followed. Among the first state park sites acquired by Minnesota, for instance, were Birch Coulee (1889), an 1862 battleground of the Sioux War, and Camp Release (1895), where Sioux Indians once besieged white settlers. Beginning in 1891, Illinois acquired several forts, Indian-white battle sites, places associated with Abraham Lincoln, and Indian mounds. Some of these were purchased as state park or monument sites; others were donated to the state. Likewise, in Texas, the first state parks included battlefield sites. North Dakota's state park system began with the creation of historical parks, which were first administered by the state historical society. In Massachusetts, New York, Ohio, Alabama, and Tennessee, historical and archaeological societies acquired properties that were later transferred into state park systems.[75]

The emphasis on preserving military sites, particularly those places in the East where colonists vanquished the British or where, in the trans-Mississippi West, Euro-American settlers gradually forced American Indians onto reservations, has been viewed as another manifestation of the resurgence of nativism and nationalism that swept the United States in the post–World War I era.[76] There is some truth in this observation: certainly patriotism was both a motivating force and a goal, and the park movement was, to a certain degree, the benign side of the same "national character" that gave rise to immigration restriction and the resurgence of the Ku Klux Klan. But it is also true that battlefields, forts, and places associated with great political leaders were places that most people could agree represented American history uniquely and, equally important, should be preserved. The Iowa Board of Conservation, like its counterparts across the nation, had to make choices. Those choices were limited by a narrow view of history and culture, to be sure, but they also were limited by funding and by the board's ability to persuade private owners to part with desired properties.

Lack of money kept the board from realizing the goal of restoring Fort Atkinson and probably from acquiring a good many historic properties in need of major repair or expensive upkeep. Limited funding also encouraged the board to retain a number of old buildings, particularly farm buildings, that came with land acquisitions; many were converted to new uses. An old log cabin in Bixby State Park, for instance, became a picnic shelter. Farm houses or log cabins were remodeled for the use of custodians in Clear Lake, Dolliver, Flint Hills, and Pammel state parks. Materials from more than one farmhouse were recycled into the golf lodge at Lacey-Keosauqua. In Palisades-Kepler, the custodian made his home in an old tavern overlooking the Cedar River.[77] Thus, structures that would be valued for their historical interest today survived for a time. However, this was merely historic preservation by convenience. As historian E. R. Harlan once remarked, referring to the several farm buildings that came with the Lacey-Keosauqua land, "the objection to their remaining on the park would be their atrocious architecture."[78] In the 1920s, farm buildings held little aesthetic, let alone historic, value. Consequently, there was little hesitancy to replace the converted farmhouses and the aging log cabins when funds became available to build modern facilities. The same New Deal programs that enabled development in state parks on a much larger scale in the 1930s would obliterate these vestiges of the agricultural past that preceded many state parks.

Forestry

The board's authority with regard to forestry work generated controversy that ultimately drove deeper the wedge between conservationists and the State Horticultural Society. By virtue of the 1917 State Park Act the Board of Conservation had authority to investigate and recommend acquisition of forest reserves as well as "investigate the means of promoting forestry" for resource conservation purposes.[79] However, the SHS, a state-chartered entity, had come to view forestry matters as its exclusive province, in large part because the 1906 Forest and Fruit Tree Reservation Act designated the secretary of the SHS to act as state forestry commissioner.[80] The secretary's duties were to see that landowners complied with provisions of the act specifying the species of trees (fruit bearing and forest types) and the minimum acreage required to qualify for property tax exemptions. In reality, the title of State Forestry Commissioner carried little weight, inasmuch as the law was designed to encourage private landowners, through tax exemptions, to plant trees in the hope that the state could thereby replenish timber lost to commercial use. Legislators had no intent to establish state forests, and, as of 1920, fewer than 15,000 acres of timber had been "reserved" in widely scattered woodlots and groves. Nonetheless, the 1906 law had given the SHS some official responsibility over forestry matters.

The 1917 State Park Act did not repeal the 1906 act, but its language was certainly broad enough to allow the Board of Conservation to impinge on the duties the 1906 act conferred on the State Horticultural Society. Moreover, it was not long before Pammel and others sought to expand the board's authority with respect to forestry. The reasons were obvious. Lands acquired for parks frequently included timbered portions that required maintenance. Other park areas were eyed for possible reforestation or conversion to woodlands. Had the membership of the SHS still been dominated by a mix of experimentalists and professors of horticulture, botany, and forestry, cooperation between the board and the society might have been relatively easy to attain. But by this time, commercial nurserymen were a strong voice within the SHS. Therein lay the rub.

In late 1920 G. B. MacDonald proposed to Pammel that the State Park Act be amended to appoint a state forester, to serve without salary, who would assist the board in carrying out its responsibilities with respect to forestry. The Iowa Conservation Association generally supported his proposal in a rambling

Gilmoure B. MacDonald, 1883–1960. *Courtesy Iowa State University Library, University Archives.*

resolution that urged the state "to inaugurate a policy for state forestry including the planting of timber along streams and roads and in its state parks and wherever the acreage of timber may advantageously be increased."[81] Pammel clearly understood the implications of MacDonald's proposal, for he immediately discussed the matter with SHS officers, including the secretary, R. S. Herrick. Actually, Pammel took MacDonald's proposal a step further as he worked with legislators to redraft another proposed senate bill that would have transferred from the State Horticultural Society to the Board of Conservation the duties connected with tax exempt acreage on private lands that were covered with forest trees; under the proposed changes, tax exempt fruit orchards would remain under SHS aegis. Neither the ICA nor the SHS claimed to have initiated this proposal, but it was already moving through the legislative process. On January 12, 1921, Pammel presented an early version of the proposed bill to the SHS for discussion at a meeting the executive committee held specifically for this purpose. After airing "considerable difference[s] of opinion," the executive committee agreed to the transfer on the condition that the bill would be amended to make the secretary of the SHS an ex officio member of the Board of Conservation and provided "that nothing contained in the provisions of the law . . . shall be interpreted as authorizing the Board to grow trees and plants for sale."[82]

Although Pammel did not, perhaps could not, foresee the implications of his subsequent actions, they caused an irreparable crack in relations with the State Horticultural Society. He readily assented to including the SHS secretary on the Board of Conservation, but, for reasons not explained, he proceeded to ignore the second condition of approval. To Senator Foskett he wrote, "Personally, I do not think that this should be in the bill and therefore, have not added it."[83] It may have been that he simply felt it was enough to ensure representation of SHS interests on the board. More likely, he saw the measure as one that might limit board authority; and, if anything, Pammel was bent on expanding the board's powers, even if it meant alienating influential members of the SHS. Toward that end, he began to view the few thousand acres in "forest reserves" as potential new parks. He urged Senator B. J. Horchem to introduce the bill with yet another provision that would allow the board to enter into agreements with private landowners in order to use such tax-exempt reservations for park purposes.[84]

Pammel's deliberate snub precipitated a small storm of protest from a commercial nurseryman in the SHS. "Now Dr. Pammel," wrote E. M. Sherman, president of the Sherman Nursery Co. in Charles City and a member of the SHS board of directors, "I do not wish to seem insistent, but we have been caused no end of trouble during the past ten years by State Nurseries [elsewhere in the United States] growing and selling stock to the general public and I cannot consent to seeing such a state of affairs put under head way in the State of Iowa without my protest." To the chair of the SHS Legislative Committee, W. P. Dawson, Sherman speculated that Pammel had "some intentions which he may wish to carry out in the future which are contrary to the spirit of the above recommendations."[85] When Pammel shared the contents of Sherman's letters with G. B. MacDonald, the latter replied that he was "not at all surprised" and expected that "others of a similar nature" would be forthcoming considering "that many of the nurserymen do not understand a strictly forestry nursery proposition."[86]

Sherman's protest effectively killed any support for the bill from the SHS. Pammel tried to explain his way out of the predicament, but his ego prevented him from going so far as to make overtures of appeasement.[87] By the time Senator Horchem introduced the bill in mid-February 1921, Pammel appears to have sensed it would go nowhere, since by now some Board of Conservation members felt they should move cautiously in seeking additional changes to the State Park Act.[88] Nonetheless, Horchem kept moving his bill forward. By the time it came up for a vote on the floor, the bill contained no provision for adding the secretary of the State Horticultural Society to the Board of Conservation, but it would have authorized the board to arrange for using privately owned timberlands for park purposes. Additionally, the bill called for a $3,000 appropriation to establish a state nursery.[89]

As Pammel had suspected, the reforestation bill was defeated, which left him bristling. In a "personal and confidential" letter to Senator Newberry, his long-time friend from Strawberry Point, he aired his frustration. "The reason it failed," he stated bluntly, "is because Professor Beach and Mr. Herrick and perhaps Mr. Dawson double crossed us. I can't see why it was any affair of the Horticultural Society." Pammel then detailed a history of uneasy relations between the SHS and the Iowa Conservation Association, which he felt had influenced the attitude of the SHS executive committee. In fact, the ICA had played no part in drafting or amending the 1921 reforestation bill, but some members of the SHS apparently thought otherwise. Pammel's harshest words were reserved for SHS secretary, R. S. Herrick: "Apparently Mr. Herrick does not distinguish between the Conservation Association and the State Board of Conservation." A few days later, he wrote as an afterthought, "I am more determined than ever that the State Horticultural Society will not have its way in having this reforestation come under the Horticultural Society." [90] To Herrick, Pammel wrote that "the conservation interests in the state will hold you, in part, responsible" for poisoning relations between two organizations that once were allies.[91] But whatever role Herrick may have personally played in defeating Horchem's reforestation bill, it merely exposed a deepening rift between the State Horticultural Society, which now accommodated diverse commercial interests, and those people who felt that the Board of Conservation should have authority over forestry matters.[92]

Pammel and MacDonald kept pressing for more action on the state level. At the 1921 summer meeting of the Iowa Conservation Association, MacDonald delivered a pointed address in which he compared Iowa ("practically nothing along the line of extensive reforestation") to Pennsylvania ("state nurseries supply millions of trees not only for planting on the state forest lands but these are supplied free of cost to land owners . . . desiring to undertake reforestation work") and Michigan (where "in . . . state nurseries, the trees are produced at a small cost per thousand"). It was not for want of effort on MacDonald's part that Iowa was doing "practically nothing." The state had several small reforestation projects underway in northeast Iowa and on the Mesquakie settlement near Tama. But a few hundred acres were a paltry sum compared to the possibilities. MacDonald proposed a "system of forest parks for every section of the state," acquired and administered by the Board of Conservation, and the establishment of state nurseries "to stimulat[e] reforestation projects in every corner of the state" on private land at nominal cost.[93] Precisely what Mr. E. M. Sherman feared.

His hackles apparently still up, Pammel pounded the same theme in a less-focused speech before the same body. "I hope the next session of the legislature will pass a good reforestation bill. I am going to take my coat off and back it with all my power," he assured his audience. Recalling a recent visit to inspect a proposed state park area near Anamosa, Pammel said that he found "forty or fifty acres of that once fine timbered tract had been cut long ago," reducing the land to gullied waste. "Mind you," one can almost see the finger pointing, "some people said in Des Moines 'It is not your [the Board of Conservation's] problem to go ahead and plant trees on these areas.' Does this not sound like a strange philosophy?" Just whose problem was it, then? Pammel called for nothing less than a law that would give the state control over all timbered land within its borders, on private land as well as public, justifying such drastic action as "a matter of economic necessity" to preserve the fertility of agricultural land and to keep streams and navigable rivers from choking on silt.[94]

The battle was rejoined during the next legislative session. MacDonald drafted a new reforestation bill which, with Pammel's help, was introduced in both houses in February 1923. The new bill would have empowered the Board of Conservation to establish and operate state nurseries. Presumably in an attempt to head off opposition from commercial nurserymen, MacDonald worded the bill to limit the use of state-produced nursery stock principally to state parks. Supplying trees to private landowners would have been permitted, too, but subject to considerable restriction.[95] The nurserymen, however, were not to be swayed. Shortly after the bill was introduced and circulated, R. S. Herrick reported to "Mack" that, once again, "the nurserymen, headed by Mr. E. M. Sherman, are fighting this bill." He was quick to add, though, that he was "remaining neutral." "I do not intend to do anything that will cause anyone to say that I double crossed them in the matter." Pammel's rebuke had struck a nerve.[96]

While MacDonald and Pammel had embarked on a determined mission to establish forestry as a legitimate function of the Board of Conservation, political realities forced a retreat. The Iowa Conservation Association officially endorsed the bill this time around, but it was not enough to influence passage.[97] Other legislation adopted by the legislature in 1923 signaled waning support for the Board of Conservation in general. A hotly debated bill to reorganize the Department of Agriculture carried a provision that would have abolished the board, as well as the Geological Survey and the Forestry Commission. Conservationists probably would have been happy to see the latter go, since the State Horticultural Society had by now made it clear that forestry should not interfere with the profits of commercial nurseries. The threat of abolishment turned out not to be serious, though, since friends of conservation in the Senate easily preserved all three agencies. However, the legislature did reduce the Board of Conservation's annual appropriation from $100,000 to $75,000.[98]

If the House Appropriations Committee had had its way, the annual sum would have been slashed to $25,000. Pammel took quick action to head off this effort to squeeze the board dry. He dashed off a letter to the chair of the Senate Appropriations Committee, B. M. Stoddard, asking legislators to consider the high public demand for state parks within the state. "Do we want to send these people to Minnesota?" he implored. "I am told on good authority [who knows?] that the state of Minnesota plans to catch all of the tourists from Iowa by establishing a system of state parks across the southern part of that state." His appeal, which other conservationists buttressed with similar telegrams and letters, netted $50,000 in reinstated funds.[99]

Failing to secure legislation that would have placed forestry in Iowa on a par with that of other progressive states, Pammel and MacDonald quietly did what was possible working from their respective capacities at Iowa State College and within the Board of Conservation. In effect, what could not be accomplished through the legislative process would be attempted through voluntary cooperation. As early as 1920, MacDonald had initiated a practice of supplying the board with trees, free of cost, from the nursery stock produced at Iowa State; and in 1922 the board asked him to draw up plans for reforesting denuded areas with native species.[100] At about the same time, MacDonald began to promote forest conservation among farmers through the Forestry Section of the Iowa Experiment Station.[101] After the 1923 reforestation bill fizzled, Pammel simply asked the Board of Conservation to "extend to the Forestry department at Ames the privilege of using our state parks" to grow the native trees and shrubs needed to carry out reforestation within the parks.[102] It was a much less ambitious program than either of them wanted, but it was also one that could be implemented quickly and with no political bickering. Thereafter, MacDonald became the board's unpaid "consulting forester" and he worked closely with the board and with park custodians to establish and maintain plant nurseries in selected state parks.

MacDonald not only took a keen interest in the forestry programs of other states, but he followed federal legislative developments closely. In 1924, he worked through the Iowa Conservation Association to mobilize support for passage of the Clarke-McNary Act, which appropriated $2.5 million to establish cooperative federal-state programs of fire control and a much smaller amount, $100,000, to produce and distribute trees for reforestation purposes and to encourage farmers to plant woodlots, shelterbelts, and windbreaks.[103] Iowa stood to benefit very little from the 1924 act, inasmuch as forest fires were unheard of in the state, but MacDonald rightfully saw the act as an important shift in federal forestry policy. He joined other state foresters in urging the U.S. Department of Agriculture to allocate the bulk of the $100,000 appropriation to "farm states" because, as he put it, "many people in the prairie sections feel that the shelterbelt, windbreak and reforestation work are just as important for these states as the fire protective work is for the timbered sections of the country." [104]

MacDonald also worked to get anticipated Clarke-McNary funds distributed either through the Board of Conservation or the Extension Service at Iowa State instead of through the State Horticultural Society. In this he achieved a measure of success. Iowa's share of the first $100,000 appropriation was only $2,000, but it all came to the Extension Service through the State Department of Agriculture. MacDonald used the money to launch "some real reforestation work," as he put it, which meant boosting the output of the Experiment Station nursery and expanding demonstration plots. With continued funding under the Clarke-McNary Act, MacDonald was able to supply nursery stock to the Board of Conservation for reforestation in state parks and to initiate a program of selling, at a nominal cost to private landowners, trees for windbreaks and shelterbelts.[105]

Commercial nursery owners had thwarted MacDonald and Pammel in their efforts to bring forestry under the purview of the Board of Conservation, but MacDonald nonetheless astutely judged the direction national forestry policy was headed and used this to good advantage in Iowa. Between 1924 and 1930, the federal government revamped what had become a disjointed forestry program that was conserving very little of the resource to a policy of federal-state cooperation in order to expand forestry work beyond public lands.[106] On the national level, the strategy of decentralizing forest policy helped to master extreme positions on both sides of a highly controversial debate on the need for federal forest regulations. Importantly, MacDonald employed the new federal policy to build a stronger state program. It was folly to think that Iowa or any other prairie state would ever claim much of the federal pot of money meant chiefly for preventing and fighting forest fires, but MacDonald spent every cent he got. Working with what was available through Iowa State, supplemented by federal funding, he built a modest, but solid, forestry program that jointly served the needs of both farmers and the Board of Conservation.

Unbeknownst to him or anyone else at the time, his federal connections would, with the New Deal, place him in a key position to advance both forestry and state park development in the 1930s.

Water Resources: Natural Lakes and Meandered Rivers

Just as lake preservation was the catalyst for enacting state park legislation, so too did water resources—lakes and streams—become focal points of controversy over resource protection versus public access for recreation and private use for commercial profit. In slightly different form, the same controversy raged over national parks. Forced to defend parks against those who were concerned chiefly with the economic value of resources in the West, the National Park Service, supported by preservationists, formed a "pragmatic alliance" with railroads to promote the economic value of national parks as tourist attractions. This placed the National Park Service in the awkward position of promoting an ever-increasing number of park visitors in order to *save* the parks from utilitarian conservationists and a variety of private interests bent on exploiting mineral, timber, and water resources to their fullest. "Given a choice," though, "preservationists clearly preferred roads, trails, hotels, and crowds to dams, reservoirs, powerlines, and conduits."[107] The automobile quickly overshadowed the railroad in terms of preferred visitor conveyance, but the die had already been cast. Within a few years of its inception, the National Park Service was in the business of managing visitors rather than managing resources. In later decades the National Park Service would begin to think in terms of "complete conservation" in national parks, trying to buck the tide it had created. The Iowa State Board of Conservation, in contrast, started out thinking of parks as "complete conservation" areas, then slowly gave ground throughout the 1920s.

Vehicles and people eventually posed a dilemma for state park management. Nonetheless, like the National Park Service, the Board of Conservation initially looked at visitors as a source of justifying its own existence and, equally important, its annual appropriation. That parks were for the "common man" was an oft-repeated sentiment. As a case in point, when the legislature threatened to cut state park funding drastically in 1923, Pammel defended the budget by hastily assembling statistics that showed more than 232,000 people had visited state parks during the previous year.[108] As important as automobiles were to increasing public demand for recreation in state parks, though, public access to water was the key to expanding the recreational use of parks.

Between 1919 and 1931, the board established eleven state parks principally to provide lake or river access. At least a dozen more were situated along major rivers or contained bodies of water suitable for recreational development. Well over half the state's parks, then, provided access to some of the sixty-three lakes and 800 miles of rivers under board jurisdiction as of 1931. Fishing remained the primary water recreation throughout much of the 1920s. The board actually spent very little money to encourage water recreation, although park custodians maintained swimming beaches at Clear Lake, Eldora Pine Creek, Lewis and Clark, Lost Island, and Palisades-Kepler. Bathhouses were provided at Eldora Pine Creek, Lewis and Clark, and Lost Island.[109]

Regardless of how little the board spent, the magnitude of water recreational use increased to the point at which the board felt moved to impose use restrictions. In May 1929, all "high power boats" were ordered off Twin Lakes, and the board set a ten-mile-per-hour speed limit for all other motorized boats. When boat owners protested, the board listened to their complaints but made it clear that rules and regulations would be set. Early the next year, the board announced general rules for power boats on lakes and streams and new provisions for inspection and licensing. The new rules set both operating times and boat equipment requirements. The board also imposed a fifteen-mile-per-hour daytime speed limit, dropping to ten miles per hour at night. This time, the speed limits provoked much greater protest from individual boaters as well as from the National Outboard Association and the Mid-West Outboard Association. When it became clear that the governor and the Executive Council would not stand behind the speed limits, the board finally agreed to raise the limit to twenty miles per hour both day and night. Outside pressure also forced the board to allow the testing of racing boats on certain rivers. It held the line on seaplanes, though. In June 1931, the board flatly refused to grant permission for seaplanes to land on Spirit Lake.[110]

Speedboating, like golf courses, was an indication of things to come. Compared to other water-related issues in the 1920s, however, boating restrictions were a minor matter. Like forestry policy, water resource policy also reflected the degree to which the state would centralize jurisdiction over natural resources. During the 1920s, hydroelectric power dams and water pollution posed a much greater threat to Iowa's lakes and rivers. At first, the state moved toward centralizing authority over water resources, a promising sign for conservationists. It was not long, however, before the state began to reverse itself, and by the end of the decade, water resource policy was fragmented.

As of 1920, the Board of Conservation still had no specific jurisdiction over lakes and rivers. The plausible explanation for this, as previously noted, is that most legislators saw lakes and streams as inherently connected with the work of the Fish and Game Department. However, Fish and Game was not really concerned with lakes and rivers per se, only with the fish and aquatic mammals that thrived there. When fish failed to thrive in lakes and streams, the accepted solution was to restock the waters. As a result, the state's water resources were almost totally unmanaged, although it was clear from the beginning that the Board of Conservation, by the very nature of its duties, would have to deal with lakes and streams. Thus it was that the board began working with W. E. Albert in 1919 to establish a policy of cooperation between the two agencies. This, in turn, led to the 1921 changes in the State Park Act that gave the board much greater authority over water resources.

In order to justify the proposed 1921 amendment that would expand the board's purview, Pammel emphasized the recreational potential of lakes and streams. Appearing in February 1921 before a joint meeting of the Fish and Game committees and the Conservation committees of the House and Senate, Pammel spoke of recreational use in terms of economic benefit. Estimating that 1.5 million people had visited state parks during the summer of 1920, a wildly inflated figure based on no data whatsoever, Pammel nonetheless proceeded to place a dollar amount on the value of recreation. Pulling another figure out of the hat, he judged each visit to average three hours, which he arbitrarily valued at $.20 per hour. In this manner, he figured that the state provided $500,000 worth of recreation annually, "certainly a good investment for its citizens and future citizens."[111]

The "lake proposition" also was seen as a way to win greater support among sport hunters for the overall goals of the board. Public access to still-viable lakes was one issue; restoration of several partially drained lakes was another. Both, it was argued, would benefit the recreational sports of hunting and fishing in particular. Pammel thus buttressed his case for the 1921 amendments by arguing that drainage laws had produced some notable failures; Rice Lake was the example most often cited, where the drainage ditch had silted in.[112]

Shortly after he spoke before the joint committee meeting in February 1921, Pammel outlined his legislative suggestions in a letter to Representative J. C. Sterling (Hamilton County), chair of the House Conservation Committee. Most of his suggestions subsequently turned up, almost verbatim, in committee-sponsored bills. Pammel called for several amendments to the law, including the controversial forestry provision opposed by commercial nursery owners, but the "lake proposition" had equal priority. Pammel, speaking for the board as a whole, felt strongly that Iowa's lakes and streams should be turned over to the Board of Conservation as *de jure* state parks. In a now-familiar refrain, he appealed on behalf of the "common man, the man who cannot go out of the state for an extended vacation."[113]

The bill, which passed both houses of the legislature and was approved in April 1921, authorized the Board of Conservation "to assume control and management of all meandered streams and lakes belonging to the State which [were] not already under some other jurisdiction." [114] This amendment conferred an extension of authority that was both broad and, as would shortly be revealed, vague. Meandered streams under board jurisdiction would, in time, be defined as the Des Moines, Raccoon, Nishnabotna, Cedar, Iowa, Upper Iowa, Skunk, Maquoketa, Little Maquoketa, Wapsipinicon, and Turkey rivers, although none of these rivers was specifically identified in the law. [115]

Proposed hydroelectric power dams posed the first major threat to the board's control over streams and to those state parks located along such bodies of water. Passage of the Federal Water Power Act in 1920 paved the way for large-scale hydroelectric dam construction under federal regulation. Lesser provisions in the act also permitted private power companies to develop water power on navigable streams, grants of privilege that had been suspended since 1913 as Congress struggled to reach a compromise water resource policy that would appease a variety of interests. [116] It was not long before power companies in Iowa applied for federal permits to build dams in several locations along Iowa's major rivers. The first state park to be threatened by construction of a power dam was Ledges. During much of 1922 and 1923, the fate of Ledges State Park hung in the balance after Iowa Traction Company of Cedar Rapids [117] proposed to erect a series of hydroelectric dams on the Des Moines River, one of which was to be located in Boone County near the park. In May 1922, the company requested permission from the Board of Conservation to overflow about 150 acres of the park. [118]

Although the 1921 amendments gave the board jurisdiction over streams and lakes, it was not clear just what legal standing this gave the board to stop federally sanctioned water power development on those navigable rivers that the board ostensibly "controlled." Consequently, the board took its time studying the issue, requiring Iowa Traction to supply detailed maps and to flag the line of inundation. Between July 1922 and February 1923, several members visited the site, probed local sentiments, and solicited outside opinion from landscape architects and engineers. [119] The board also requested that the attorney general render an opinion regarding its legal rights to represent the state in matters pertaining to the protection, maintenance, improvement, or development of the Des Moines River. [120]

Local sentiment ran hot and cold for Ledges, another complicating factor. In 1919–1920, citizens had contributed a total of $16,000 toward land acquisition in order to protect from despoliation the limestone ledges that gave the park its name. Two years later, many of the same people enthusiastically supported a dam that would inundate the very heart of the park, flooding seventy-five percent of the existing recreational land as well as a scenic view of the ledges. The editor of the *Boone News-Republican* proclaimed "that if it comes to a show down between the park and the dam, 'We're all for the dam first.'" The park had its local protectors, to be sure, who urged the board to stand firm against encroachment. Nevertheless, more than a few people found the idea of a large lake with opportunities for resort and amusement development much more enticing than a 600-acre scenic woodland river valley with a few camping areas, picnic spots, and hiking trails. "What is the sacrifice of a tree or two to the wonderful opportunity for cheap power for factories and the added pleasure of a water resort," the *Boone News-Republican* asked rhetorically.[121]

With local support divided, the board proceeded cautiously, although Pammel seems to have spent more time than usual communicating with the attorney general's office, with various legislators, and with the governor. He alludes to many private conversations in his letters, particularly those written to fellow board member May McNider, who was up in arms over the power dam issue. Inevitably, compromise positions cropped up. Landscape architect Francis A. Robinson prepared an equivocal report in which he admitted that raising the water to the level proposed would result in "an irreparable loss" of park value, but pointed out the added benefit of "a body of water large enough to afford recreation for a large group of people in the torrid climate of our summer."[122] In his report and in a personal appearance at a conference held on February 9, 1923, Robinson suggested that if the power company would agree to maintain the water at a lower, fixed level, the loss of existing recreational and scenic lands might not be too great and the park would have the added value of a lake.[123]

Robinson's suggested compromise was politically attractive, and several local citizens urged the board to adopt it. However, it was equally unacceptable to the power company and to members of the board. On the one hand, the power company could not commit to any fixed water level and also generate electric power without interruption. On the other, as Pammel put it, "Parks are created for scientific, recreational and historic values. The point is, is the company going to destroy these three values?" May McNider framed the answer quite clearly: "I think that if the dam ever goes in there as planned you might as well say good-bye to the park."[124]

Pammel's and McNider's comments made their sentiments clear, although the board's official action was limited to a motion asking the power company

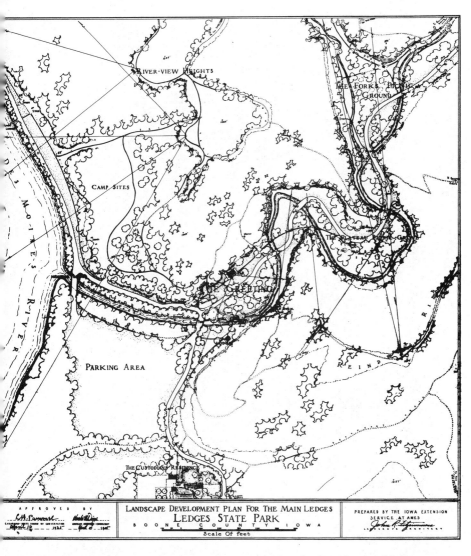

Ledges State Park, 1925 landscape development plan, Charles Diggs and John Fitzsimmons. *Courtesy Iowa Department of Natural Resources.*

Ledges State Park, probably "The Greeting," c. 1931. *Courtesy Iowa Department of Natural Resources.*

Ledges State Park, footbridge across Pea's Creek, c. 1931. *Courtesy Iowa Department of Natural Resources.*

to prepare draft legislation for the board's consideration.[125] Draft legislation was indeed forthcoming, but it was not submitted to the Board of Conservation. Shortly after the February 1923 hearing, Representative Criswell of Boone introduced, "largely at the request of the people in Boone County," a measure designed to give the Executive Council the power to enter into compromise settlements on behalf of the Board of Conservation in order to assess damages against corporations going through state-owned land. Criswell's bill would have effectively stripped the board of its control over streams whenever the board opposed private development, such as power dams. Criswell and his bill supporters were of the opinion that the Executive Council looked favorably upon granting the power company's request for a dam that would overflow Ledges State Park, although Pammel, who had queried the governor, was confident this was not the case.[126] Pammel attempted to quash the bill in committee, but Criswell nonetheless managed to bring it to the floor of the House for a vote, where it was soundly defeated.[127] The board's authority was thus upheld, and Ledges was spared the insult of a power dam.

The victory may have been firm, but May McNider feared this was only the beginning. She preferred to see a "no power dams in parks" policy embedded in legislation and therefore suggested to Pammel that a pending conservation bill be amended to include a provision stating that "neither power dams nor

Ledges State Park, park lodge, c. 1931. *Courtesy Iowa Department of Natural Resources.*

Ledges State Park, registration kiosk, c. 1931. *Courtesy Iowa Department of Natural Resources.*

back waters of power dams controlled by private corporations shall be permitted in public parks." Pammel, however, would not go so far. "I am satisfied that the Conservation Board will always have idealists on the Board like yourself and that there is no immediate danger," he wrote. More to point, he felt such an amendment would be politically unpalatable. Humiliated by his unsuccessful attempt to secure a reforestation bill in 1921, Pammel was more willing to submit to the dictates of politics in 1923. He would not jeopardize pending conservation measures by adding a provision that was sure to bring out the opposition. "I do not want to endanger our park legislation at this time. I am afraid that when we introduced the reforestation bill *which was a real conservation measure* [author's emphasis] that I made a mistake." [128]

Both of them were right. On the one hand, Pammel managed to keep some conservation legislation reasonably intact and to avert a drastic cut in the board's appropriation. In 1923, the board lost its second bid to establish plant nurseries, but the legislature did reorganize the board by expanding it to five members with staggered terms, all appointed by the governor, and by replacing the Curator of the Historical Department with the Secretary of the Executive Council, the latter change designed to establish closer ties between the board and the council. Pammel had long wanted a larger board, but he wanted one that gave the scientific community at least one assured seat. This, like forestry legislation, he did not get. What he did get was legislation that assured continuity among board members from one year to the next—"absolutely essential for the success of the park movement" was Pammel's rationale.[129] On the other hand, the threat of power dams did not go away. The Board of Conservation would face this issue over and over in the coming years.

In late 1924 and early 1925, the board studied the probable effect of a proposal to divert water from the Cedar River for water power development at Muscatine. The proposed plan, one in a long string of such proposals, called for three low-head storage dams between Cedar Rapids and the Cedar-Muscatine county line, which would control the flow of water to a much larger dam (100-foot fall) near Muscatine, the latter creating an 8000-acre pool. When there was no immediate threat to state park lands, Pammel was more inclined to accept the utilitarian conservationist point of view. Noting that a dam at Muscatine "might prove valuable for power development" and conserve coal resources, he was principally concerned that the state's jurisdiction over meandered streams be protected through the powers of the Board of Conservation and the Executive Council.[130]

Controversy erupted again in 1927, when Central States Electric Company proposed placing a dam twenty-six feet high across the Des Moines River near Dolliver State Park in Webster County, a short distance up the river from Ledges State Park. Although this proposal also had plenty of local support, many people who had donated money to create the park in the first place stepped forward to oppose the dam, just as had happened in Boone County. Conservation Board member Mary C. Armstrong of Fort Dodge helped to mobilize the opposition, and she was joined in the effort by Fort Dodge Mayor C. V. Findlay. After Governor John Hammill stated that the people of Webster County could have whatever they wanted, Armstrong predicted that he would "get some red hot letters" from Fort Dodge because construction of a power dam was "a State matter and not at all a matter for the county to decide."[131]

The 1927 controversy fixed on Woodman Hollow, an area north of the park containing several rare plant species, which would have been almost completely flooded. In October 1927, the Board of Conservation went on record as opposing the dam, stating that it would threaten "all the features which make this park valuable and worth conserving" and, in particular, would destroy most of Woodman Hollow, thus "imped[ing] the scientific study for which said tract was acquired." Armstrong arranged to have the board's resolution published in an open letter to the "citizens of Webster County," adding that "to agree to the construction of a dam in Dolliver would be breaking faith with the people of Webster County who contributed $11,000" toward park acquisition.[132]

Louis Pammel, who retired from the board in the midst of this conflict, continued his activism as a private citizen, working to defeat the dam project. He praised the board's resolution, wrote a letter of protest to Governor John Hammill as well as other high officials in state government, and sent an "interview statement" to the Des Moines Register. When the governor agreed to a special hearing before the State Board of Conservation, Professor Pammel attempted to round up as many fellow scientists as he could to appear and testify against the project.[133] The first hearing, held December 6 before the Board of Conservation, elicited a small crowd of opponents and proponents. Interestingly, the opposition included former boosters from Boone who, four years earlier, felt they had "lost" a dam and recreation lake because of Ledges State Park. They now raised fears that a dam in Webster County would cause a "water famine" downstream.[134] During the next two days, hearings continued before the Executive Council on two alternate proposals, one for a dam near Kalo, approximately four miles from Dolliver State Park, and the other for a dam immediately north of Woodman Hollow. After three days of listening to both sides, the Board of Conservation recommended that the low-head dam at Kalo be approved, the dam near Woodland Hollow denied. The Executive Council concurred, thus ending the threat to Dolliver State Park and Woodland Hollow.[135]

Dolliver, like Ledges, had been spared, but the battle over Dolliver had a sobering effect. By now it was clear that state parks were not immune to private development if enough pressure could be brought to bear on state officials. Moreover, this time a tract highly valued for scientific purposes, Woodman Hollow, had been seriously threatened. This prompted the board, for the first time, to distinguish among parks based not just on their size or location but on their scientific value. Buttressed by demands from the Iowa Academy of Science that it set aside plant and animal "sanctuaries," the board declared Woodman Hollow to be a "reserve for the full protection of plant and animal

Woodman Hollow, c. 1931. *Courtesy Iowa Department of Natural Resources.*

life therein." Woodman Hollow thus became the first "preserve" within the state park system, a move clearly calculated to send a message that the Board of Conservation considered certain areas inviolate.[136] Still, the Board of Conservation had no blanket policy concerning power dams. The situations at Ledges State Park and Dolliver State Park merely set precedence. As a result, the board continued to deal with such proposals on a case-by-case basis, each time rejecting them or postponing a decision until applicants dropped their plans.[137]

The board's position on dams was complicated by the fact that it did not oppose all dams, only power dams that would impound a large pool of water and inundate areas prized for their scientific or scenic qualities. The board routinely approved the construction of low-head dams at lake outlets in order to restore and control water levels, and no one was opposed to building dams for the purpose of creating swimming and fishing areas. As a case in point, at the same time the board was skirmishing with Iowa Traction over the proposed hydroelectric dam in Ledges, it was considering the construction of a low-head dam across Prairie Creek in Dolliver State Park in order to create a thirty-acre swimming hole.[138] Likewise, in Backbone State Park, where the Maquoketa River had been dammed in the nineteenth century to power a mill, the board left the existing structure intact, eventually replacing it and improving the lake with swimming and boating facilities.

Rejecting power dams but embracing "beauty dams," as they were called, left a murky policy area concerning low-head dams for small power-generating stations or other purposes. Here the line was much less clear. For instance, in October 1931, the board considered three permit applications for low-head dams, one along the Cedar River at Waterloo, another along the Des Moines River near Boone, and a third along the Des Moines River at Wallingford. After discerning that none of the proposed dams would raise water levels by much, the board indicated to the Executive Council that it would "not disapprove" of the applications—an equivocal position the board tried to clarify further by advising the council that as a matter of policy it did not favor the construction of any dams except "beauty dams" in meandered streams under its jurisdiction.[139] "Beauty dam" became a convenient metaphor for any dam that restored, enhanced, or created a water recreation area.

Jurisdiction over lakes and streams also brought the issue of water pollution before the board. Part of the "lake proposition" was to clean them up. Since the early part of the century, various members of the Iowa Academy of Science had worked on the problems of municipal sanitation, and both the University of Iowa and Iowa State provided leadership in the areas of sewage treatment as well as water pollution control. As early as 1911, Macbride had argued that it should be a "criminal offense . . . to allow any species of filth, from the hog-lots, barnyards, privies, dead animals, or anything of the sort to drain into or find exit in the waters of any lake or stream."[140] State legislators responded by reorganizing the Board of Health in 1913—it had been around since 1880—and granting it the power to enforce local sanitary regulations if so petitioned by local citizens.[141] These powers, however, were so insufficient to deal with the growing severity of water quality problems that in 1919 the Iowa Conservation Association passed a resolution calling upon the state legislature to get tough with polluters, asserting that Iowa's streams and lakes "should not be made the dumping grounds for disease-laden, filthy sewerages, trade waste, etc."[142]

Although Thomas Macbride and the Iowa Conservation Association may have viewed water pollution as an environmental issue, this ran counter to the direction the country as a whole was headed at the time. By the turn of the century, cities had ceased to think of sewage and waste accumulation as strictly a nuisance problem. The links between refuse and disease were now recognized. With this shift, new professions of public health and sanitary engineering emerged. As it turned out, though, public health professionals tended to focus on identifying and treating the bacterial causes of disease rather than cleaning up the environmental messes that spawned disease. Likewise, sanitary engineers focused on devising technological solutions to master the logistical problems of waste collection and disposal. Waste management thus became a municipal engineering problem rather than an environmental issue, and sanitary engineers concerned themselves with the problems of disposal rather than the environmental problems caused by waste generation. The growth of these two professions effectively divorced the issue of water pollution from the issue of water quality.[143]

In addition, pollution was still perceived as an urban problem, even though it was known that wells in small towns and rural areas were showing signs of increasing contamination. By the 1910s, a few cities in Iowa had hired sanitation engineers and asserted some control over waste discharge and disposal, but state politicians displayed little inclination to challenge home rule. Sanitation was considered a municipal prerogative, even when pollution problems flowed beyond city limits. Sanitary engineers therefore responded to the dictates of municipal governments and followed the trends of their profession, which called for applying technological solutions to the problem of waste removal. Not only did this approach ignore the waste-generation half of the equation, but by and large, sanitary engineers sanctioned the use of streams and lakes to dilute raw sewage. They justified this position by claiming that a certain level of contamination was tolerable as long as human health was not seriously threatened.[144]

In 1915, Anson Marston, dean of engineering at Iowa State organized the first meeting of what would become the Iowa Water Pollution Control Association. World War I interrupted formal organization, but in 1920 Iowa State engineering faculty, consulting engineers, and municipal sanitation officials resumed annual conferences to share information about effective sewage treatment, plant technology, and operations. Still, as of 1920 only about 120 towns and cities in Iowa had any sewage treatment facility, and virtually nothing was being done to control pollution from industrial sources or from animal waste.[145] Iowa's lakes and rivers were in danger of becoming open sewers.

It was not until 1923 that the state legislature took any significant steps to combat water pollution. That year the legislature passed several bills pertaining to public health, among them the Stream Pollution Act. The provisions of the act, however, completely ignored any jurisdiction the Board of Conservation had over state waters. Rather, it gave the State Board of Health authority to investigate polluted waters and order polluters to cease discharging sewage or other befouling substances into lakes and streams.[146] Even so, its operations were severely limited by budget constraints. It took a pollution crisis in the Shell Rock River basin below Mason City to convince the legislature to get serious. The Iowa Division of the Izaak Walton League demanded that legislators do something to clean up the streams. Not only was the stench unbearable at certain times of the year, but municipal sewage augmented by waste discharge from meat packing and sugar beet processing plants had caused fish kills as far as ninety miles downstream.[147] As a result of this outcry, the state legislature appropriated a modest sum in 1925 so that the state sanitary engineer could actually begin studying stream pollution problems.[148]

The small sum appropriated in 1925 enabled the Board of Health to investigate the problems of only one river basin. After Alfred H. Wieters took over as state sanitary engineer in 1926, however, the Board of Conservation called upon the Board of Health to investigate the extent of pollution in rivers, streams, and lakes *throughout* Iowa, noting that the pollution of the Cedar River through Palisades State Park was so "excessive and flagrant" that it would consider no improvements to the park until the pollution abated.[149] Shortly thereafter, Wieters began investigations on the Cedar, Des Moines, and Iowa rivers. Routine water sampling at fixed stations revealed, over the next few years, zones of moderate to heavy pollution below nearly every sizable city along these rivers. Eventually, the department's investigations lead to a major cleanup campaign, but this did not happen until the 1930s, when federal funds became available for such purposes.[150]

Meanwhile, water quality in general continued to deteriorate. Moreover, the Board of Health's investigations were directed at only a handful of rivers; lakes and smaller streams were not included. In September 1930, the Board of Conservation convened a special meeting with the governor, the fish and game warden, and members of the Board of Health to discuss the mounting problem of stream and lake pollution, specifically to consider means of exterminating lake algae without harming fish and other aquatic life. All agreed that lake dredging was considered the only safe means of treating the increasing incidence of lake algae. Following this conference, the conservation and health boards jointly prepared draft legislation to address stream pollution and algae through a separate Water Sanitation Commission.[151] This proposal never was enacted, but not for lack of interest. Rather, in 1931 the state embarked on an ambitious new undertaking to develop a long-term plan for the conservation of all natural resources.

With the 1921 amendments to the State Park Act, the legislature handed the Board of Conservation what appeared to be sweeping authority over state waters. A decade later, however, the board rightfully observed that it had only "nominal" jurisdiction.[152] Its actual power to restrict public use was hampered by public demand and the propensity for elected officials to shift ground when pressured from the outside. Equally important, authority over lakes and streams remained divided. The Fish and Game Department restocked lakes and streams with fish and, in order to minimize fish losses, regulated lake water levels. In addition, the Fish and Game Department operated fish hatcheries in Backbone, Dolliver, and Palisades-Kepler state parks. However, the Fish and Game Department remained a separate agency, and coordination with the Board of Conservation depended on cooperation from the fish and game warden. With respect to boating regulations, the board recommended and the Executive Council approved licensing and inspection regulations, but the power to appoint inspectors rested with the governor. On matters of water quality, the board could only appeal to the State Board of Health for cooperation. It could not play the role of lead agency in order to address water pollution problems. With respect to power dams, the board managed to hold a tough line, but it had no authority to make binding decisions and the federal government could, at any time, assert primary jurisdiction over rivers considered to be navigable.

In retrospect, the 1921 amendments constituted an important step inasmuch as the state acknowledged, for the first time, that it must assert control over its lakes and streams, but subsequent actions revealed that the state would not go so far as to centralize authority over water resources in one agency. Likewise, the mandate to create forest reserves and establish a comprehensive forestry program remained largely an unrealized goal. The state had no real forestry policy as of 1931, although thanks to G. B. MacDonald's voluntary service to the Board of Conservation, the board was able to extend its orbit to encompass reforestation and maintenance of timbered areas within state parks. While many conservationists wanted authority over all natural resources consolidated in the Board of Conservation, there is little indication that either Pammel or the major organizations of the conservation constituency were prepared to make this a political issue and risk losing support from the legislature. The failed efforts in 1921 and 1923 to establish forestry under the Board of Conservation effectively tested the limits of political support for centralized authority. Fiscal constraints also placed limits on what the board could do. To the extent that this precluded recreational development in state parks, limited funding was a boon to resource conservation interests. However, the same constraints also precluded the board from taking seriously its mandate to preserve historic places.

The one thing that remained unchanged during the 1920s was the sense of mission that drove the Board of Conservation. In some respects this, too, was part of Louis Pammel's legacy. He may not have achieved what he perceived as the ideal mix of members, but the system worked to combine citizens who were committed to finding solutions. The board did not lose its momentum when Pammel left it in 1927, nor was it ever troubled by internal dissension. The strength of the board was perhaps the greatest asset of the conservation movement in Iowa during the 1920s, and that strength would serve the state well in the coming decade.

4

Toward a Resource Agency
The Twenty-five-Year Conservation Plan

We cannot move on to new frontiers. There are none. . . . The trails that led to the choicest fishing grounds, the wooded hills and deeply shaded pools where the big bass lie, where once we trudged with our entire equipment for a month's cruise in a 40 pound sack, are now main-traveled highways, cluttered with empty tomato cans, pop bottles, and road signs to lend speed to the heedless motorist.

We cannot bring the old conditions back and we must not resent nor despair of a situation which is now history and cannot be rewritten. But we are the biggest fools in the world and unworthy of the intelligence with which we credit ourselves if we do not immediately act to preserve what there is left of our outdoor natural endowment . . . not only for our own enjoyment and well being but as a heritage to those who shall follow after us.

—J. N. "DING" DARLING[1]

The economic depression that devastated one-third of the nation's people during the 1930s was also an unquestionable boon to state, as well as national, parks. New Deal relief programs, augmented by increased state appropriations, put thousands of people to work in parks. CCC, CWA, PWA, WPA, and NYA crews constructed buildings, roads, trails, bridges, and dams; planted forests; dredged lakes; and riprapped shorelines.[2] In Iowa alone, work relief programs funded the construction of at least one thousand park buildings and structures, ranging from sturdy stone-and-timber lodges to inconspicuous stone culverts under footpaths. Most of these structures are still in use, giving our parks their distinctive rustic appeal. In addition, several thousand acres were set aside as forest reserves, forming the genesis of Iowa's state forests. This is the tangible legacy that is chiefly associated with park development in the 1930s. The legacy park visitors so greatly enjoy today, however, probably would be much less rich had it not been for visionary state planning that preceded President Franklin Roosevelt's New Deal.

Jay N. Darling, 1876–1962. This photograph of "Ding"
appeared in *Iowa Conservationist* with his 1945 article
"Poverty or Conservation Your National Problem."
Courtesy J. N. "Ding" Darling Foundation.

Jay N. "Ding" Darling and the *Twenty-five Year Conservation Plan*

Jay N. Darling, best known to Iowans as the Pulitzer Prize–winning cartoonist
whose works held a spot on the front page of the *Des Moines Register* for over
forty years, also became Iowa's voice of conscience for conservation. Not con-
tent to be merely a social and political commentator, Darling, like Pammel
before him, was an activist, although these two important personalities operated in different realms and their paths rarely crossed. The only time they are
known to have shared the same stage is at the 1930 dedication of Pammel State
Park. Pammel, the academician-turned-administrator, learned how to work
behind the scenes but within the political system in order to influence change.
Darling, the artist-as-pundit, jabbed and poked at public attitudes and the
political process from the pages of newspapers and magazines. As a cartoonist,

he played the role of an "outsider," employing an acerbic wit to shame witless hunters as well as scheming politicians. But as his fame and fortune grew, Darling gained access to power circles. By the late 1920s, he had earned a reputation as an environmental critic, and he parlayed that reputation into effective political action. During the early 1930s, Darling played an important role in refocusing resource conservation policy in Iowa and in restructuring the administrative apparatus for policy implementation.

Darling's zeal for conservation came through his love of the outdoors. His position in the world of journalism just happened to provide a convenient platform from which he could broadcast his views. The two sides of his personality were evident very early. In college he studied biology, earning a bachelor's degree in 1900. After graduating from Beloit, he went to work for the *Sioux City Journal*, first as a reporter and then as a cartoonist. In 1906, he moved to the *Des Moines Register*, where his political cartoons began to attract the attention of other newspaper editors. The *New York Globe* lured him East in 1911, but the New York sojourn was short-lived. He liked the exposure to East Coast audiences and the proximity to political power, but big-city life did not entirely suit him or his wife, Genevieve, so he negotiated a return to the *Register* in 1913. In the end, he got the best of both worlds. From 1917 until his retirement in 1949, Darling lived and worked in Iowa, and his cartoons were syndicated through both the *Register* and the *New York Tribune*.[3]

During the first twenty years of his career, journalism was the true focus. He never associated himself closely with the Iowa Conservation Association or any of the other groups instrumental in establishing the Board of Conservation. Moreover, until the late 1920s he seems to have had only passing interest, at best, in the board's work. When he finally did take greater interest in the board, Darling often referred to it as "the Park Board," not the Board of Conservation, betraying his personal conception of the agency. His ideas about resource conservation, and many of his goals, were shaped by his association with hunters and sportsmen's organizations. It was a world decidedly masculine. Parks, for Darling, were linked with aestheticism, not "conservation," as indicated by his cartoons for the Outdoor Good Manners campaign, which denounced the despoliation of natural scenery and wildflowers by thoughtless park visitors. He did not think of parks as laboratories for resource conservation in the same way as Macbride, Pammel, and the other academy-based scientists who had been leaders in the park movement in Iowa. Where they saw parks as a means to establish conservation as a legitimate function of government, Darling reflected a new generation that increasingly viewed parks as publicly held outdoor recreation spots. For him, resource conservation was a complex societal problem, and he did not see parks as strategic tools for solving these problems. Moreover, even though Darling understood that societal values were at the root of resource issues, he placed great faith in the ability of professional experts to manage resources wisely and the power of education to reshape society.

By 1930, Darling had become an articulate spokesman for the protection and scientific management of wildlife. He made his entrance into conservation politics with the aid of the Izaak Walton League. As the Iowa Conservation Association faded during the 1920s, the Izaak Walton League quickly gained recognition and strength in numbers. The league emerged from a now-legendary meeting of fifty-four prominent Chicago men, all gentleman anglers, who gathered at the Chicago Athletic Club on Saturday, January 14, 1922. Their aim was to unite fisherman across the nation into a fraternal, even militant, organization that would, on the one hand, promote "good sportsmanship" among anglers, and, on the other, fight to save the waterways and wildlife species that were threatened by pollution, drainage, or other forms of aquatic desecration.[4] Iowa proved to be extremely fertile ground for organizing. When the league held its first national convention in 1923, in Chicago, fifty-one delegates representing several Iowa chapters were in attendance. Shortly thereafter, Iowa formed the first state division of the Izaak Walton League of America. Membership in the Iowa division quickly rose from 2,000 in 1923 to 4,000 in 1924, then to 7,500 in 1926. Jay Darling was among its most active members.[5]

In part, the rise in membership can be attributed to the 1923–1924 fight to save the Winneshiek Bottoms of the Mississippi River, an area between Lynxville, Wisconsin, and Lansing, Iowa, frequented by hunters and fishermen from both states. The fight began when the Army Corps of Engineers proposed to remove timber from the Winneshiek (islands that became overflow lands during flood stage), riprap the shores, and convert the land to tillable acres. The Izaak Walton League, with national president Will Dilg leading the effort, drafted a bill to create a wildlife refuge extending for nearly three hundred miles along the Upper Mississippi River, from Rock Island, Illinois, to Wabasha, Minnesota. The Federation of Women's Clubs, both the national and Iowa organizations, joined the league in support of the bill, as did the Iowa Conservation Association, and a persistent group of conservationists in northeast Iowa who, since 1909, had been promoting the creation of a national park in the Upper Mississippi River Valley. The annual Wild Life Protection School even became the scene of a special conference on aquatic resources. In August 1923, conservationists from Minnesota, Wisconsin, Illinois, and Iowa met at McGregor with representatives from federal and state agencies to discuss clam production, fish farms, water resources, and the need to protect the Winneshiek.

This sudden and vibrant coalescing of interests drew sportsmen, park advocates, wildlife conservationists, politicians, and federal bureaucrats together in a common cause: to protect a stretch of the river that was variously, and passionately, prized for its spectacular scenery, its abundant fisheries, its migratory waterfowl, and its mystical backwaters and sloughs. In February 1924, an impressive array of conservation advocates and scientific specialists gathered in Washington to deliver strong statements of support at the Congressional hearings that preceded passage of the Upper Mississippi River Wildlife and Fish Refuge Act. Among them were Cora Whitley, representing the Iowa and General Federations of Women's Clubs; botanist Bohumil Shimek, representing the Iowa division of the IWLA as well as allied conservation groups; plant physiologist A. L. Bakke of Iowa State; and soils expert W. G. Baker from the Iowa Experimental Station.[6]

As politics go, the War Department succeeded in amending the 1924 act in order to maintain a measure of control over the river for navigation purposes. This amendment eventually allowed the Corps of Engineers to dredge a nine-foot navigation channel and construct a series of locks and dams through the Upper Mississippi River from St. Paul to St. Louis, right through the Winneshiek. Nonetheless, the refuge gradually materialized. The 1924 act carried a $1,500,000 appropriation for acquiring private property, since much of the land adjacent to the river was privately owned. After the Iowa legislature passed corollary legislation authorizing the U.S. government to acquire land from the state, land purchases began. By 1940 the refuge contained nearly 25,000 acres, and it eventually grew to encompass approximately 200,000 acres. Various provisions of the state enabling legislation gave the state game warden broad authority over land use within the refuge. The warden, for instance, had the power to establish and control fish hatcheries and game farms. Whenever land was transferred from the refuge into a public park, the warden also could establish game refuges or other wildlife sanctuaries and could forbid hunting and trapping.[7]

The 1924 Upper Mississippi River Wildlife and Fish Refuge Act represented a turning point in the history of federal wildlife legislation. Congress authorized the first federal wildlife refuge in 1903, and by 1924 nearly half of the states had federal refuges. However, the Upper Mississippi authorization was the first to carry an appropriation, and it set the tone for two additional pieces of federal legislation in the 1920s. The 1928 Bear River Migratory Bird Refuge Act also carried an appropriation for land acquisition, as did the 1929

Norbeck-Andresen Migratory Bird Conservation Act, which laid the foundation for systematic development of private lands for wildlife refuges throughout the United States. Between 1903 and 1925, the number of acres in federal wildlife refuges grew from three to more than 450,000. From 1926 through 1930, the figure gradually increased to about 744,000. During the ensuing decade, however, the acreage increased exponentially, reaching an astounding 9,617,713 acres by the end of 1941.[8] Jay Darling, as it turned out, was instrumental in securing the initial federal funding that led to this remarkable growth.

The Izaak Walton League made its mark with the Upper Mississippi refuge issue, and by 1924 it claimed more than 100,000 members nationwide.[9] The federal government's failure to take the lead in wildlife conservation during the 1920s may account, in part, for the phenomenal growth. Donald Swain argues that the volunteerism policies of the Harding, Coolidge, and Hoover administrations severely hampered scientific research within the Bureau of Biological Survey and the Bureau of Fisheries. In addition, both of these agencies tended to identify with utilitarian conservation ideals and often catered to commercial interests. As a result, the principles of scientific game management developed largely outside federal purview.[10] A more immediate reason for the league's early success no doubt was Dilg's decision to take on the Mississippi River wildlife refuge campaign. It was an issue that could draw support from many states; and, in fact, much of the organization's strength came from the Midwest, where membership was highest.[11] The fight to establish the Upper Mississippi River Wildlife and Fish Refuge demonstrated that a highly focused, well-financed grassroots campaign could move Congress to adopt conservation legislation that challenged the Corps of Engineers' supremacy over navigable waterways and, additionally, challenged the preeminence of states in regulating fish and game matters.[12] It was more than an important victory for the Izaak Walton League; it propelled the cause of wildlife conservation in general during the 1920s.

Emboldened by the successful fight to save the Winneshiek Bottoms, the Iowa division maintained its political momentum for a time by pressing the State Board of Health to investigate pollution along the Shell Rock and Cedar rivers. Many Waltonians, though, appear to have been drawn into the organization more for its fraternal aspects than its advocacy positions. By 1925, "activism" had taken on a volunteer spirit. Waltonians worked with the Fish and Game Department, sometimes as "complimentary deputy wardens," assisting with fish rescue work and raising pheasants for release. Local chapters sponsored annual clean-up days and picnics. Others sponsored outdoor activities for "orphan boys" from the city, tree- and game-cover planting projects, or winter bird-feeding programs. In this respect, the work of the Iowa Ikes resembled that of local women's clubs, although the latter operated within a more sophisticated organizational framework and a top-to-bottom educational program. This dissipation of energy seems to have been the source of bitter controversy that erupted in 1927, when then-president Bohumil Shimek staged a secession movement, charging that the league had become little more than a clique for wealthy sportsmen who had no real interest in conservation. Shimek's Rebellion, as it has been called, coincided with Dilg's removal from the national organization, and the upheaval cost the Iowa division more than half its membership.[13]

In perverse fashion, the schism also seems to have rekindled a sense of mission among Iowa Ikes, who initiated a new campaign to eliminate political corruption in the Fish and Game Department. In 1928 and again in 1929, the Iowa division sponsored legislation to establish a nonpartisan Fish and Game Commission to oversee the affairs of the department. The goal was to curb what was widely perceived as a rampant disregard for fish and game regulations by W. E. Albert and some of his deputy wardens.[14] Recalling the history of that campaign many years later, Jay Darling noted that "fish hatcheries and restocking of private waters for political benefits to favored candidates" triggered resentment among organized sportsmen, especially as they watched game fish populations steadily decline in lakes and streams throughout the state. Recreational anglers began to question what the money they paid each year for hunting and fishing licenses was really funding. As Darling remembered it, "The appointment of political fence-menders to jobs as Game Wardens and the consequent violations of game laws and excessive bag limits finally blew up a storm of public sentiment, but it was not until Dan Turner, a newly elected Governor, was willing to adopt personally a policy for taking conservation out of politics that the long campaign of the members of the Izaak Walton League burst into bloom."[15]

The Iowa Ikes' effort was successful in 1931, when the state legislature established a five-member Fish and Game Commission. In effect, the law created a new authority, on par with the Board of Conservation, and to whom the state game warden thereafter reported. Additionally, the new commission was authorized to "acquire lands and waters for hunting, fishing, and trapping purposes" and to establish "fish hatcheries, game farms, and protected bird refuges."[16] This authority overlapped that of the Board of Conservation. Moreover, the Fish and Game Commission now had the Fish and Game Protective Fund to finance land acquisition as well as pay for departmental operating costs; it did not have to rely on legislative appropriations, as did the Board of Conservation.

The same legislature also passed a joint resolution calling on the State Board of Conservation to engage a professional "park expert and regional planning engineer to provide, within two years, a comprehensive, budgeted, State-wide park, fish and game program" that could be implemented over a twenty-five-year period.[17] Darling and the Iowa Division of the IWLA were behind this action, too. Shortly before the law establishing a fish and game commission was adopted, Darling spoke at the Iowa division's annual convention, where he urged the Ikes to unify in support of a specific conservation program. "We are 100 percent for conservation," he noted, "but we are divided into a hundred different Chapters, each bent on its own local projects." The specific program that he proposed was a long-term plan for conservation and state park development. It would engage engineers, landscape architects, scientists, and planners in a survey of the state's lands and waters, study the feasibility of selected areas for both conservation programs and outdoor recreation, and devise a plan for land acquisition and the development of additional parks, recreation areas, refuges, preserves, fish hatcheries, and game farms. "When completed," Darling calculated, "such a plan should accurately locate and specify the use and character of every state park, game refuge, fish and game nursery, scenic highway, reclaimed lake and marsh which will be needed for the future recreation and enjoyment of the population of Iowa."[18] Darling also proposed that the Board of Conservation and the Fish and Game Department be merged into one agency. Duplication of effort and expenditures, he believed, would dictate as much. He also saw that wildlife would benefit if game refuges and state parks were integrated under one agency. The Iowa division not only supported his plan; a committee helped draft the legislation that put the plan in motion.[19]

Jay Darling was among those Governor Dan Turner appointed to the Fish and Game Commission.[20] His presence strongly suggests that this body was created chiefly to shape a state agency that would complement the envisioned state conservation plan. There probably was little intent that the commission would have anything but a temporary existence, since Darling had already announced his idea of merging the Board of Conservation with the Fish and Game Department. The "statewide conservation survey," in fact, was the only topic of discussion at the commission's first regular meeting in May 1931.[21] However, the legislation specified that the Board of Conservation, not the Fish and Game Commission, initiate the survey. To get the ball rolling, Darling appeared before the board in June 1931, advising its members that he had already asked several landscape architects to appear before a (not-yet-formed) joint committee of the board and the commission.[22]

How the board reacted to Darling's announcement, as well as the legislature's directive, can only be guessed because the evidence of what transpired is limited to official records. There is no indication, for instance, that a joint committee was ever formed, although in November 1931 the board and commission adopted a practice of holding joint meetings on a fairly regular basis, a practice that continued for two years. In August 1931, however, the Board of Conservation agreed to pay all expenses connected with the survey and bill the Fish and Game Commission for half. A month later the board selected Jacob L. Crane, Jr., a recreation planner from Chicago, to coordinate the survey work and prepare a report.[23] Neither of these business items is reported in the Fish and Game Commission minutes. It is perhaps revealing that, ten years later, Jay Darling would misremember events in terms that clearly subordinated the board's role. Commenting on a draft of a magazine article Crane had sent him for review, Darling attempted to write the Board of Conservation out of the historical picture. "I am under the impression that Margo Frankel's Board was at that time known as the State Park Board and I am also of the very definite conviction that not until after you came into the picture and began to coordinate all the interests[,] the State Park Board was hardly aware of the proposal to make a state survey under technical supervision." On this point, Darling may have been essentially correct; what followed, however, does not square with the facts. "That proposal and all the momentum which finally made it effective, including the legislation and the appropriation of $25,000 from the Fish & Game funds to finance the Survey were generated from the Fish & Game Department and it hardly seems a fair presumption to name the Park Board first, if at all, in the formative period of the 25-Year Program."[24] At the very least, the survey, and, more important, what lay behind it—creation of the Fish and Game Commission—initiated a period of apprehension. The board and the commission *did* cooperate until the two bodies were merged into a new State Conservation Commission in 1935, but the minutes of both reflect an alliance that was at times uneasy.

Whatever the politics behind the state survey and Crane's resulting report, the *Iowa Twenty-five Year Conservation Plan* was a pioneering effort. It was the first comprehensive, statewide conservation study that attempted to relate state park planning and development to broader resource conservation needs.[25] Crane and his assistant, George Wheeler Olcott, assembled an impressive team of experts to assist them in conducting the survey. Drawing from the ranks of college and university faculties, they enlisted the aid and advice of many, including John Fitzsimmons and P. H. Elwood for landscape architecture and planning, George Kay and J. H. Lees for geology, E. R. Harlan and

Benjamin Shambaugh for history, Charles R. Keyes for archaeology, and G. B. MacDonald for forestry. When state expertise failed, Crane went elsewhere for talent. Most notably, he turned to Aldo Leopold, who had surveyed Iowa's game populations in 1928 when he was writing his *Game Survey of the North Central States* (1931), a seminal work in scientific game management theory. Darling, who had met Leopold in 1928 when he was conducting his first survey, undoubtedly was responsible for bringing Leopold on board. Among consultants, Leopold was a dominant figure, but Crane and Olcott also consulted with the Bureau of Biological Survey, the Bureau of Fisheries, the Bureau of Agricultural Economics, and the National Conference on State Parks. In addition, they turned to local organizations and individuals for much information: county farm agents and county engineers, game wardens, sportsmen, ornithologists, women's clubs, business and civic organizations, and the Farmers' Union.[26]

The *Iowa Twenty-five Year Conservation Plan* was everything the 1919 Board of Conservation report was not. A sense of urgency pervaded the work of the board in the early years, and need correspondingly dictated a hastily compiled report in 1919. Throughout the 1920s, however, there was a growing realization that resource conservation problems could not be solved simply through parks and preserves or by the Board of Conservation alone. The state conservation plan thus addressed far more than parks and preserves. Crane himself ranked the state's outstanding conservation problems, in order of priority, as soil erosion control, lake and stream restoration, woodland conservation, and park expansion, including the provision of more hunting and fishing areas.[27] Nonetheless, in its final form, the thrust of the plan was toward "map[ping] out a program for the economical and orderly" development of "more and better recreation facilities." Conservation programs were integral aspects of the plan, but the capstone was recreational development. As the reasoning went, "every phase of public recreation . . . is dependent upon three major correlative factors—erosion control, the conservation of surface waters, and the conservation of forest and small cover on the lands."[28]

In a sense, the state conservation plan rationalized a host of ideas, dreams, and half-realized programs that had been stirring around together, but never really mixing, for more than a decade. Crane's genius was that he consulted practically everyone, spent enough time in the field to understand the territory, and had a great ability to translate ideas into feasible goals. At the center of the plan were parks and preserves, with the existing system to be expanded and reordered. A separate system of sixty to eighty preserves would include prehistoric and historic sites, unusual geological phenomena, areas containing rare plants, forest tracts, sites of outstanding scenic beauty, three water-power

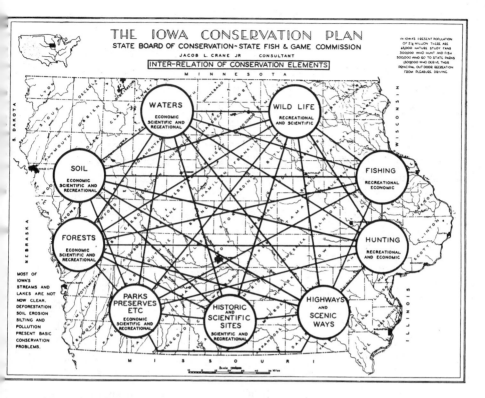

"Inter-relation of Conservation Elements" from *Iowa Twenty-five Year Conservation Plan*. *Courtesy Iowa Department of Natural Resources.*

ponds, and at least one large prairie tract. Preserves would admit some recreational use, but a half-dozen "sanctuaries" would not. The latter would be inviolate ranges for sharptail grouse, wild turkeys, prairie chickens, ruffed grouse, beaver, or deer. They also would provide the ultimate protection for rare plants, trees, and songbirds. Many state parks were to be reclassified as preserves in order to distinguish them from areas "intended to fill the demand in all parts of the state for recreation facilities." State parks, in other words, would include state holdings large enough "to accommodate intensive recreation by large crowds in a setting of relatively unspoiled natural landscape." All together, the plan called for seventeen state parks (ten existing and seven new) distributed across the state so that no Iowan had to drive more than two hours from home to enjoy "large scale outdoor recreation." [29]

In addition to the park and preserve systems, the plan envisioned a wildlife management system on a scale never before imagined. It called for constructing as many as thirty artificial lakes in the southern half of the state, dredging twenty-five existing lakes to enhance fishing as well as water recreation, and providing hundreds of public access points along streams and lakes. More than one hundred sites were designated as refuges for upland game and migratory waterfowl. All existing lakes and marshes were incorporated into the waterfowl management program, with separate areas set aside for refuges and public hunting grounds. Tying together all these parks, preserves, wildlife refuges, fishing spots, and public hunting grounds was a planned system of scenic highways with scores, if not hundreds, of wayside parks. The envisioned result: a "finished" countryside that would rival the northern European landscape once familiar to the ancestors of so many Iowans. To implement this vastly expanded recreational, preserve, and wildlife management system, Crane's report recommended, as had Jay Darling, consolidating the Board of Conservation and the Fish and Game Commission, and possibly the State Geological Survey, into one agency.[30]

At about the same time the state conservation survey was thrust on the Board of Conservation, Herbert Evison, secretary of the National Conference on State Parks, met with the board and then toured Iowa to gather information as part of a nationwide survey of state parks. The board subsequently asked him to provide "his frank opinion on the park system of Iowa."[31] Evison obliged with a lengthy report, which Crane surely studied, since he acknowledged Evison's contributions in the *Twenty-five Year Conservation Plan*. Evison's suggestions, however, had little apparent influence on the board's operations, no doubt because the recommendations of the Crane report were far more sweeping. Nonetheless, in retrospect the Evison report is an interesting document because it amplifies diverging opinions about what state park systems should encompass. Evison's report contained a lengthy critique of the fish hatchery at Backbone State Park. His remarks focused on the architectural design of the hatchery rather than its use, but they left no doubt that he considered fish hatcheries to be incidental uses that interfered with the recreational and aesthetic functions of state parks. Crane, in contrast, had directed a survey designed to expand and revamp the existing park system in order to make a greater place for fish and wildlife.

The twenty-five-year plan did indeed become the State Conservation Commission's primary reference, and remained so well into the 1950s. By 1941, however, the plan's instigator, Jay Darling, had also become its greatest critic. A Republican and self-proclaimed fiscal conservative, Darling viewed Roosevelt's New Deal programs with suspicion. He judged that more than half of the projects recommended in the plan had been completed or were in the process of completion as of 1941, but the means by which these projects had been realized—federal aid—had undermined the basic aim of comprehensive, coordinated, and continuous planning. "Public attention became captivated by the struggle for ccc camps," he complained, "and the background of unified planning on a continuous basis was lost sight of, and the general tendency now is to look for outside aid for specific projects instead of implementing the general program of state restoration." Without naming names, Darling also faulted lack of leadership from the governor's office, politics-as-usual, and, most important, a lazy citizenry. "The one lesson to be learned from this whole picture is that the public is interested in the benefits but will neither exercise themselves to see that proper methods are used to achieve those elements of restoration which provide benefits nor will they contribute any personal effort toward their accomplishment. There is still a complete lack of understanding of the fundamentals of conservation of natural resources in the state of Iowa." [32]

Darling might also have added that he took himself out of state affairs just when the state conservation plan was implemented. In 1934, he accepted a presidential appointment to what became known as the Beck Committee, so named for its chair, Tom Beck, editor of *Collier's* magazine. Despite his personal antipathy toward Franklin Roosevelt, colored by a deep and abiding friendship with Herbert Hoover, Darling would not refuse a president's call to national service. The three-person Beck Committee, which included Aldo Leopold as well as Darling, was charged with helping the Roosevelt administration devise a wildlife program that would complement its programs to take submarginal land out of agricultural production. Darling's performance on the Beck Committee brought unanticipated results. When Paul Redington chose to resign his position as chief of the Bureau of Biological Survey (forerunner of the U.S. Fish and Wildlife Service) rather than implement the recommendations of the Beck Committee report, Roosevelt tapped Leopold to replace him. When Leopold declined the offer, Roosevelt turned to Darling. To the amazement of almost everyone who knew him, Darling accepted, only, however, after extracting promises from the administration that he would be given a free hand to run the agency. During the eighteen months that Darling served as chief, he completely reorganized the BBS in order to give professional wildlife biologists greater authority. Equally important, with a hard-won initial appropriation of six million dollars under the 1934 Duck Stamp Act, Darling launched a program of developing national wildlife refuges based on biological research. [33]

These were impressive achievements that probably saved the Bureau of Biological Survey itself from extinction, but it meant that for all of 1934 and most of 1935 Darling was absent from Iowa. Coincidentally, these were two very critical years in terms of implementing the *Twenty-five Year Conservation Plan*.

Had Darling remained active in state conservation administration during this time, things might have gone very differently. But he did not. Although Darling had fulfilled an important role as protagonist, it was another prominent conservationist, Gilmoure B. MacDonald, who chiefly influenced the way federal moneys were spent to implement the state plan for resource conservation and outdoor recreational development.

The Transition Years: 1931–1935

From April 1931 to May 1935, the Board of Conservation and the Fish and Game Commission operated as separate entities, although their functions increasingly overlapped. In its first year of operation, the Fish and Game Commission spent more than $300,000, three times the amount appropriated to the Board of Conservation. Some of this money was spent on land acquisition to provide access to existing lakes and on various construction projects to enhance their fishing attributes, particularly dams to raise the water level. This, of course, brought the commission directly into the Board of Conservation's orbit, which is one of the reasons they often met jointly.[34]

During 1931 and 1932, Jacob Crane kept both agencies apprised of progress on the state conservation survey. Aldo Leopold also appeared in person from time to time in order to share the findings of his game survey and to promote recommendations for wildlife protection and habitat restoration. As a result of cooperating on the state conservation survey, and based on recommendations submitted by Leopold, the two agencies jointly adopted, in December 1932, a policy for establishing wildlife refuges. The policy defined two classes of refuges: "protective" refuges, which included state parks as well as other areas established "to safeguard" certain rare species, and "productive" refuges, areas intended "to produce an outflow of other species," meaning desired game species such as pheasants. Following adoption of the refuge policy, the Board of Conservation agreed to designate several state parks as game refuges. Even so, the board cautioned the commission not to use state parks for game restocking purposes unless so requested by the superintendent of parks.[35]

At Crane's recommendation, the board and the commission also began to consider sites for constructing artificial lakes, particularly in the southern part of the state. The rationale was twofold, designed to benefit the interests of both agencies. First, artificial lakes would expand the park system into areas that held few public recreational spots. Second, artificial lakes would enhance the fish and game potential in southern Iowa. As the conservation plan neared completion, both agencies agreed that, in the case of development projects, neither would deviate from recommendations in the plan without consulting

Crane first.[36] Until New Deal programs started in 1933, the Fish and Game Commission had considerably more money to spend on land acquisition, thus it took the lead in developing artificial lakes. However, federal funds greatly accelerated the pace of construction. By mid-1934, Lake Wapello in Davis County was nearing completion. This was "the first of the new lakes," a combination fishing spot and wildlife refuge. Nine more artificial lakes were in various stages of completion, six of them as joint projects with the Board of Conservation.[37]

For a time, the board and the commission cooperated to study means of controlling lake pollution. In this effort they were frustrated by the Department of Health's lack of leadership; and they deplored the irresponsible actions of communities, particularly those around Iowa's Great Lakes, who dumped their sewage into the state's waterways in utter disregard of fish life. When federal aid became available to build public works projects, the Board of Conservation heartily endorsed using such funds to build municipal sewage treatment plants, a subject this chapter will return to.[38] Conversely, the annual algae problems of the Iowa Great Lakes, aggravated by sewage pollution, caused the Fish and Game Commission to rethink its policies on water pollution. It finally went on record as "not favoring the expenditure of public funds on water areas where extreme pollution is being carried on by communities."[39] Stream and lake pollution became, instead, the commission's chief rationale for spending money on artificial lakes. "There are numerous streams all over southern Iowa," the commission's 1934 report stated, "but they carry all the silt and pollution from farther north and the possibilities of ever bringing them back to, much less maintaining them in, any degree of satisfactory fishing condition were remote."[40]

Beginning in 1933, the availability of federal money for conservation and park programs forced even greater coordination between the board and the commission. Their individual and joint programs to create artificial lakes, restore existing lakes, and develop state parks were greatly enhanced by federal aid. However, before the state conservation plan was unveiled and before federal aid started flowing, at least one local community raised money to put relief workers in state parks. In 1932, the City of Estherville provided funds to hire unemployed men, who worked under the supervision of Park Superintendent M. L. Hutton and State Landscape Architect John Fitzsimmons, to build a custodian's residence and amphitheater in Fort Defiance State Park. After the Iowa State Emergency Relief Committee was established in early 1933, the board conferred with it regarding a regular program of using relief workers in other state parks.[41]

It was a stroke of immense good fortune that the *Iowa Twenty-five Year*

Conservation Plan was completed at the same time President Roosevelt authorized the Civilian Conservation Corps. To receive federal aid, states first had to submit a plan outlining how such aid would be used. Iowa was among the first states to qualify. G. B. MacDonald, appointed state director of the Emergency Conservation Work program, was the federal government's point man in Iowa with principal authority over all Civilian Conservation Corps camps. In April 1933, sixteen camps were authorized for Iowa. MacDonald conferred with both the board and the commission to establish procedures for using some of these camps in state parks. It was decided that M. L. Hutton would take charge of all engineering work on ECW projects under the board's jurisdiction, and both bodies would appoint members to a joint executive committee that would coordinate ECW projects with MacDonald. In February 1934, the two agencies approved, jointly, development plans for thirty-two projects in state parks. They also approved the proposed organization of a State Planning Board, providing that it would have no direct control over the actions of the board or the commission.[42]

The Iowa State Planning Board played a secondary role in park and conservation planning during the New Deal era, the main tasks already having been laid out in the *Twenty-five Year Conservation Plan*. Nonetheless, the planning board's studies and reports provided an opportunity to reemphasize and augment the goals of the conservation plan. For instance, the State Planning Board report placed even greater emphasis on the recreational value of parks than had the state conservation plan. The "human conservation" argument of two decades earlier, which had justified parks as oases of pleasure for working-class families, now gave way to a new sociologic rationale that stressed the need for recreational outlets to absorb increasing amounts of leisure time. To be sure, high unemployment accounted for some short-term increase in "leisure" time, but labor reform legislation and changes in the workplace also led to dramatic declines in the average number of hours workers spent on the job. The planning board thus predicted steadily increasing public demand for recreational areas. To meet this demand, the board reiterated its support for the state park and recreational program set forth in the state plan. Additionally, the planning board strongly recommended enabling legislation to foster the acquisition and development of county parks. Visualizing county parks as the "connecting link" between municipal and state parks, the planning board plotted out a system that filled the gaps between state parks. The idea behind this "infill system" was to place most Iowans within ten miles of a "well-preserved and managed" county park.[43]

Likewise, the State Planning Board reiterated support for acquiring, restoring, and preserving both prehistoric and historic sites. Acting on information compiled by archaeologist Charles R. Keyes, with assistance from many, the planning board identified sixty-two sites worthy of state acquisition. The list of sixty-two marked Iowa's first real statewide survey of historical remains and the first attempt to establish a benchmark for what should be considered historically significant. Adding its recommendations to a long line of others, the planning board once again called for federal protection of the prehistoric Indian mounds in northeast Iowa. Even so, the report made it clear that many of the sites on the list were neither of national nor statewide importance. Prefiguring the emergence of professional historic preservation standards and guidelines in the 1960s, the planning board declared the selected sites as important because "they are of value in themselves as typical and worthy objects marking milestones of progress and growth in Iowa. . . . Each site, each object and each location are the physical remains of the age in which they were created. They are the illustrative material to accompany that story of Iowa's growth and development." Notably, the planning board departed from earlier attempts to establish preservation of historic material culture as a function of the Board of Conservation. Instead, it recommended site acquisition either through the Board of Conservation, the State Historical Department, or the State Historical Society.[44]

In 1935, the legislature acted on the central recommendation of the *Twenty-five Year Conservation Plan*, merging the Board of Conservation and the Fish and Game Commission into one entity. The new agency became known, variously, as the State Conservation Commission or the Iowa Conservation Commission, legally constituted of seven members appointed by the governor, plus an administrative and field staff.[45] Margo Frankel was the only member of the Board of Conservation, and the only woman, to be appointed. As senior representative from either parent board, she also served as the first chair of the State Conservation Commission. Two members of the Fish and Game Commission were appointed to serve on the new commission: A. E. Rapp of Council Bluffs and Dr. W. C. Boone of Ottumwa. The other four members were Dr. Frank J. Colby of Forest City, Dr. E. E. Speaker of Lake View, Logan Blizzard of McGregor, and W. A. Burhans of Burlington. By design, it was both a bipartisan and a geographically representative group.

Organizationally, the identity of the two parent bodies was preserved by assigning state park functions to a new lands and waters division, fish and game functions to a division of the same name. G. B. MacDonald's shadow position as state forester was made official, a move that recognized forestry as a legitimate function of the conservation commission. Forestry activities were officially reported with those of Lands and Waters, but MacDonald actually functioned as an autonomous figure, often consulting directly with the commissioners. A new administrative division had some responsibility for coordinating the work of the entire agency. However, funding provisions of the 1935 law reinforced the distinct functions of Lands and Waters and Fish and Game by reserving the fish and game protection fund as the exclusive financial resource of the latter division, minus a portion sufficient to pay one half of the agency's administrative expenses. Operating funds for the Lands and Waters Division continued to come from biennial appropriations. In a move that foreshadowed the relative status of state parks versus fish and game functions within the agency for the remainder of the decade, the commissioners selected M. L. Hutton, former superintendent of state parks, rather than I. T. Bode, former secretary of the Fish and Game Commission, to be the new director of the State Conservation Commission.[46]

Expediency dictated that the work of the professional staff be minimally disrupted by reorganization; New Deal programs were fully operational, and no one had time to figure out how to do a new job. Further, the organizational divisions and their separate funding sources would hinder the kind of integrated resource management envisioned in the state conservation plan. Coordination at the top, however, was initially strengthened by the commissioners, who assigned themselves each a set of counties to supervise, continuing a long-standing practice of the old Board of Conservation.[47] To some observers, the commission seemed to be holding to an outmoded procedural method. In practice, however, the district system of supervision meant that every commissioner had the opportunity to see, first-hand, how the resource management functions of the agency were interrelated in the field. Parks, lakes and streams, wildlife, and forests were now under one umbrella agency, and that decision would not be reversed or compromised.

Toward a Resource Agency: New Deal–Style Conservation in Iowa

"The first time I saw the light on some finances that would make possible the execution of the Twenty-five Year Plan was this," John Fitzsimmons recalled in 1941. "I had been looking over Lake Wapello for the Fish and Game Commission, and was in the dining room of the hotel at Ottumwa. Mr. Darling came and rapped on the window and said, 'Fitz! What do you think! Uncle Sam has just given us all the money we need. We can go ahead with the Twenty-five Year Plan.'" The long-range conservation plan was precisely the kind of state commitment that President Franklin Roosevelt was looking for in the spring of 1933 in order to launch his Civilian Conservation Corps. "Professor MacDonald ... was given a very definite sign of approval," Fitzsimmons continued, "when the President of the United States said, 'Give Iowa all it wants.' We had the first ccc camps west of the Mississippi River."[48]

With sixteen CCC camps in place and thirty-two projects approved, the New Deal came to Iowa in mid-1933. Federal aid would enable much of the state conservation plan to materialize in less than a decade, although progress would be uneven. As in the past, personalities would shape plan implementation, although no one would ever hold so firm a hand as Louis Pammel had in the 1920s. Still, G. B. MacDonald, who directed the flow of federal funds and who had a long history with conservation in Iowa, would have a lasting influence.

Beginning in 1918, MacDonald had served as deputy state forester under the Iowa Department of Agriculture. In this capacity he had worked closely with the State Board of Conservation, helping to regenerate woodlands on newly set aside parks and preserves. Even though he and Louis Pammel had failed in the 1920s to secure legislation that would transfer forestry from the Department of Agriculture to the Board of Conservation, MacDonald poured more personal energy into the cause when Jacob Crane was hired to produce a state conservation plan. Only the preliminary findings of his forestry survey were incorporated into the *Twenty-five Year Conservation Plan*. He completed the work under the auspices of the State Planning Board, taking the time to produce detailed information about tree species and soil conditions for forestry across the state.[49]

Importantly, MacDonald undertook the forest and wasteland survey anticipating federal involvement, whereas the rest of the state conservation plan envisioned no such partnership. His preliminary recommendations contained in the 1933 state conservation plan attracted attention among Robert Fechner's staff at Civilian Conservation Corps headquarters in Washington, D.C., particularly because MacDonald articulated the relationships among forestry and soil erosion and farming. New Dealers fixed on MacDonald as their man to carry out the CCC mission in part because his vision meshed with Roosevelt's goals for the CCC and in part because he was open to working with federal agencies. As a result, MacDonald served as state director of the Emergency Conservation Work program, the CCC's bureaucratic home from 1933 to 1938. Because of his special concerns for forestry and soil conservation in an agricultural state, many of the CCC camps operating in Iowa were, at one time or another, assigned to reforestation or soil erosion control work. Even so, the Civilian Conservation Corps, and other New Deal relief work programs, are primarily remembered for their work in Iowa's state parks.

State Parks

The CCC contributed more to state park development than any other federal relief program. Between 1933 and 1942, 2.5 million men between the ages of seventeen and twenty-eight "enrolled" in the Civilian Conservation Corps nationwide, where they were put to work planting trees, improving national and state parks, restoring historic sites, fighting forest fires, assisting flood victims, and generally caring for the nation's resources. Of this number, nearly 46,000 CCC enrollees worked on projects in Iowa. At peak construction in mid-1935, forty-six camps were operating in Iowa, although the average number of camps was closer to thirty throughout most of the program's life.[50] The bureaucracy developed to implement projects remains one of the most fascinating case studies in the history of intergovernmental relations; and the CCC is often cited as the most popular and most effective of all the New Deal programs. Its twin goals of employing young, single men and engaging them in meaningful work to restore the public domain enjoyed widespread public support. The U.S. Department of Agriculture and the U.S. Department of Interior provided technical services through several agencies, including the National Forest Service, the National Park Service, the Soil Conservation Service, and the Biological Survey. Technical services included planning and supervising work projects, furnishing equipment, and transporting enrollees to work projects. The Department of Labor made the final selection of enrollees, and the Department of the Army housed, fed, and clothed them. In addition, the Army supervised all CCC camp activities, including programs of physical exercise, counseling, and educational training.[51]

As director of the ECW in Iowa, MacDonald shepherded proposals through the chain of command. Other duties included recommending camp assignments, supervising technical staff, and administering federal and state funds. Federal funds paid for supervisory and technical staff, equipment, and construction materials. Just how much federal aid Iowa received is an elusive figure, but as of 1940, the State Conservation Commission estimated the value of "artificial construction" in state parks at $5 million, most of it coming from federal sources.[52] Factoring in all the related costs that went into planning and construction would raise this figure substantially. For instance, during the calendar year 1935, the highest level of CCC operations, the combined total of funds expended in Iowa was nearly $6.1 million for labor, CCC camp construction, food, direct project expenses, and operational expenses.[53] Direct project expenditures from July 1935 to June 1938 were reported at approximately $3.9 million for land acquisition, lake dredging, surveys, and construction. Of this figure, $3.7 million came from federal sources. State expenditures covered a much smaller portion of the total cost, although the legislature did augment State Conservation Commission funds in 1934, 1935, and again in 1937, with supplemental appropriations totaling about $1 million, which went into a special Emergency Relief Administration Fund administered solely by G. B. Mac-

Donald. State dollars were used variously to acquire land, purchase equipment and materials, and pay technical staff workers who were not covered by federal funds. Importantly, the state legislature also began to appropriate smaller sums from time to time for special projects such as lake dredging and pollution control; this practice would become a pattern in subsequent decades.[54]

In operation, the CCC was a flexible organization, and camp assignments frequently shifted. Of the total forty-six camps organized in Iowa, forty-one were assigned, at one time or another, to state park projects. CCC enrollees built an untold number of buildings, dams, bridges, water systems, sewer systems, fish-rearing ponds, trails, roads, and fences. Approximately seven hundred of these structures remain standing. They also excavated artificial lakes, rehabilitated drainage ditches, built terraces and earthen dikes, collected seeds, and planted trees, reshaping as well as restoring the natural landscape. As projects changed, so did camp assignments and camp locations. During the life of the CCC, anywhere from fifteen to twenty-five camps were assigned, off and on, to Soil Conservation Service projects. Another two or three camps generally were on assignment in state forests at any one time. Occasional assignments included working on drainage levees and conducting biological surveys.[55]

As important as the CCC was to park development in Iowa, the contributions made through four other New Deal agencies cannot be overlooked. The Works Progress Administration, the Public Works Administration, the Civil Works Administration, and the National Youth Administration also put people to work in parks. WPA workers, for instance, constructed buildings, fabricated park signs, dug water and sewer systems, and built roads in twenty-seven state parks. Larger civil works projects, such as concrete bridges and dams, generally fell within the purview of the CWA and the PWA, although WPA or CCC crews might also be assigned to such projects. It is not clear what criteria, if any, determined which agencies would be involved in which projects, but the overall ad hoc character of the New Deal kept day-to-day operations in a state of constant flux. Many decisions undoubtedly turned on the funding and personnel available at any given time. The speed at which work progressed was simply amazing. It has been estimated that in the first two years of operation, New Deal programs had advanced park development at the national, state, and municipal levels by at least a decade.[56] This was certainly true in Iowa.

State park development also depended upon close cooperation with the National Park Service in order to meet federal design standards. In mid-1933, Ames B. Emery, an NPS architect assigned to Iowa, appeared before the board to present the rustic architectural design principles that NPS had adopted for park buildings, furniture, and other features.[57] His appearance was part courtesy call, since John Fitzsimmons had already implemented the rustic style in Iowa's parks. Judging from the few park buildings that had been constructed as of 1933, Fitzsimmons's interpretation of rustic architecture lacked the formulaic quality that came to be associated with the style during the 1930s, but he adhered to the use of native materials and the philosophy of blending manmade structures into the natural landscape. In any case, Emery had the board's assurance that it would cooperate with the Park Service; and, indeed, as the hammers began to fly, the board came to rely more and more on standardized building plans supplied by NPS designers.

In practice, all development plans for state parks originated with the board (after July 1935, the State Conservation Commission). While the board approved overall park development plans, Margo Frankel was chiefly responsible for approving all design and construction plans. By this time, she had been recognized by the American Scenic and Historic Preservation Society as one of the outstanding leaders, nationwide, in park development.[58] Thus, other board members deferred to her judgment. Initially, much of the design work took place in the Central Design Office of the Extension Service at Iowa State College under Fitzsimmons's supervision. After plans had been approved by the board/commission, they were submitted to the National Park Service for approval. Iowa first reported to the Indianapolis Branch Office of Planning and Design, later to the Omaha Branch. The use of standard building plans enabled projects to proceed swiftly. By early 1936, the pace of park development was such that the inspector for Iowa moved his office from Omaha to Ames in order to keep up with the workload. Inspectors checked and approved plans, participated in the design planning, and visited parks to inspect work in progress.[59]

Scattered evidence indicates that Fitzsimmons and the Central Design Office sometimes chafed under the restrictions imposed by federal guidelines. For instance, NPS inspectors felt that the Central Design Office had a tendency to overdo its designs—unnecessarily large bathhouses, interior ornamentation, decorative light fixtures, extensive landscaping, that sort of thing. Fitzsimmons and his group were asked more than once to keep designs simple, with construction demands that were within the skill level of CCC enrollees. The Board of Conservation, however, did not want simplicity to slip into "cheap." In July 1934, it advised G. B. MacDonald not to approve any more shed-type roofs on state park buildings, nor to approve buildings with rolled roofing. The board also protested the hiring of CCC foremen who were unqualified to supervise construction.[60]

Since Margo Frankel handled park design matters for the board, presumably it was she who insisted on reasonably high design and construction stan-

Left to right: Walter H. Beall (Northeast Iowa National Park Association), Margo Frankel (Board of Conservation, 1927–1937), unidentified man, Roger W. Toll (superintendent, Yellowstone National Park), Mabel Volland (Board of Conservation, 1929–1932), somewhere on the Upper Mississippi River, 1931. *Courtesy State Historical Society of Iowa.* In the 1920s and early 1930s the Board of Conservation supported a multistate movement to establish a national park on the Upper Mississippi River. After Congress passed the Upper Mississippi River National Park bill in 1930, the National Park Service sent Roger Toll on a five-day investigative trip, after which he recommended against a national park. Effigy Mounds National Monument, however, was later established because of this broad-based effort.

dards. This may have been one reason that the board, and later the State Conservation Commission, utilized WPA workers, typically more skilled, so extensively on park projects. Another reason may have been that the National Park Service did not work fast enough for MacDonald. In July 1935, at MacDonald's urging, the Conservation Commission notified officials in Washington that if the NPS regional office in Omaha did not start moving the paperwork faster, it would send a delegation to D.C. in order to "adjust" matters.[61] Delays and ruffled feathers aside, five New Deal programs made it possible for Iowa to improve twenty-five existing state parks between 1933 and 1942. New construction included everything from drinking fountains to toboggan slides. Table 4 summarizes the extent to which New Deal conservation

Table 4.

New Deal Work in Existing State Parks, 1933–1942

Park	County	Date Acquired	Improvements under New Deal Program(s)
Backbone	Delaware	1918	CCC
Bixby	Clayton	1926	CCC
Ambrose A. Call	Kossuth	1925	PWA, WPA
Clear Lake	Cerro Gordo	1924	WPA
Dolliver Memorial	Webster	1920	CCC, WPA
Farmington	Van Buren	1919	WPA
Flint Hills	Des Moines	1925	WPA
Ft. Defiance	Emmet	1923	CCC, WPA
Lacey-Keosauqua	Van Buren	1919	CCC, WPA
Lake Manawa	Pottawattamie	1932	CCC, WPA
Ledges	Boone	1920	WPA
Lewis and Clark	Monona	1924	WPA
Maquoketa Caves	Jackson	1921	CCC, WPA
Okamanpeden	Emmet	1923	CCC
Palisades-Kepler	Linn	1922	CCC, WPA
Pammel	Madison	1923	CCC, WPA, CWA
Pillsbury Point	Dickinson	1928	CCC
Pilot Knob	Hancock / Winnebago	1921	CCC, WPA
Pine Lake	Hardin	1920	CCC, WPA
Rice Lake	Winnebago & Worth	1924	CCC
Springbrook	Guthrie	1926	CCC
Twin Lakes	Calhoun	1923	CCC
Walnut Woods	Polk	1925	CCC, WPA
Waubonsie	Fremont	1926	CCC, WPA
Wildcat Den	Muscatine	1926	WPA, CWA, NYA

CCC = Civilian Conservation Corps, CWA = Civil Works Administration, WPA = Works Progress Administration, NYA = National Youth Authority, PWA = Public Works Administration.

Table 5.

New State Parks, 1933–1942

Park	County	Date Acquired	Improvements under New Deal Program(s)
Beeds Lake	Franklin	1934	CCC, WPA
Black Hawk Lake	Sac	1934	CCC, WPA
Brush Creek Canyon	Fayette	1936	CCC
Echo Valley	Fayette	1934	CCC, WPA
Geode	Henry	1937	CCC
Gull Point	Dickinson	1934	CCC
Heery Woods	Butler	1935	WPA
Lake Ahquabi	Warren	1934	CCC
Lake Keomah	Mahaska	1934	CCC, NYA
Lake Macbride	Johnson	1933	CCC, WPA
Lake of Three Fires	Taylor	1935	CCC, WPA, PWA
Lake Wapello	Davis	1932	CCC
Mill Creek	O'Brien	1935	WPA
Mini Wakan	Dickinson	1933	CCC
Pike's Peak	Clayton	1936	CCC
Pike's Point	Dickinson	1936	CCC
Preparation Canyon	Monona	1934	CCC
Red Haw Hill	Lucas	1936	CCC, WPA, CWA
Stone	Woodbury	1935	CCC
Swan Lake	Carroll	1933	WPA
Union Grove	Tama	1940	WPA
Wanata	Clay	1934	CCC

and relief programs contributed to development in these state parks. Between 1933 and 1942, Iowa also added and improved more than twenty parks and recreation areas (table 5). Many of these were designed around new artificial lakes.

The magnitude of federal aid established the National Park Service as arbiter of park design. Rustic buildings constructed of native materials and sited in naturalistic settings became *the* park aesthetic of the 1930s. Federal aid also stimulated an unprecedented level of park development throughout the United States. Iowa was but one of thirty-seven states that acquired new lands and expanded their park systems. The ccc program prompted another eight states—Colorado, Mississippi, Montana, New Mexico, Oklahoma, South Carolina, Virginia, and West Virginia—to establish their first state parks. Because of its achievements in the 1920s, Iowa was among a handful of states considered to be in the "mainstream park movement" during the 1930s. Others included Indiana, Pennsylvania, New York, Ohio, and Illinois.[62]

New Deal work relief and conservation programs presented an unprecedented opportunity for the National Park Service to promote much wider adoption of the rustic design principles that had been evolving both in national and state park systems since the late 1910s. In 1933, Dorothy Waugh began compiling, for the NPS, designs and technical information for administration, community, and service buildings; privies; picnic tables, shelters, and fireplaces; park benches and trail steps; entranceways, lookouts, and guardrails; and various other types of park structures. The NPS then circulated these designs in loose-leaf portfolios for use on CCC and PWA park construction projects. In compiling these portfolios, Waugh relied heavily on existing examples, drawing on a body of design work already developed for several state, county, and municipal park systems, Iowa's system included. As construction progressed in parks throughout the United States, the NPS highlighted outstanding examples in two additional publications, *Park Structures and Facilities* (1935) and the three-volume *Park and Recreation Structures* (1938), both edited by Albert Good. These two publications also contained Iowa examples. Good's 1938 work, in particular, featured photographs or architectural drawings of stone barriers, picnic shelters, cabins, hearth rings, camp stoves, picnic tables, and trailside benches located at Backbone, Dolliver, Swan Lake, Springbrook, Pine Lake, Ledges, and Gitchi Manitou state parks.[63]

Backbone State Park

Nothing demonstrates the ascendancy of recreation in the 1930s better than Backbone State Park in Delaware County. A long, irregularly shaped tract that stretches approximately four miles from north to south, Backbone is situated between the towns of Strawberry Point on the north and Dundee on the south. The park's borders encompass one of the most interesting geologic formations in Iowa: a small area cut through otherwise agricultural fields where the Maquoketa River loops along a narrow, winding limestone ridge. The summit of the ridge rises from 90 feet to 140 feet above the river, and time has carved the exposed surfaces into rugged, picturesque forms and deep ravines. When the park was dedicated in May 1920, it covered 1,280 acres. Small parcels were added throughout the 1920s and 1930s, and by 1940 Backbone had grown to 1,415 acres. Between 1950 and 1974, another 368 acres were added. Today, state holdings cover approximately 1,784 acres, 186 of which have been designated as a state forest.

Backbone was the first state park, an honor more than happenstance. To many people, Backbone was the paradigm. Natural scientists, who frequented the place locally known as the Devil's Backbone during the late nineteenth century, marveled at the beauty of this craggy canyon along the Maquoketa River, tucked into gently rolling farm country. As the park movement gained strength, so grew the idea of preserving the Devil's Backbone area for public benefit. Thus it was that even before the Board of Conservation was organized, a local group initiated the process of securing land in the Backbone region for state park purposes. When the board convened for the first time in December 1918, it followed through by voting to acquire at least twelve hundred acres.

In keeping with board policy, comparatively little development took place until 1933. G. B. MacDonald oversaw a limited reforestation program; a stand of white pines was planted in 1922, and in 1928 the Iowa Daughters of the American Revolution donated stock for an additional stand of six thousand trees. Landscape architect E. A. Piester worked with Delaware County engineers and the State Highway Commission to design a park road system that was completed in December 1924. Sometime during the early 1920s a stone barn and a stone pumphouse also were constructed, both without board approval. The appearance of these buildings went hand-in-hand with repeated efforts to keep local citizens from constructing private summer cottages in the park or near the entrances.[64]

The year 1925 marked the real beginning of development in the park, although the extent of improvements between then and 1933 was modest compared to the development that followed. After several years of discussions with the Fish and Game Department, the board authorized construction of a trout fish hatchery on a sheltered grassy meadow not far from the Maquoketa River. This was the hatchery that Herbert Evison later found so objectionable. In 1925, the board also authorized funds to convert the "unauthorized" stone barn into a custodian's residence and to construct a new barn adjacent to it, the work done by Anamosa prison inmates. John Fitzsimmons drew up designs for both buildings, handsome structures with lower walls of stone construction, reflecting the rustic architectural style then coming into vogue for park buildings. Additionally, the board authorized construction of stone portals at the north entrance. Landscape architect Francis A. Robinson designed the portals, simple rectangular columns with hipped caps. Local citizens from Strawberry Point provided the stone and sand for the portals; Anamosa inmates constructed them. All of these structures remain standing.[65]

The board authorized construction of one other building before the extensive CCC building projects that began in 1933. This was an auditorium, still standing, located near Richmond Springs at the north end of the park. A lovely, open-wall stone structure designed by Fitzsimmons, the auditorium was constructed in 1931. By this time Pammel had been gone from the board for four years, and the policy of limited development in state parks was giving way to new ideas based on the recreational value of parks. The auditorium was to have been but one building of a large nature-study complex linked to Richmond Springs by foot trails. Conceptual plans show a proposed complex of no fewer than three lodges, twenty-one cabins, a dining hall, and three service buildings.[66] In one form or another the nature-study complex was incorporated into subsequent park development plans. Conceptual drawings turned into actual work plans after the CCC was authorized in March 1933.

For reasons that are unstated in the record, Fitzsimmons sent Thomas Macbride some of the material detailing plans for developing Backbone with the nature-study complex, multiple camping and picnic areas, and an expanded network of roads and trails. Macbride, the aging dean of conservation, was aghast. "I am frightened when you talk of landscape-development," Macbride confessed to Fitzsimmons. "For the wild things, the birds and trees, we fain would save from threatened absolute destruction, *Nature's landscape* is exactly right. The wild woods and thickets undisturbed, grasses, sedges, composites, hazels, cedars are precisely right. To these we can add nothing."[67] In another letter, Macbride opened with two pages of anecdotal introduction recalling the extraordinary effort it had taken to arouse public and political interest in a park system, then pass the state legislation that enabled conservationists to bring the Backbone under state ownership. With that preface, he expressed dismay that Backbone State Park seemed to be going the way of Mt. Rainier and other national parks. "Is all the quiet beauty of our park, like Mt. Rainier at Seattle, simply a hot kitchen ... where we get chicken dinner?"[68]

Macbride's outpourings elucidate the shift in fundamental concepts underlying the mission of Iowa's state park system between the late 1890s, when parks were just a hazy dream, and the early 1930s, when Iowa embarked on a building program that would create the park system we recognize today. The streak of moralism that fired the first generation of conservationists had cooled. To protect natural resources, conservationists had cut an epistemological pathway between resource preservation and public access. Macbride could give lip service to the multiple-use concept, which was the middle way, but he was steeped in the old school. In the end, he had great difficulty accepting the practical results of this approach. His exchange with Fitzsimmons signifies the tension between those who saw the state park system as a vehicle for protecting splendid natural areas and those who saw parks as public playgrounds in splendid natural, or even just naturalistic, settings.

During the 1930s, recreational development became the driving force in park design, eclipsing Macbride's philosophy. The trend acknowledged public demand, and it proceeded under the guise of "multiple use," the catchword of conservation in the 1920s and 1930s. In simple terms, landscape architects, also known as landscape engineers in the parlance of the day, had become the professional experts of park design and planning. For states that initiated state park development in the 1930s, or in those states where development had proceeded with aesthetic and recreational considerations uppermost, the shift was less noticeable. However, in Iowa, where decisions about park acquisition and development had long included some consideration of ecological values, the trend strained relations. People like Macbride, Jay Darling, and G. B. MacDonald, who were trained in the sciences, were apprehensive about transferring the reins of park development to landscape "designers."

The tension can be explained in the way the two sides defined the term "multiple use." On the one hand, for Macbride it meant protecting native species in their habitats and, at the same time, providing quiet places for human respite and spiritual regeneration. For Darling, it meant restoring wildlife species and their habitats and, at the same time, expanding the recreational outlets the public demanded. For MacDonald, it meant restoring woodlands to curb soil erosion and provide wildlife cover and, at the same time, accommodating public recreation. Landscape architects, on the other hand, tended to define the term much more narrowly. For them, "multiple use" meant chiefly "multiple recreational use." This led directly to the design theory of "use areas," that is, concentrating points of recreational use within parks so as to minimize the human impact to natural scenery.[69]

By 1942, Backbone was Iowa's premier example of a multiple-use state park as defined both ways. Not only was it one of the largest parks in the system, it was also the most extensively developed. From 1923, when the road system was built, to 1931, when the auditorium was completed, Backbone had been "improved" with perhaps as many as fifteen major structures. During the period of CCC construction, 1933 to 1942, two camps, SP2 (at Dundee) and SP17 (in the park), built at least seventy additional structures, approximately half of which could be considered major, including picnic shelters, cabins, boat- and bathhouses, overlooks, fish hatchery buildings, bridges, and latrines.[70]

The areas developed within the park during the 1930s were differentiated by recreational function, "use areas" in the vocabulary of landscape architects. In the southern area of the park, CCC crews developed an overnight cabin and recreational complex around a new artificial lake. During the late nineteenth century, a low-head dam had been built across the Maquoketa River to power a mill, creating a small pool of water. This was expanded with the construction of a larger dam that created a 125-acre lake. Construction of a bathhouse, a swimming beach, a boathouse, several overnight cabins, and an overlook marked the lake as a resort area for swimming and boating rather than fishing. The center of the park was developed as a picnicking, hiking, and camping area. Recreational facilities here included walking trails and trail steps, stone benches, picnic shelters, and a concession building. In the north end of the park, Richmond Springs was "enhanced" with rock work and linked to the auditorium with trail steps, although, as Macbride had wished, the area around the auditorium itself was left in a natural state. There would be no lodges, no dining hall, no service buildings for a nature-study center, although CCC crews did build hiking trails and latrines. However, the trout hatchery, also in the north area, was expanded considerably with a custodian's residence, new service buildings, new fish rearing ponds, and trout raceways.[71] The hatchery represented multiple use in the other, more inclusive, sense. Moreover, the stand of white pines located within park boundaries eventually would be redesignated as a state forest.

Macbride died in 1934, so he did not witness the actual transformation of his beloved park into recreational "use areas." Nor did he live long enough to see the state park north of Iowa City that bears his name. Among the eighteen artificial lakes created during the 1930s was one near Solon, in Johnson County. The lake itself started as a Fish and Game Commission project, with $32,000 in local support.[72] In keeping with the recommendations of the *Twenty-five Year Conservation Plan*, the initial purpose was to provide an area for fishing, hunting, and waterfowl and game production near two major urban centers: Iowa City and Cedar Rapids. In late 1933, CCC crews began constructing a dam that would impound runoff water to form a 138-acre artificial lake. The project quickly took on an expanded recreational character, however, a bow to public demand. In addition to the lake, CCC crews built a dining hall and a bathhouse as well as custodial and maintenance buildings.[73] By early 1937, the park was ready for public use, and a private concessionaire was on hand to sell refreshments as well as rent boats and beach equipment to the more than 50,000 visitors who came to Lake Macbride during its first full year of operation. Annual park use tripled within as many years, proving the extent of public demand for water-based recreation areas.[74] Whatever the merits of Lake Macbride State Park, however, it is inconceivable that Thomas Macbride would have been flattered to be its namesake.

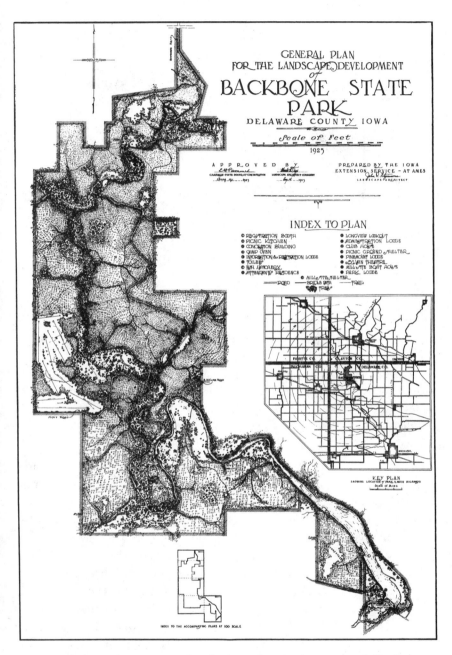

Backbone State Park, 1925 general park plan, John Fitzsimmons and Charles Diggs. *Courtesy Iowa Department of Natural Resources.* The index lists several planned facilities, some of which were never constructed. Stippling indicates the extent of planned reforestation, much of which was carried out. A historic millsite was located at the extreme southern tip of the lake.

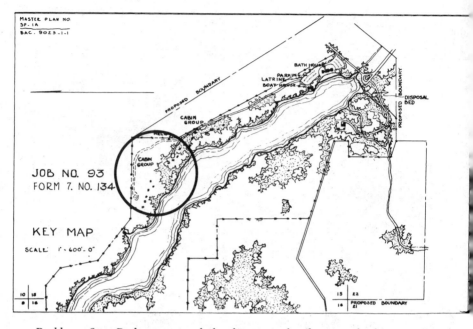

Backbone State Park, c. 1937 park development plan for second cabin group at
southern end of Backbone Lake. *Courtesy Iowa Department of Natural Resources.*
Note the expanded boundaries. The millsite has been replaced by "disposal
beds." The shore area has been developed with cabins, boathouse, bathhouse,
parking areas, and (unmarked) an assistant custodian's residence (now a
campground).

The focus on recreational development tended to make Iowa's state parks,
and state parks everywhere else in the country, look more and function more
like scaled- down versions of national parks catering to tourists. When the New
Deal came to a close in 1942, the system contained eighty-seven parks, preserves, forests, historic sites, waysides, and parkways. There were seventy-
three well-equipped overnight cabins in nine state parks, and public demand
for many more. Lake Ahquabi and Dolliver Memorial each had group camping facilities, including cabin clusters, dining and assembly halls, and shower
rooms. A third group camping complex was under construction at Springbrook State Park. Tent camps were available in thirty-two parks. Three natural
lakes had been dredged to enhance boating and fishing; eighteen new artificial
lakes were open for public use.

Private concessionaires ran boat liveries, dining halls, refreshment stands,
and bathhouses in twenty-seven state parks. Steadily increasing numbers of
private docks and commercial launches further pointed to the popularity of

CCC boys relaxing in front of the new bathhouse at Backbone Lake, c. 1935.
Courtesy CCC Museum, Backbone State Park, Iowa Department of Natural Resources.

motor- and sailboating. To meet the recreational demand, some parks even stayed open year-round. For winter sports enthusiasts, Stone State Park near Sioux City offered an ice rink, and toboggan and ski runs. Portable toboggan slides were deployed at Fort Defiance, Pine Lake, and Pilot Knob. Skaters could enjoy cleared and smoothed ice on numerous lakes; and at some lakes, concessionaires provided night lighting, rented skates, and sold hot drinks and sandwiches.[75] The State Conservation Commission, however, drew the line at tennis courts, swimming pools, and playground equipment. Despite constant requests for such facilities, the commission would not countenance these as fitting uses for state parks.[76]

F. T. Schwob, who was named director of the State Conservation Commission when M. L. Hutton died in 1941, admitted in retrospect that Iowa had been shortsighted to focus park development so heavily on recreation. "In all of our planning and development work," he noted, "there never was any attention or thought given to any phase of the conservation program except the provision of recreation." Among the mistakes, he noted that the state had failed to achieve effective agency coordination in order to "utilize natural resources wisely." For instance, in carrying out the artificial lake construction program, the commission rarely had acquired sufficient land in the surround-

Iowa's State Parks and Preserves, 1937, center map. Courtesy State Historical Society of Iowa, Iowa City.

ing watershed to control siltation from soil erosion. Likewise, none of the dams to impound water for artificial lakes had been designed for flood control, even though the vast majority of them were located in the headwaters of streams. Moreover, in its rush to expand the park system, the commission had committed funds disproportionately to building artificial lakes, leaving little in the budget for dredging natural lakes to improve their water quality.[77]

No matter how extensive the recreational development, though, it did not begin to meet demand. By the end of the decade, 3.6 million people a year were visiting Iowa's state parks, and the problem of maintenance had assumed "gigantic proportions." Visitors didn't want to experience nature unspoiled. Grass had to be planted, then mowed, weeds sprayed; there was firewood to cut and distribute to campsites; roads and bridges needed maintenance; deer herds required feed; picnic tables had to be painted. Above all, the piles of garbage that visitors left behind had to be removed. In its biennial reports to the legislature, the State Conservation Commission hinted broadly at greater costs looming on the horizon. Although the widespread use of stone construction had the added advantage of limiting future maintenance costs, all those new appurtenances would need routine repair and occasional replacement.[78]

If public demand for recreation became the driving force behind land acquisition and development, and Schwob's criticisms contained more than a kernel of truth, the State Conservation Commission nonetheless gave more than lip service to resource protection. In 1936, sixteen state parks and preserves were designated as game refuges.[79] The camping, picnicking, and lake access areas of Black Hawk State Park, for instance, existed side-by-side with designated wildlife refuges, fish hatcheries, and a tree nursery. By 1942, the State Conservation Commission was responsible for approximately 80,000 acres of public lands and waters, plus 800 miles of rivers and streams. Approximately 20,000 acres of land were located in state parks or preserves, another 13,000 acres were classified as state forest reserves. Public waters included approximately 42,000 acres of natural lakes. Designated wildlife refuges, public hunting grounds, game bird hatcheries, and fish hatcheries occupied another 5,000 acres, although many of the state parks, preserves, and natural lakes were managed, either in whole or part, as wildlife refuges. Even so, these 80,000 acres represented less than one percent of Iowa's total land area. The vast majority of Iowa's land remained in private hands, with few governmental controls over land use.[80] At best, the conservation commission could only provide environmental leadership by example. This is precisely the approach that G. B. MacDonald took with the state's forestry program.

Forestry

The *Twenty-five Year Conservation Plan* identified woodlands as the state's most pressing conservation problem because it was integral to controlling soil erosion, decreasing the rate of lake siltation, and restoring wildlife habitats. MacDonald's complete plan for establishing a system of national and state forests, however, appeared in the Iowa State Planning Board's 1935 report. By this time, the National Forest Reservations Commission had already approved the purchase of more than 800,000 acres in southeast and southwest Iowa. MacDonald proposed federal purchase of approximately 500,000 additional acres of rough submarginal land in northeast Iowa, along the Des Moines and Cedar rivers, and throughout the Loess Hills region. In addition, he tagged for state purchase more than 290,000 acres in smaller submarginal tracts scattered throughout Iowa. Soil conservation was the long-standing rationale for seeking forest reserves in Iowa, although MacDonald was keenly aware of the need to sell forestry in terms of its economic benefit. He was among those who promoted the multiple-use concept, although by this he meant more than controlled logging. MacDonald's idea of multiple use included managed timber production, but he also advocated the use of woodlands to propagate fish and game resources, and to provide greater recreational opportunities.[81]

Once MacDonald completed the forest and wasteland survey, he lost no time implementing a forestry program. The basic obstacle to achieving a program of forest conservation in Iowa, however, was that woodlands outside state parks were on privately owned land, chiefly on farmland. Any feasible conservation plan, therefore, had to address land use by private owners, affecting some 2.5 million acres of land. Accordingly, he adopted a two-pronged strategy: first, government purchase of tax delinquent lands whenever possible, and, second, a forest management program designed to secure and implement cooperative agreements with private landowners.[82] The latter goal received a big boost during the 1935–1936 legislative session when the legislature amended the 1906 Fruit Tree and Forest Reservation Act, transferring the administration of approximately 45,000 acres of private land in tax-exempt forest reserves from the Department of Agriculture to the State Conservation Commission. During the 1935–1936 biennium, MacDonald also set out to establish state forest reserves as "areas that demonstrate true conservation in the sense of proper use of the land and water."[83] He spent $12,000 from the State Emergency Conservation Fund to purchase White Pine Hollow, a 390-acre stand northwest of Luxemburg in Dubuque County. Originally acquired by the Dubuque County Conservation Society for eventual preservation under state ownership, White Pine Hollow marked the first state forest acquisition. Additional tracts of land were added to several existing state parks and preserves for reforestation purposes.[84]

The accomplishments of 1935–1936 signaled an important *new* beginning, but the first solid achievements in forestry came during the next two years. Again using special state conservation funds, MacDonald acquired more than 11,000 acres of cutover woodlands, rough brush-covered tracts, and worn out fields in Lucas, Monroe, Lee, Van Buren, Allamakee, and Clayton counties. In keeping with the principles of multiple-use management, these areas were slated for a combination of timber production, wildlife production, and grazing; limited recreation; and experimental areas for the study of forestry problems. Two CCC camps were assigned full-time work surveying the southern woodlands, building erosion control structures, grading and surfacing roads, replanting trees, and preparing game food and shelter areas. MacDonald also managed to establish a state tree nursery on a newly acquired tract of land near Ames, despite renewed opposition from commercial nursery operators. By spring 1938, enrollees at a third CCC camp had four million seedlings ready for planting in state parks and forests as well as for distribution to farmers under the provisions of the 1924 Clarke-McNary Act.[85]

By 1940, the state nursery operation was in full swing. CCC, WPA, and NYA workers collected seeds from trees and shrubs in state parks, from which seedlings were grown. At the main nursery in Ames, CCC crews constructed warehouses, an underground cellar for storing nursery stock, an irrigation system, and a drainage system. To augment the supply of trees, auxiliary nurseries had been established at Walnut Woods, Lake Manawa, Black Hawk Lake, Lake Ahquabi, Lake Wapello, and Wapsipinicon state parks, and the Josh Higgins Parkway near Cedar Falls. Much of the stock produced at these locations was planted in state parks and forests, but more than half went to farmers for erosion control and reforestation purposes. A gift of 330 timbered acres near Boone overlooking the Des Moines River gave MacDonald an opportunity to establish a laboratory for experimenting with various forestry practices in a natural environment. The Holst Forest Area, named for the donor, additionally served as a demonstration area.[86]

MacDonald also initiated the first timber stand improvement and planting demonstrations on private land during the 1939–1940 biennium. These outings developed into a cooperative farm forestry project established under the provisions of the 1937 Norris-Doxey Act. In 1940, the State Conservation Commission hired a farm forester to assist MacDonald with this project. Working in cooperation with various agencies—the Soil Conservation Service, the Extension Service, the U.S. Forest Service, the Iowa Agricultural Experiment Station, and the Bureau of Agricultural Economics—the farm forester, Manford Ellerhoff, began introducing the principles and techniques of "woodlot economy" to a handful of cooperating farmers in Allamakee County. In this manner, MacDonald and his new assistant hoped to convince farmers throughout the state that proper woodland management could augment farm income as well as provide wildlife cover and control soil erosion.[87]

In addition to the state forest program, the U.S. Forest Service also began to acquire land for national forests in Iowa, a move that MacDonald encouraged and the State Conservation Commission welcomed.[88] After Iowa passed the National Forest Enabling Act in 1933, authorizing land acquisition for the purpose of establishing national forests in Iowa, the National Forest Reservations Commission outlined its plan to purchase 829,000 acres of submarginal and marginal farmland in thirteen southern Iowa counties. These acres were referred to collectively as the Hawkeye Purchase Unit. In keeping with recommendations in the state conservation plan, the Forest Service agreed to adopt a balanced management policy of timber production, fish and game protection, and recreational use.[89]

Despite ambitious plans, however, the federal purchase program moved with ponderous slowness. To a degree, limited funds and administrative regulations hampered implementation, but a persistent mindset that identified national forests with western states also meant that farm states were typically last in line for funding consideration. In 1936, as a show of good faith, the National Forest Reservations Commission authorized one purchase: one hundred acres near Keosauqua for use as a plant nursery, which soon was turned over to the state for operation. In addition, U.S. Forest Service personnel assisted the state in establishing priorities for land acquisition, regardless of whether the money came from federal or state sources. In this manner, the objectives of both programs began to mesh, and at one point the State Conservation Commission even began to make plans for eventually turning over state forest reserves to the U.S. Forest Service.[90] No substantive federal action occurred, though, until 1941, when the U.S. Forest Service purchased 4,400 acres of distressed, tax-delinquent lands, all the tracts located in proximity to state reserves in four southeastern counties. Subsequent purchases brought total federal reserves in Iowa to 4,700 acres.[91]

By the close of the New Deal, state forest holdings had reached approximately 13,000 acres, with another 56,000 acres in private forest reserves. This represented a tiny fraction of the 2.5 million acres of woodlands still standing in Iowa, and it was a far cry from what MacDonald had hoped to achieve. At one time, when federal and state moneys were flowing, MacDonald had visions of placing no less than 915,000 acres of woodlands in state forest reserves alone.[92] His expectations far exceeded anything possible. The reality of federal and state funding simply did not match MacDonald's vision of large forest reserves in Iowa. When it became clear that the U.S. Forest Service would never have a real presence in Iowa, MacDonald and his colleagues at Iowa State asked the State Conservation Commission to take over all the forestry responsibilities being carried out by the Forestry Department and the Agricultural Experiment Station. In December 1941, the commission agreed to terms of the transfer, which effectively placed forestry administration in the hands of the assistant state forester, Harold Bjornson, located in Des Moines. This action once again raised fears among commercial nursery operators that the state would compete for their customers. After lengthy negotiations, the commission met some of their demands, adopting a policy that limited the size of stock sold to the public and discontinuing the practice of distributing trees for Christmas tree farms.[93]

No matter how small the gain, MacDonald's efforts firmly established forestry as a function of the State Conservation Commission, realizing a goal that stretched back as far as the founding of the Iowa Park and Forestry Association in 1901. Considering the setbacks that MacDonald and Pammel had experienced during the 1920s, the successes of the 1930s, though limited, were substantial. Of everything that had been accomplished—the CCC program, the state reserves, the federal reserves, the state nursery operation, and the cooperative work with farmers—MacDonald felt that the farm forestry program gave the most promise of long-term success. By the end of the decade, farmers were requesting more trees than the program could supply, and MacDonald "anticipated that the state nurseries [would] be a real factor" in the conservation programs that were being set up all across Iowa under the auspices of the Soil Conservation Service.[94]

Piecemeal Water Policy: The Iowa Great Lakes Sewer System

As early as 1930, at the urging of the Board of Conservation, the governor, the fish and game warden, and members of the State Department of Health met with the board to try and figure out the best procedure for addressing lake and stream pollution problems. In the end, they proposed a new water sanitation commission to coordinate the necessary investigations and make recommendations, a proposal that never made it through the legislature.[95] Rather, surface water problems were incorporated into the statewide survey that led to the *Twenty-five Year Conservation Plan*. Accordingly, cleaner waters became part of the rationale for recommendations to control soil erosion on private lands, to restore drained lakes and marshes, to improve natural lakes by dredging, and to acquire more land adjacent to waterways. Additionally, the state conservation plan recommended some form of state assistance to towns and cities for abating water pollution from sewage, noting that sewage pollution had reached the serious stage at two natural lakes—Storm Lake and East Okoboji—and that sooner or later municipal sewage, industrial waste, and farm waste would impair all lakes and streams.[96]

Jurisdiction over water affairs, however, was vague and scattered in early 1930s. It would take a determined effort just to coordinate the activities of existing state agencies, let alone establish definite water policies. Between 1933 and 1935, the State Planning Board attempted to pull together information from several sources—the Board of Conservation, the Fish and Game Commission, the State Hygiene Laboratory, the State Board of Health, the Iowa State Geological Survey, the U.S. Geological Survey, and the Iowa Institute of Hydraulic Research—in an effort to establish a base for planning. The effort was largely wasted. Discussion of surface water problems occupied much of the State Planning Board's 1935 report, but in the end it could only recommend continuing investigations and tackling as many local problems as feasible through the use of relief labor.[97]

Throughout the 1930s, important investigations did take place, although their scope was limited and funding was never adequate to the task. In 1932, the Board of Conservation and the Fish and Game Commission jointly funded a study conducted by Iowa Lakeside Laboratory to determine the relationship between lake and stream pollution and the growth of blue-green algae. Two years later, the board agreed to fund algae control experiments conducted by the University of Iowa.[98] Noting the failure of chemical treatment as a long-term solution, the Board of Conservation favored lake dredging, while the Fish and Game Commission tended to favor building artificial lakes where related environmental factors in the surrounding watershed, such as soil erosion and livestock effluent, could be more easily controlled. Consolidating the Board of Conservation and the Fish and Game Commission into the State Conservation Commission certainly reduced the number of agencies handling larger or smaller pieces of the entire water resource puzzle, but no real water-quality policies emerged. The Conservation Commission continued the lake construction and improvement projects initiated under its parent agencies. In addition to creating eighteen artificial lakes, the commission dredged parts of Storm Lake, Black Hawk Lake, North Twin Lake, and Lake Manawa.

Neither the studies nor the lake construction and dredging projects got to the heart of water pollution problems. There was just too much human, animal, and industrial waste, and too much soil, going into lakes and streams. As it turned out, the problem itself would force some sort of solution. The result was a political accommodation known as the Iowa Great Lakes Sewer System. It was a solution encouraged by a desire to keep Iowa's premier lakes clean enough for fishing and other recreational uses, by the availability of federal funds for public works projects, and by a political climate that increasingly favored special state appropriations for specific projects. Building the Iowa Great Lakes Sewer System launched the state into a long-term venture with consequences far beyond those envisioned at the outset. Not only did the state underwrite part of the construction cost, the State Conservation Commission ended up operating the system for many years.

The beginning of this unintended state-local partnership is hard to pinpoint, since the size of the lake complex, and the unwillingness of private property owners to part with much land for public access, meant that in the 1920s and early 1930s the Board of Conservation was constantly butting heads with landowners and private businesses over dock permits, commercial boat operations, livestock in the water, and similar issues concerning shore usage. In any case, when the Board of Conservation learned, in May 1932, that a private hotel and restaurant was dumping raw sewage into West Lake Okoboji, it asked the Department of Health to investigate and report its findings. If the Department of Health ever did investigate, there is no evidence that it reported its findings to the board. Two years later the board decided to act on its own, requesting that the City of Okoboji take proper steps to dispose of the effluent draining into the lake from septic tanks, cesspools, and outhouses. At the same time, the Fish and Game Commission resolved not to spend any more public funds "on water areas where extreme pollution was being carried on by communities," and sent letters protesting the pollution in East Okoboji, Minnewashta, and Gar lakes.[99]

William P. Woodcock of Spencer, who was appointed to the Board of Conservation in 1932, spent his three-year term trying to initiate a positive response on the local level. Although details are sketchy, Woodcock reportedly began to promote the idea of a trunk-line sewer system that would serve cottages around all the lakes. Eventually, the Okoboji Protective Association, a private group of lakeside landowners, gave him support. It spearheaded an effort to form a new organization, the Iowa Great Lakes Sanitary Commission, with representatives from the lake towns. Together, the protective association and the sanitary commission raised money for an engineering survey.[100] Woodcock was not appointed to the new State Conservation Commission in 1935, but he continued his involvement in the sewer issue as a member of the sanitary commission. The commission, for its part, continued to press for pollution control around the lakes. After Iowa Lakeside Laboratory issued a 1936 report on the condition of the lakes, the Conservation Commission condemned the lake towns for "using the lake system as a dilutent and conveyance" and justified the need for state action because it was "beyond the point of debatable issue" that sewage was ruining one of the state's best public recreational areas.[101] It implored the Executive Council to find surplus funds somewhere and to give *some* state agency the authority to take corrective measures.

The availability of federal relief funds for public works projects, and pressure from the State Conservation Commission, prompted the Iowa Great Lakes Sanitary Commission to seek financial aid from both the state and federal governments. The state legislature responded in 1937 by appropriating $125,000 to initiate construction of the sewer system, contingent upon additional federal funding, which ultimately came through the Works Progress Administration. The State Conservation Commission, not the State Department of Health, was charged with implementing the project. Having been largely responsible for getting the project off the ground, the commission willingly accepted this responsibility and immediately proceeded to secure easements, select an engineering and construction firm, and draw up an agreement for cooperative local operation and maintenance.[102]

The sewer project moved along smoothly for just about one month, when, suddenly and without warning, the City of Spirit Lake refused to sign the long-term operation and maintenance agreement. The towns of Arnolds Park, Okoboji, West Okoboji, Orleans, and Wahpeton all signed the agreement, which vested administrative authority with the Dickinson County Board of Supervisors and called for each participating community to pay a proportionate share of the operating and maintenance costs. Spirit Lake's refusal to sign placed federal funding in jeopardy and brought the project to a temporary halt. Governor Nels Kraschel was furious. He wrote a chastising letter to the mayor of Spirit Lake and sent a copy to the *Spencer Daily Reporter* for publication. The newspaper then entered the fray and demanded an explanation from the Spirit Lake City Council. After a week's delay, the council publicly responded, charging that the State Conservation Commission had been vague and arbitrary in figuring future costs and that Spirit Lake's proportionate share was too high. The governor and the press continued to demand a favorable resolution, but in the end Spirit Lake forced the State Conservation Commission to meet its demands. Following an acrimonious three-day conference of state officials, engineers, mayors, and city council members, Spirit Lake agreed to new terms that vested administrative authority in a reorganized Iowa Great Lakes Sanitary Commission and that placed a cap of $6,000 on annual operating and maintenance costs with the proportionate community shares to be determined by the sanitary commission.[103]

Once the new agreements were signed, the WPA released $166,000 for the first phase of construction, and work on the sewer project began in November. During the ensuing year, approximately six hundred WPA workers excavated a nine-mile trench and laid tile for the trunk line, which extended from the south shore of Spirit Lake to the site of a proposed treatment and disposal plant at the southern end of Lower Gar Lake. This section was intended to handle sewage from Spirit Lake and Arnolds Park, which were then pouring treated sewage directly into East Okoboji and Minnewashta lakes.[104] Early in 1938, the Okoboji Protective Association and the sanitary commission began the process of applying for additional funds to construct a treatment plant and lateral lines that would connect the towns around West Lake Okoboji into the system. WPA officials promised an allocation, provided that local sponsors paid for the engineering survey. The difficulty sponsors experienced raising a mere $2,500 for engineering work gave yet another hint of the local jealousies that would plague administration of the sewer system into the 1950s. This time, it was the Dickinson County Board of Supervisors that refused to make anything more than a small contribution, and local backers were forced to go begging for private subscriptions to augment the shares already paid by cooperating towns. It took a campaign of several months duration to raise the $2,500, but the effort paid off with another $400,000 WPA grant and an additional $7,800 from the state. The main trunk began operating in October 1939, disposing of wastes from Arnolds Park and Spirit Lake. Eight months later, in July 1940, the sewer project was completed.[105]

The haggling over which towns and how much each would pay to operate the sewer system began immediately. As it turned out, engineers had vastly underestimated ongoing operating and maintenance costs. In July 1940, Wahpeton, Orleans, and West Okoboji gave notice that they were withdrawing from the project; and a large delegation from Spirit Lake, Okoboji, and Arnolds Park appeared before the State Conservation Commission to demand that it pay for all costs above the agreed upon $6,000 annual figure. To quiet the immediate situation, the Executive Council ordered the commission to pay the deficiency.[106] Realizing that it was being forced into an ongoing financial commitment, the Conservation Commission took steps in 1941 to draft enabling legislation that would establish a sanitary district with power to levy user fees or taxes in order to make the sewer system self-supporting. However, just when the commission was on the verge of getting a sanitary district bill introduced, State Representative W. A. Yager of Spirit Lake intervened to kill the measure. The commission ended up subsidizing the Iowa Great Lakes Sewer System another year.[107]

In 1942, the Conservation Commission again tried to get a sanitary district bill before the legislature, and once again Representative Yager stopped it. More ominous, Yager, speaking on behalf of the Great Lakes Sanitary Commission, proposed that the State Conservation Commission take over operation of the sewer system and then collect, or attempt to collect, proportionate shares from the benefited towns. The Conservation Commission continued to push for a sanitary district bill, but by now the issue was so politicized that it was perennially buried in committees. To make matters worse, the condition of the lakes did not improve, actually, could not improve, because only half the communities were using the system and farmers adjacent to the lakes were doing little to prevent livestock waste and eroding soil from washing into the water. By 1945, even the communities that *were* hooked into the sewer were "not very much concerned about fulfilling their commitments on funds," and so the state legislature conveyed to the State Conservation Commission exclusive control and operation of the Iowa Great Lakes Sewer System.[108]

The Great Lakes Sewer was now unofficially known as "the state sewer," which adequately expressed commissioners' sentiments about the decade-long saga. With no alternative, the State Conservation Commission proceeded to cajole communities into new cost-sharing agreements. After the usual posturing, all the communities except Orleans agreed to pay some part of the cost; Arnolds Park and Okoboji were the only willing participants, and Spirit Lake would not agree to pay its full share.[109] And so the situation continued. By 1948, the annual operating costs were pushing $18,000, but the towns that benefited "seemed to contribute toward the sewer what and when they wished." Having failed to eliminate all the sources of municipal sewage emptying into the Iowa Great Lakes, and having failed to establish a local entity with responsibility for controlling pollution in the entire lakes complex, the State Conservation Commission now sought legislation that would give it sufficient police power to curb lake pollution. This effort, too, failed, leaving the commission completely boxed into an untenable situation.[110]

By 1950, the commission seems to have adopted a "hold the line" attitude. When local citizens from Clear Lake sought state funds for a sewage disposal system to abate lake pollution, the Conservation Commission politely told them that it had no jurisdiction to conduct studies, hold hearings, or approve plans, and that the group should seek cooperation from the Department of Health. Trying to extricate itself from the Great Lakes Sewer business and to prevent similar entanglements, the commission shifted political ground by redefining the issue as a public health problem, not a lake pollution problem.[111] This represented a fundamental shift. Contrary to public health officials and sanitary engineers, who commonly viewed waste disposal as a legitimate use of public waters, the commission had consistently maintained since the 1920s that sewage disposal inherently degraded the quality of lakes and streams. But in 1950 the commission retreated and seemingly gave in to the notion that public health concerns should dictate acceptable levels of water pollution. In effect, prolonged and unresolved conflict caused the commission to rationalize away one of its most important principles.

An opportunity for escape from the Great Lakes Sewer debacle came in 1951. The state legislature appropriated $350,000 to pay a portion of the costs of a new sewer system at Clear Lake and another $350,000 for sewer improvements in the Great Lakes area.[112] When the State Comptroller automatically credited the funds to the State Conservation Commission for control, the commission deftly responded "no thank you," pointing out that under the provisions of the act, the Budget and Financial Control Committee was the agency authorized to manage the funds. The commission then initiated steps to transfer ownership and operation of the system built in the 1930s over to a new Iowa Great Lakes Sanitary District, a transfer that was completed in May 1952.[113] The saga did not exactly end there, however. The commission eventually agreed to manage funds for the Clear Lake project, and it did hire an engineer to protect the interests of the state during planning and construction, and as late as 1961 the state legislature was still channeling through the State Conservation Commission special appropriations for sewer improvements in the Iowa Great Lakes region.[114] Nonetheless, in 1952 the State Conservation Commission succeeded in extricating itself from the sewer management business.

Surely, the Iowa Great Lakes Sewer System was not the model of "state assistance" envisioned by the *Twenty-five Year Conservation Plan*. The experience had been a disaster. It had exposed the depth of local intransigence when it came to controlling water pollution. Local citizens were not uniformly opposed to taking responsibility for disposing of their own sewage, of course. Some cared very much about water quality and the lakes environment, but not enough people to overcome the influence of penurious attitudes and petty local politics. The most that could be said for the unintended partnership was that the State Conservation Commission had succeeded in reducing the amount of effluent pouring into the lakes. In the process, it had learned a sobering lesson. Clearly, the state could not build and operate sewer systems to clean up every lake and stream in Iowa. It was equally clear that few local governments would take the initiative to do so.

Sanitary engineers liked to paint the 1930s as the "golden era" of waste treatment in Iowa, and, to be fair, some of the larger polluters were no longer befouling rivers by the end of the decade. Acting under pollution abatement orders issued by the Department of Health and approved by the Executive Council, sixteen cities along the Des Moines, Cedar, and Iowa rivers constructed treatment facilities between 1934 and 1940. The Great Lakes Sewer System was by far the largest sewer project funded with federal aid in Iowa, but all together, at least 125 treatment plants were either constructed or improved thanks to the Works Progress Administration and the Public Works Administration. These constituted a substantial advance in water pollution control during the 1930s, but progress slowed to a snail's pace when federal funding ran out. Less than half of the state's municipalities had *any* type of sewage treatment facilities as of 1940, and only eighty or so communities built new facilities during the next two decades.[115] Environmental degradation from sewage in Iowa's lakes and streams would continue until the late 1960s, when the federal government began to set national water quality standards and then allocated massive amounts of aid to help cities construct or upgrade their sewage disposal systems.

Assessing the *Twenty-five Year Conservation Plan*

When war mobilization replaced relief spending and federally funded projects began to end, it was inevitable that a period of assessing the *Twenty-five Year Conservation Plan* would begin. Jay Darling, the true mastermind behind the plan, easily was the most critical. In 1941, he could see few positive results.

Hoping for a groundswell of public support, in 1933 the Board of Conservation and the Fish and Game Commission had sent copies of the plan to every public library. Sadly, most of the copies just sat on library shelves, unthumbed and rarely checked out. "Public pressure," Darling noted, "has been from the start entirely lacking. Had it not been for a few of the ardent promoters of the program and their persistence in the early years of its functioning, it is doubtful if the survey and program would have served any good purpose. The tendency," he continued, "was always to cling to the old method of state expenditures or unsystematic and desultory projects with no coordination or unity in mind. With the passing out of the picture of the original group of commissioners who organized the program, all evangelism and publicity along the program lines ceased." Still, there had been "profit in the undertaking," if one judged by increased public landholdings, increased public use of recreational areas, the restoration of certain natural areas, and increasing revenue to the State Conservation Commission.[116]

G. B. MacDonald left no reflective comments for the historical record, which is unfortunate considering his long tenure in the conservation movement and his pivotal role during the 1930s. John Fitzsimmons, however, who had nearly as long a tenure in the movement as MacDonald, assessed the *Twenty-five Year Conservation Plan* from a very different perspective than did Jay Darling. In a special conference with the commission, Fitzsimmons readily acknowledged that "there ha[d] been a change of policy," but he did not see public apathy or politics-as-usual behind that change. What Fitzsimmons understood best were the changes that had taken place within the Board of Conservation and then the State Conservation Commission since the early 1920s. "The first Conservation Commission I was acquainted with," he recalled, "was very strong on conserving anything that could be conserved, and would take anything. Dr. Pammel was so absorbed in preserving nature that he never saw the other side of the picture. He would take a single tree in the middle of the road if somebody would give it to him. . . . From that ultra side of conservation as personified by Pammel, Armstrong, McNider, it has changed to a form of recreation within natural environment." As a landscape architect, Fitzsimmons himself personified the "other side," and so it is not surprising that he thought the change in policy direction was "a healthy growth with the times." More to the point, though, he perceived that the State Conservation Commission had generally accommodated both sides of the picture. "I do not think the Commission has ever lost track of good preservation policies," he explained, "at the same time providing recreation."[117]

Given his perspective, Fitzsimmons's critique emphasized the need for more thoughtful project-specific deliberations without abandoning the basic goals of the overall plan. On one level, this criticism excused the 1930s focus on state park development in the rush to put people to work. Recalling the madcap days of early CCC projects, when staff at the Central Design Office would toil through the night in order to have construction plans on site for work crews the next morning, he acknowledged that lodges and extensive nature trails had been built on the assumption that people would use them, which, visitor surveys later revealed, had not necessarily been the case. "Let the demand build itself rather heavily ahead of any additional work you do," he counseled. On another level, though, Fitzsimmons's words echoed Darling's observations about lack of unity and purpose among commissioners. In couched but unmistakable language, Fitzsimmons spoke of a "rubber band policy" that more and more often allowed local demands to dictate where state parks would be located and how they would be developed. Pioneer, Cold Springs, and Lake Macbride state parks were cited as particular examples where the commission had lost control over development. Moreover, the commission had not always known when to stop. "We have areas where we have just kept on with the physical development," Fitzsimmons pointed out. "For instance, Backbone. We have had a camp there since CCC started. Now we find that we have plenty of physical development and have really neglected the natural resources." [118] Fully, though mistakenly, anticipating that the federal government would resume public works projects after the war, Fitzsimmons urged the commission to begin laying out a postwar program to take up those parts of the state conservation plan in which progress had been slow or modest, such as stream improvement, land acquisition along streams and lakes, more marshes for migratory waterfowl, and more forests.

While the assessments offered by Darling and Fitzsimmons measured achievements against the expectations that had prevailed in the early 1930s, equally interesting and ultimately more telling assessments came from the commission's staff. In May 1942, Director Fred Schwob asked employees to "reappraise" the state conservation plan, and he gave them a three-page outline to follow in organizing their comments. Two lengthy responses survive, one by E. B. Speaker, superintendent of fisheries, and the other by James R. Harlan, superintendent of public relations. [119] What is remarkable is not so much the assessments they offered, although each document reveals considerable difference of opinion among professional staff regarding appropriate policies, programs, services, and techniques. Rather, there is a richness of detail that reflects the many ways in which broad policies of the State Conservation Commission were now institutionalized and implemented. In less than a decade, Iowa had created a resource management bureaucracy. The implications cut both ways. On the one hand, the commission now had a professionally trained and dedicated staff of people who understood the complexities of resource management. The transition from commission-driven to staff-driven policy had begun. On the other hand, it was becoming increasingly difficult to maintain that unity of purpose Jay Darling felt was completely missing. Indeed, an element of drift was evident as early as 1938, when the commission first decided it was time to bring the state conservation plan up to date. Louise Parker, in particular, who succeeded Margo Frankel in 1937, noticed that commission meetings were almost entirely devoted to specific cases, and it was she who requested that the commission set aside time to discuss general problems and long-range plans.[120]

Unified departments of conservation or natural resources, such as Iowa's, were a sign of increasing modernization in state government. By 1941, Arizona and Colorado were the only states without a state park agency of some kind, though the administrative apparatus varied widely from state to state. Iowa, which had been a leader in the trend toward unifying parks, forestry, and wildlife in one agency, was now one of sixteen states so organized. Most of them were located east of the Mississippi River. In the trans-Mississippi West, only Iowa, Minnesota, Oklahoma, and California administered their state parks as part of a unified resource system.[121] However, Iowa no longer could claim leadership in terms of landholdings or any other standard measure of public benefit. Even in comparison with other states in the agricultural North Central region, Iowa ranked somewhere in the middle. Minnesota, with approximately 1.4 million acres in state parks and related areas, ranked far above the others. At the other extreme was North Dakota, with only 5,150 acres.[122] Iowa's public lands and waters totaled approximately 80,000 acres.

Relative size of landholdings aside, Iowa still had a remarkably diverse system, one that reflected the integrative approach to conservation and park development that had prevailed since the beginning. Iowa's holdings incorporated forests and a whole range of areas that no longer fit the evolving definition of a "state park." Along with reassessing the state conservation plan, the State Conservation Commission authorized staff to draw up a classification system along lines suggested by the National Park Service. By now, "real" state parks were seen as "areas large enough in extent to provide recreation and service for large crowds and still have sufficient area to preserve the natural landscape character with all that it contains."[123] The emphasis had shifted to large-scale recreation with nature as a backdrop. Fewer and fewer people saw parks as a vehicle for preserving natural resources; preserving the "character" of the natural landscape was enough. Narrowing the concept of a state park created, accordingly, a need to define everything else in the system. In this regard, the multiple-use approach went hand-in-hand with categorizing areas by function. Thus, the predominant use of each property now determined its classification. Accordingly, the Iowa state park system, as of 1941, consisted of fourteen state parks, twenty-seven state recreation reserves, twelve lake reserves, seven state forests, four historical-archaeological areas, four geological-biological reserves, and five wayside parks. In addition, the Conservation Commission managed ninety-one wildlife refuges, eighty-six public shooting grounds, and an assortment of game bird and fish hatcheries.[124]

Unified by agency did not, however, mean unified management. The Lands and Waters Division and the Fish and Game Division still operated under different budget setups. Moreover, although there was considerable overlap of activities and responsibilities in the field, each division had its own management style and priorities. In this respect Jay Darling was right. In the federally funded development rush of the 1930s, Iowa had compromised the interconnected approach to developing state parks, preserves, wildlife refuges, and recreation areas that was envisioned in the *Twenty-five Year Conservation Plan*.

5

Seeking Balance

Wartime Interlude

The State Conservation Commission, like everyone else, curtailed new construction, conserved materials, and economized on travel during World War II. Federal relief aid vanished when the nation went to war. Dozens of commission employees were called for military service, park attendance dropped by about one-third, and the pace of conservation work slowed accordingly. The practice of hiring seasonal park naturalists was discontinued in favor of cheaper, self-guided nature trails. Commissioners and administrative staff debated closing outlying parks, but in the end did not, because rural communities objected and because the cost of maintenance was nearly the same whether parks were open or closed. In some cases, local groups or communities agreed to take over or assist with maintenance. More often, park custodians simply put in extra hours and improvised on maintenance when the cost of labor and materials went up. Although deferred maintenance would catch up with the commission, "making do" was not entirely without benefit. Economy measures provided a convenient excuse to remove the toboggan slides, skating rinks, and other winter sports facilities, which had attracted only light use since their construction. Floodlights salvaged from the Stone State Park toboggan slide even went to the war effort, illuminating the new air base at Sioux City.[1]

Gas rationing and the inability to buy tires dampened park attendance, but the commission opened up several parks for military use, a move that concessionaires welcomed. Resident tent camps were allowed in Lake Macbride, where Air Cadets from Iowa City lived, and at Oak Grove, which the Army Air Force used for military field training. Cadets stationed at the Ottumwa Naval Air Base had privileged use of a section set aside for recreation at Lake Wapello. Cabins at Palisades-Kepler were offered as housing for personnel stationed at the Naval Flight Preparation School at Mt. Vernon. Women in the WACS practiced truck driving at Lake Ahquabi. Ledges, Pammel, and Stone state parks were used intermittently for overnight hikes, special maneuvers, and compass training. Additionally, the commission made all armed services personnel welcome in state parks by waiving the usual charges for camping and lodge use.[2]

When the commission's assistant forester, Harold Bjornson, left for military service, and the state nursery superintendent resigned, the forestry program slowed to a crawl. Likewise, land acquisition slowed dramatically when conservation and relief appropriations ended. The state nursery curtailed tree propagation, although it continued to supply farmers with seedlings and saplings from existing stock. To meet the high wartime demand for lumber, the U.S. Secretary of Agriculture ordered state agencies receiving Norris-Doxey allotments to develop forest products marketing services for farmers. As a result, the commission assisted the Extension Service at Iowa State College by paying the salary and expenses of foresters who doubled as timber cruisers to assist farmers in processing trees for market. No timber was taken from state lands, but approximately 7.5 million board feet of lumber were harvested from farms and privately owned forest reservations.[3]

Although the commission hunkered down "for the duration," progress did not stop. Anticipating the resumption of development after the war, the commission kept a list of projects for postwar attention. Its August 1944 report to the Postwar Rehabilitation Committee outlined nearly $29 million in land acquisition and development projects.[4] The commission also finished projects that were underway. To complete the group camp at Springbrook State Park, for instance, three former CCC barracks were remodeled to serve as an administrative building, a recreational hall, and a dining hall. In deference to its employees serving in the armed forces, however, the commission rejected an offer to have prisoners of war work in state parks and forests. At the urging of Louise Parker, who supported conservation education wholeheartedly, the commission engaged a local film company to produce a series of movie shorts on parks, hunting, and conservation. Despite reduced travel allowances, the commission sent its administrative staff to professional conferences and on investigative trips whenever it could. The commission also continued to acquire new lands, mostly for parks. Even so, land acquisition reflected the new austerity mode. Between June 1942 and June 1944, slightly fewer than 800 acres were added to the system.[5]

When gasoline rationing was lifted in 1945, park attendance immediately rebounded. Overnight cabins were full to capacity in the ten parks where they were available, and the demand kept increasing. Now, however, visitors were crowding into areas where makeshift maintenance had been the rule for four years. Custodians were still shorthanded until the end of the 1945 season; equipment had been repaired to the point of needing replacement; tools were worn. Buildings had been patched with whatever material was handy; their furnishings were getting shabby. The parks, in short, needed attention.[6]

A brief, but unsuccessful, effort in 1945 to establish a permanent source of federal funding for state park acquisition and development prefigured the financial straits that began to plague state parks.[7] The 1940s saw the commission move in several directions trying to carry out recommendations of the *Twenty-five Year Conservation Plan* and, as the associated costs mounted, searching for ways to shift responsibilities and trim expenditures. On the one hand, the commission renewed a long-standing commitment to historic preservation, embarked on prairie preservation, resumed construction of artificial lakes, and began acquiring land for wildlife refuges. On the other hand, the commission began to divest the system of smaller units with only local use.

Louise Lange Parker and the Historical Program

During the 1940s, the Conservation Commission attempted to tackle plans that had been ignored or underfunded in the 1930s. Among other concerns, Louise Parker made historic sites and historic preservation one of her priorities. Additionally, the approaching celebration of the state's centennial in 1946 stimulated localized interest in preserving sites important in Iowa history.[8] This combination of circumstances produced a short-lived "historical program," as it was presented in the commission's biennial reports from 1942 through 1948. It would be wrong to characterize the historical program as solely Mrs. Parker's creation or even as her primary interest; however, she carried out much of the administrative work that accompanied the acquisition and care of historic sites.

Born in Michigan and educated at the University of Illinois, Louise Lange came to Des Moines after earning a bachelor's degree in library science. In 1910, she took a staff position with the Iowa Library Commission. Two years later, she married Addison Parker, a Des Moines attorney, after which time she focused her energies on raising a family and participating in the types of organization to which upper-middle-class women of her generation gravi-

Louise Lange Parker (1886–1980), photographed in 1954. *Courtesy Addison M. Parker, Jr.*

tated.[9] She belonged to the Des Moines Women's Club, served on the Conservation Committee of the State Federation of Women's Clubs, and was a founder and one-time president of the Des Moines Garden Club. She shared with Margo Frankel the same social status and outside interests, and the two women developed a close friendship though their work with the women's and garden clubs. The Parkers and the Frankels also traveled in the same social circles as the Jay N. Darlings. Thus, it was almost natural that Louise Parker would be selected to complete Margo Frankel's term on the State Conservation Commission after the latter retired in 1937. Mrs. Parker was then appointed to the commission, ultimately serving two six-year terms. When the Natural Resources Council was formed in 1949, Governor Beardsley appointed her to a four-year term on that body.[10]

An exchange of letters that took place in 1947 reveals much about her dedication to the cause she took up a decade earlier. Joe Kautsky, owner of Kautsky Sporting Goods in Fort Dodge, had written to commend her leadership on an issue that was of particular importance to sportsmen; and, in the course of his remarks, he confessed that, "When I first met you, it seemed to me that a woman on a commission of hard-boiled hunters and fishermen did not just ring true." Parker took this as an opportunity to educate him a bit. In a return letter she confided that, although she did not "qualify as a 'sportsman,'" she grew up in a family of sportsmen and liked to fish. "There are however, in addition to the sportsmen who are concerned with the protection of wildlife," she continued, "those who are interested in the aesthetic values; those who go to the woods to study all forms of wildlife; those who go because of the love of the out-of-doors." Lightheartedly making the point that parks had not been created solely, or even chiefly, for sportsmen, she closed by noting that "wild-life belong to the State. It is the province of the State to so manage them that they may be enjoyed by all the people of the State—Women, too!" [11]

Those words were written in 1947. When she came to the commission in 1937, though, Parker's sense of its mandate was much less clear. For a year or two, she demonstrated no discernible pattern of support through her voting record. In this sense, she reflected a state of increasing confusion that characterized the commission as a whole, due in part to numerous changes on the commission itself between 1935 and 1940, and due in part to the enormous workload of the 1930s. Proposals and issues often came before the commission in haphazard fashion: one minute it would deal with personnel matters, the next with the issue of playground equipment in state parks, or sand and gravel leases, or predator control, or proposed land acquisitions. Split votes became much more common. The director, M. L. Hutton, was the only person with a view of the whole situation, and he often did not have authority to act on routine matters.

Nonetheless, having accepted the charge of serving on the commission, she then devoted herself to serving the best interests of the public. She proved to be a quick study, and, judging from commission minutes, soon became a dominant personality. Indeed, it was Parker who, in 1938, sought to refocus the commission on setting broad policies and programs for carrying out the goals of the *Twenty-five Year Conservation Plan*. [12] Parker commanded respect from staff and fellow commissioners alike, although she never displayed the domineering presence one associates with Louis Pammel. Much of this respect no doubt came from her willingness to devote considerable time conducting the commission's business. But she also developed strong opinions about what the commission should be doing, and she had the ability to articulate her positions clearly and cogently. When she found an issue or a project that appealed to her personal interests, she never hesitated to give it as much attention as she could. [13] In 1941, the commission elected her as vice chair, a position she held until her second term ended in 1949. During her twelve years of service, Parker was the driving force behind the Iowa Teachers Conservation Camp at Springbrook State Park, behind the first prairie acquisitions, and behind the historical program. One of her last contributions was a tribute to

Margo Frankel, who died in 1948. Parker consulted with her husband, Henry Frankel, to settle on an appropriate memorial for his wife somewhere in the park system. This led to his donating a small tract of timberland on the (then) outskirts of Des Moines, which, in 1949, was named Margo Frankel Woods.[14]

For the commission's historical program, the centerpieces were Fort Atkinson in Winneshiek County and Plum Grove in Iowa City. After acquiring Fort Atkinson in the early 1920s, the Board of Conservation, in keeping with its policy of delaying improvements, spent very little to restore or maintain the site. Other than appropriating small sums for rebuilding one cannon house and erecting a flag staff, the commission spent no money on the site. No one questioned the historical value of Fort Atkinson, but restoring the site required more money than the commission's budget would allow or the state legislature would appropriate. Consequently, the commission turned to the federal government. From late 1928 to the early 1930s, the commission worked through U.S. representatives Gilbert Haugen and Fred Biermann trying to get the federal government involved in financing restoration. None of these efforts was successful, and the old fort continued to languish.[15]

For reasons that are unclear, the commission did not take advantage of the Civilian Conservation Corps to restore the fort. Since the National Park Service and several states *did* use New Deal relief labor on a variety of historic preservation projects throughout the country, Iowa's apparent lack of effort is curious. Without doubt, Fort Atkinson was a worthy project, and no more ambitious than other notable New Deal historic preservation projects. CCC workers, for instance, restored La Purisima Mission near Lompoc, California; restored or reconstructed American Indian earthen lodges and military buildings at Fort Lincoln State Park in North Dakota; and restored portions of an early nineteenth-century village at Spring Mill State Park in Indiana.[16] Scattered correspondence indicates there actually was fleeting interest in using CCC crews at Fort Atkinson. Nelson Fardahl, connected with the Decorah CCC camp in the mid-1930s, prepared some plans. This prompted an architect working with the Bureau of Biological Survey to inspect the site and report favorably to the National Park Service on the prospects for successful restoration. The camp, however, was slated for termination in May 1936, and G. B. MacDonald was unable to persuade the NPS to extend its duration.[17] One can only conclude that local residents were not well informed about the possibility of utilizing CCC workers for historic preservation projects, and therefore did not press their case before the Conservation Commission at the right time. Likewise, no one on the commission seems to have connected historic preservation with relief work projects.

Fardahl's plans did rekindle a measure of local interest, but it was too late to take advantage of New Deal programs. Beginning in 1938, citizens from the Fort Atkinson area revived discussions with the commission, hoping to get the fort restored before its centennial year, 1940. This time they found a sympathetic listener in Louise Parker. Agreement on the level of restoration, however, was lacking. Locals wanted a completely reconstructed fort. The commission, conversely, did not want "a lot of new construction" detracting from its historical value. Mrs. Parker suggested a partial restoration, with the footprints of other buildings simply outlined. She also suggested that National Park Service historians be consulted before any work took place. Hoping for quicker action, the local delegation then turned to the Executive Council and asked that $10,000 of the commission's appropriation be earmarked for restoration immediately, but this maneuver did not work. In the end, the commission authorized $2,000 for historical research and archaeological investigations in accordance with a priority plan for restoration prepared by the National Park Service. S. S. Reque of Luther College was engaged to conduct the necessary studies. The fort's centennial year was celebrated with renewed hope for restoration.[18]

Reque produced a substantial body of important historical and archaeological data, but he was painfully slow in producing a report. He claimed that responsibilities at Luther College kept him from finishing the writing, but the tenor of communications and meetings during the next several years suggests that he was just being extremely thorough in his work. In any case, the duration of Reque's research lasted nearly as long as the active period of Fort Atkinson itself, about seven years. He appeared before the commission with progress reports in September 1940 and again in 1941. The commission anticipated his report in early 1942, but it did not materialize. Limited preservation efforts nonetheless proceeded to stabilize the existing buildings against further deterioration.[19]

After 1943 went by with still no report, the commission demanded that Reque appear for a conference, which finally took place in September 1944. By this time, the commission had spent nearly $14,000 on investigations, but with no report and no recommendations, restoration could not proceed. Once again, Reque promised a report, this time by early 1945. Only parts of it were received during the next two years. In the interim, the commission allocated another small sum to establish a temporary museum in an old farmhouse located on the fort site, so that the grounds would be presentable in time for the state's centennial celebration in 1946. But it was not until the commission threatened to turn the matter of Professor Reque's contract over to the attorney general's office that he finally submitted the balance of his report in 1947. Even then, the commission's clerical staff had to type it. The commission was thoroughly irked. As the process dragged on and on, enthusiasm for restoring the fort dissipated, although the commission continued to *preserve* the site in a state of arrested decay.[20]

As a footnote to the 1940s effort, Fort Atkinson eventually proved the old adage that good ideas never die. In 1955, the state legislature appropriated $45,000 to repair and restore the fort, one of several appropriations the legislature made that year for special projects requested by local communities. The commission once again called upon the National Park Service for an expert opinion on how the money should be spent. Regional historian Merrill J. Mattes inspected the site and forwarded feasible recommendations. Acting on Mattes's recommendations, the commission released funds to complete the partial restoration that one sees at Fort Atkinson today. The north stone barracks, only a portion of the building standing, was restored and converted into a museum. The old farmhouse was removed. A cannon house on the southwest corner and a powder house on the southeast corner, both intact original buildings, were restored for interpretive use. A replica cannon house on the northeast corner was retained as a storage building and workshop. Rather than reconstruct the remaining barracks, sutler's store, storehouses, and other buildings that had once rounded out the fort, their foundations were simply restored to delineate building outlines. To help visitors visualize the site as a fort, the former stockade wall was partially rebuilt with log palisades. By 1961, the museum was open to the public, and the site was dedicated in March 1962.[21]

Inasmuch as the commission had no professional staff expertise in historic preservation, there were also problems when the commission acquired the Robert and Friendly Lucas Home in Iowa City. At the urging of the State Historical Society and local citizens organized as the Governor Robert Lucas Memorial Association, the commission agreed, in 1940, to administer the home of Iowa's first territorial governor and his wife. A year later, through a combination of gift and purchase, the state acquired the 1844 house and 4.23 acres of the surrounding grounds.[22] Initially, plans for restoration were delayed because of the war. However, an unforeseen turn of events forced action sooner than had been anticipated. In early 1942, Henrietta Pritchard of Chicago contacted the commission offering to sell a collection of furniture once belonging to Samuel J. Kirkwood, Governor of Iowa from 1860 to 1864, for installation at the Lucas Home. Her offer aroused interest among the commissioners and staff, touching off a discussion about involving organizations such as the Daughters of the American Revolution that might be willing to contribute to the cost of interior restoration. Discussions led to an initial offer of $1,000 for the furniture, an offer than Pritchard immediately refused as "far too low." The commission increased its offer to $1,500, a sum Pritchard was willing to accept, but negotiations bogged down when she further asked to be reimbursed for shipping and storage expenses as well as attorney's fees for negotiating the sale. This the commission was not willing to do, and in June 1942 she was politely informed that because of the long delay in negotiations, the funds initially set aside for purchase were no longer available.[23]

Pritchard renewed her offer later in the year, and, although the offering price kept going up, the commission agreed to send Louise Parker on an inspection trip to Greeley, Colorado, where the furniture was stored, and to offer Miss Pritchard no more than $1,500 for all or part of the items. In July 1943, Mrs. Parker made the trip and, finding the furniture in good condition, negotiated the purchase. Once the Kirkwood furniture was safely stowed in Iowa City, in the temporary care of the State Historical Society, the local group pressed to have the home restored in time for the 1946 state centennial. As a result, the commission allocated $5,000 for exterior restoration in accordance with plans prepared by architect F. Lee Carnes of the Central Design Office in Des Moines.[24] Just as Louise Parker had been the person to negotiate the furniture purchase, she too oversaw the restoration project.

Plum Grove ready for 1946 state centennial celebration. *Courtesy Iowa Department of Natural Resources.*

Work on the house commenced early in 1944, and the project was completed in time to open the house to the public during Iowa's centennial year.[25]
In 1946, at Mrs. Parker's suggestion, the commission also formally restored the
historic name, "Plum Grove," to the Lucas home and grounds, this being the
name given to it by the original owners. A year later, Parker also led the commission to an agreement with the Colonial Dames (Iowa), authorizing the
women's group to select proper furnishings for the house and to make decisions concerning future acquisitions. This action came after the Colonial
Dames contributed $1,000 for interior furnishings and certain members vol-
unteered their services to complete the interior decorating. Parker also urged
this partnership after pointing out that "no member of the Commission had
time to give the proper amount of study" the site required. The Colonial
Dames thus assumed the responsibility for interpreting the site, a partnership
that continues to the present.[26]

These were not the only historic places considered or acquired in the post–
World War II period. In November 1941, for instance, the commission purchased an option on an 1860s building in Guttenberg, but when the war intervened it could not secure money to purchase the place. Two years later, after
everyone had settled into a wartime pattern of operations, the legislature authorized purchase of the Abbie Gardner Sharp log cabin, site of the 1857 Spirit
Lake Massacre. The cabin's location near Pillsbury Point State Park in Arnold's
Park contributed to the commission's interest in this historic structure. Once
acquired, though, the commission found that the cabin was in poor condition,
and wartime exigencies made it impossible to allocate funds for restoration.
To make the site "presentable" for the state centennial, the cabin and the
grounds were "cleaned up" in 1946. However, nothing more was done for
several years, and the cabin remained closed to the public. In 1952, four members of the Okoboji Protective Association appeared before the commission to
urge restoration, only to be mollified with assurances that "some repair work
is now being done."[27] Beyond this minor effort, the commission continued to
ignore the cabin and finally transferred administrative control to the State
Historical Society in 1960.

During the 1940s, Louise Parker, with the assistance of former commissioner Margo Frankel, also shepherded the final state acquisitions of land in
northeast Iowa for what would become Effigy Mounds National Monument.
This project represented the last major effort to materialize something of the
long-standing vision for a national park in the Upper Mississippi River Valley.
Authorization of Upper Mississippi Fish and Wildlife Refuge in 1924 afforded
a certain measure of protection for the river itself, but the scenic bordering

lands remained a source of concern. In 1931, a congressionally mandated National Park Service study rejected the national park idea, but proposed instead the creation of a national monument to preserve a portion of the remaining Indian mounds. Satisfied with a plan that was at least feasible, local supporters and the Board of Conservation (and then the State Conservation Commission) thereafter turned their efforts toward setting boundaries for the proposed monument and acquiring the necessary land. New Deal federal aid enabled archaeologists Charles Keyes and Ellison Orr to survey and document the archaeology of the area. Throughout the late 1930s, the Conservation Commission acquired land, some of it purchased as part of the Yellow River forest reserve.[28]

In 1939, Frankel appeared before the commission to urge acquiring the additional land it would take to meet the 1,000-acre-minimum requirement because the effort "might carry weight in pulling [a] proposed parkway road along the Mississippi River to the Iowa side." Frankel and Parker were dispatched to open negotiations that eventually led to the purchase of several hundred more acres, including the coveted Jennings-Liebhardt unit, which Keyes and Orr had determined held "the best of the prehistoric Indian mounds." As of 1939, the state held several hundred acres that it planned to turn over to the federal government at the appropriate time. Even though this was not enough land to meet the required minimum, the state legislature nonetheless authorized, in 1941, the conveyance of up to 1,000 acres for national monument purposes. By 1942, the commission had secured more than enough land, but the war put the national monument project on hold.[29]

When the war ended, both the proposed national monument and the proposed Mississippi River parkway were revived. The Conservation Commission favored both projects, of course, but the state legislature and the State Highway Commission were not so keen on the parkway proposal because the initial cost to the state was estimated at $625,000, and the highway commission would be responsible for ongoing maintenance. As a result, the parkway idea languished. However, the national monument plan went forward. In 1948, the state transferred to the federal government 1,000 acres in Clayton and Allamakee counties, near McGregor, for the establishment of Effigy Mounds National Monument.[30] President Harry Truman signed the official proclamation on October 25, 1949, just a few months after Louise Parker's tenure on the commission ended. In accordance with earlier agreements between the commission and the National Park Service, the state legislature authorized transfer of an additional 204 acres to the federal government in 1952. Mrs. Parker's interest in the national monument continued long after she left the commission. In 1954, she and the Des Moines Founders' Garden Club initiated a private land acquisition effort, the result of which was another 140-acre addition to the monument in 1955, known as Founders' Pond Tract.[31]

It is doubtful that the State Conservation Commission would have taken much interest in historic places had it not been for Louise Parker. Her position as a commissioner, however, afforded her an opportunity to advance a select number of worthy historic preservation projects, and she found support among her fellow commissioners. She did not initiate projects, but once proposals for historic places came to the commission, she typically expressed the greatest interest in them. As was her style, Mrs. Parker studied the subjects, cultivated community and political support, established the necessary contacts with outside experts, and facilitated interagency communication. Other commissioners and the staff supported her efforts, but the historical program was largely her creation. What she failed to consider, however, was who would carry on the work once she left the commission. She never, for instance, proposed that the commission create a staff position to administer the nascent historical program. Either the thought never occurred to her, or, if it did, she determined that there would be no support for the idea among her fellow commissioners. In any case, when she left the commission in 1949, mention of the historical program dropped from biennial reports. The commission would continue to entertain proposals for acquiring specific historic places. However, interest in administering them began to wane, even though Florence Cowles Kruidenier of Waukee, who assumed the seat Parker vacated, was a member of the Colonial Dames and had been involved in the Plum Grove project.[32]

Ada Hayden and Prairie Preservation

The drama of westward expansion was written quietly on the vast prairies of the agricultural Midwest. There was no gold or silver to excite dreams and passions. No long cattle drives or range wars contributed to the mythologizing of the West. No monumental landscapes inspired artists and invited comparisons with the cultural riches of Europe. Yet nowhere was the transformation of landscape more rapid or more complete. As native inhabitants were systematically pushed westward, then herded onto smaller and smaller reserves, the center of the nation quietly, and quickly, filled up. By mid-nineteenth century, the technologically advanced moldboard plow was ready to rip through prairie sod. Liberal land policies enticed settlers, hardy and foolhardy alike, to the hinterland. Land grants and westward fever propelled rails across the land. Capital and labor poured into the interior. Farms, towns, and cities emerged. And the ancient fertility of prairyerth made it all possible. "In short," as Aldo Leopold wrote, "the plant succession steered the course of history; the pioneer simply demonstrated, for good or ill, what successions inhered in the land. Is history taught in this spirit? It will be, once the concept of land as a community really penetrates our intellectual life."[33]

An estimated eighty-five percent of Iowa was covered with prairies before Euro-American settlement. In the space of seventy years, at most, the prairie was rolled over one furrow at a time. Between 1840 and 1880, most of the well-drained prairies were brought under the plow. During that "golden" period when agriculture reached its pinnacle in the Midwest, roughly 1890 through World War I, the wet prairies were drained for farming. By 1920, nothing but remnants remained.[34] "In one lifetime," John Madson noted, "the great tall-grass reaches of middle America had been opened, broken, and inked onto deeds. It was all deceptively swift and easy; in reality, it was one of the world's great revolutions—a vast reordering of what men felt they knew about land."[35] The next generations would ponder prairie shards and discover how fragile was the fertility of the plains. But settlers mainly saw economic opportunity in these seas of grass. In a sense, the vastness of the prairie was its undoing. There was so much of it. The prairie invited profligacy. No one lamented the disappearance while it was happening. Only afterward did literary hands begin to weave the "mystique" of the prairie.[36]

Quietly, the prairie all but disappeared, and quietly began the effort to save what remained after the "golden age of agriculture" had peaked. Prairie preservation in Iowa is distinctly linked with Ada Hayden, born on a farm near Ames in 1884. Except for brief sojourns in St. Louis, where she earned a master's degree at Washington University in 1910, and then on to Chicago to begin doctoral work, Hayden never left home.[37] Her life was circumscribed geographically, to be sure, but what might otherwise have resulted in provincialism led Hayden instead to a deeper understanding of her own microcosm. She embodied the axiom, "Blossom where you are planted." In Hayden's case, she determined to stay planted in Iowa, working alongside her mentor at Iowa State. She was one of Louis Pammel's protégés, studying under him and serving as his laboratory instructor for two years before graduating with a bachelor's degree in 1908. Pammel encouraged her to continue with graduate studies elsewhere, but she hardly charted a bold course by attending Washington University, the same institution where Pammel had earned his own master's degree. He then pushed her out of the nest again and encouraged her to go to the University of Chicago, which she dutifully did, but not for long. By 1911, she was back at Iowa State, studying under Pammel once again. In 1918, Hayden earned her doctorate, the first woman at Iowa State to do so.

After making a half-hearted attempt to seek a faculty position elsewhere, she accepted a position with her alma mater. From 1918 to 1920, she taught

Botanists Ada Hayden (1884–1950) and Charlotte King
(?–1937), photographed on the Iowa State campus,
c. 1920. King was the experiment station artist and the
seed analyst in the testing laboratory; she also assisted
Louis Pammel with his research. *Courtesy Iowa State University, University Archives.*

botany as an instructor. In 1920, she became an assistant professor of botany, a faculty rank she held until her death in 1950. Hayden's career path was not particularly unusual for women of her generation who chose the academy. Few colleges and universities would consider hiring women to teach in the sciences, and women who did secure faculty positions at institutions of higher learning rarely advanced through the academic ranks as did their male counterparts. And, like many women scholars of her generation, Ada Hayden never married. What appears in retrospect to have been timidity may have been, instead, a realization that the likelihood of making significant contributions to botany were just as great, if not greater, working on Pammel's coattails at Iowa State than they were almost anywhere else. At least there she had already earned a measure of academic respect; an equal measure would have to be hard won in new surroundings. We will never know for sure, since Hayden left few personal papers. In any case, the pattern established as student-mentor continued after she joined the faculty. She worked closely with Pammel until his death in 1934, collaborating in several of his books. From 1934 on, she was assigned to research in the Agriculture Experiment Station. In addition, she served as curator of the herbarium at Iowa State, which, in honor of her floristic studies and her role in prairie preservation, is now called the Ada Hayden Herbarium.

Hayden's own interests and personality emerged after Pammel's death. Those who knew her in the fullness of her life's work knew a woman who was "outspokenly and emotionally dedicated to the prairie effort." She became something of a legend around "Old Botany" as Agricultural Hall was known. Former colleagues described her as "commonly less than a diplomat"; "very brusque, very organized, and rather awe-inspiring"; "about as subtle as an iceberg falling through a skylight"; and, in another vein, "diversely talented and skilled." A photograph of her at about age twenty reflects an attractive woman with a soft smile and a determined gaze. Her fierce independence was evident even then, but it became her hallmark in later years. Although she is said to have been an effective team leader, in an autocratic sort of way, she preferred fieldwork alone. "In a day when women did not do some things," Duane Isley has noted, "she loaded her boat for her field work on top of the car herself, and hauled it north for the summer. And the reverse as necessary."[38]

The interest that would dominate the last decade of Hayden's life began as her doctoral dissertation on the ecology of prairie plants in central Iowa. The *American Journal of Botany* subsequently published her work in 1919, and portions of it also appeared in the *Proceedings of the Iowa Academy of Science*.[39] She also contributed a short piece on the "Conservation of Prairie" to the Board of Conservation's 1919 report, *Iowa Parks: Conservation of Iowa Historic, Scientific and Scenic Areas*. The latter may be considered the first express call for prairie preservation in Iowa, although Bohumil Shimek had been writing about Iowa's prairie ecology for some years and would continue to do so during the 1920s.[40] Hayden identified no specific locales as good candidates; indeed, it is doubtful that she had the necessary field experience at that time to recommend specific prairie remnants for preservation. She did suggest, though, that "a few acres in each county could be [preserved] without encroaching upon economic products." She had in mind prairie tracts near "the larger schools" for educational purposes, and railroad rights-of-way, which would "not conflict with private estates."[41] As it turned out, Hayden's call went unheeded in the 1920s, and the reason is probably to be found in the very couched language she used: "without encroaching upon economic products." Pammel would have been *the* person to instigate prairie acquisition, but as we have seen, even he shied away from land acquisition efforts that would be perceived as a direct challenge to farmers.

Prairie preservation resurfaced in the *Iowa Twenty-five Year Conservation Plan*, which recommended acquisition of small remnants, wherever they could be had, and at least one large tract. Apparently, enough survey work had been done by then to identify several remnants "ranging from forty to three hundred acres in area." Crane's plan did not necessarily envision the state's acquiring one of these virgin prairie tracts. Rather, it left open the possibility of purchasing submarginal land and restoring it to prairie condition, and maybe even reintroducing buffalo if the tract were large enough.[42] The survey work referenced in passing may have been conducted by the University of Iowa. An obscure document referred to as "The Shimek Plan" suggests as much. In 1934, Shimek directed a study that produced a plan for a large preserve in the Little Sioux Valley containing diverse landscapes: upland and lowland prairies, ridges, a kettle, a river floodplain, and a forest.[43] That same year Shimek appeared before the Board of Conservation to urge the purchase of a small tract of prairie near Iowa Lakeside Laboratory at West Lake Okoboji. The board respectfully declined to consider purchase of this small tract, based on lack of funds, but it did communicate with the U.S. Department of Agriculture and the Bureau of Biological Survey in an effort to interest the federal government in purchasing several hundred acres for grassland study. Nothing came of this effort, but three years later Margo Frankel made another attempt, asking Fred L. Maytag of Newton to consider donating about 1,200 acres west of Lake Okoboji for a prairie area. Apparently, he declined, since nothing came of this effort either.[44]

Margo Frankel retired from the commission in 1937, and Shimek died in 1937, events which put the prairie effort back on the shelf. Louise Parker made a brief plea, shortly after joining the commission, advocating that prairie remnants be purchased wherever and whenever possible, but except for this, prairie preservation did not begin in earnest until the early 1940s.[45] It is difficult to determine precisely the circumstances that led to a sudden flourish of activity, but one of them seems to have been a letter that Dr. T. D. Kas of Sutherland wrote to the commission in 1941 offering to sell approximately 400 acres of native prairie in O'Brien County along the Little Sioux River. The minutes reveal that Parker consulted with Ada Hayden to judge the offer.[46]

How Parker knew about Hayden can only be surmised, although it is known that Hayden gave many public presentations on Iowa's prairies, using a set of hand-colored lantern slides she produced, based on her own extensive field-work and field photography.[47] Presumably, either the two women met at one of these presentations or Parker learned of Hayden's work because of them. In any case, it was the commission's first discussion of prairie acquisition that involved Hayden in any substantive manner.

Negotiations with Dr. Kas led nowhere, possibly because no large land acquisitions were made during the early war years. In 1944, however, J. M. Aikman, chair of the Conservation Committee of the Iowa Academy of Science, instigated a wide-ranging report on the status of conservation in the state with a particular emphasis on establishing prairie preserves. Hayden and Aikman were centrally involved in preparing this report, which led the IAS to grant the committee a modest sum ($100) in order to document and evaluate vestiges of native prairie in the state as the first step toward acquisition.[48] Ada Hayden apparently directed most, if not all, of the fieldwork. The Prairie Project, as it became known, may have generated many undocumented discussions concerning tracts worthy of preservation, because in late 1944 two specific tracts came to the State Conservation Commission's attention at the same time. One was the 160-acre Walton Tract in Howard County, which the owners were willing to sell. The other was a half-section in Cherokee County, which Senator Guy Gillette thought the commission should acquire, but he could offer no assurance that it might be available.[49]

At some time during this period Hayden and Parker investigated the Howard County tract together. Parker was therefore able to speak from firsthand knowledge when the commission considered its purchase. As a result, this tract plus two smaller adjacent parcels were purchased without delay. By mid-1945 the commission held its first prairie preserve.[50] The commission also asked Hayden to prepare a report on the types of prairie areas in Iowa and to guide the commission in its prairie acquisition program. Hayden responded by organizing her material from the Prairie Project into a two-part report, subsequently published by the Iowa Academy of Science. Meticulously distilled, the Hayden report identified more than twenty specific tracts located in ten northern, central, and western counties that, in the aggregate, contained dominant species of native tall, mid, and short prairie grasses. In the introduction she noted that "although various recommendations for purchase ha[d] been made by citizens to the State Conservation Commission" over the years, unidentified members of the public had always raised objections on the basis that prairie preserves would take "potentially productive land out of cultivation."[51]

Hayden Prairie, 1949. *Courtesy Iowa Department of Natural Resources.*

Acquiring the Howard County prairie touched off a flurry of activity, but as was so often the case, the material results were less than anticipated. In 1945, the commission considered tracts in Cherokee and Pocahontas counties. However, because Hayden's report was not finished, action was deferred until an acquisition program could be worked out. When this task was accomplished in 1946, the commission once again considered land in Pocahontas County. Negotiations focused on a 160-acre "scrap of prairie," as Hayden called it, known as the Kalsow Prairie. This "scrap" according to her 1946 report "affords a clue to the wealth which made possible the prosperous farmsteads which surround this sample of the virgin prairie." It took another two years to negotiate its purchase, however, because the legislature had allocated only $24,000 for land acquisition and the asking price was $32,000. An additional appropriation allowed the commission to complete the deal in 1948.[52]

The following year, in 1949, the IAS set up an advisory committee to assist the commission with prairie management, but Hayden's death at a relatively early age seems to have undermined the Prairie Project. She carried out just one management study, with John Aikman, before she died in 1950. Aikman and J. H. Ehrenreich sporadically published prairie management studies during the next decade, but there was little in the way of actual advising to the commission.[53] Iowa held just two prairie preserves when Hayden died. The commission tried, unsuccessfully, to secure recommended tracts in Cherokee and Ida counties, but even when owners were willing to sell, state allocations consistently fell short of asking prices.[54] The next prairie acquisition did not take place until 1958, when the state purchased the 160-acre Cayler Prairie in Dickinson County, another area recommended by Hayden.[55]

Progress was disappointingly slow, yet it was enough to bring outside recognition from a number of sources. In 1945, Aldo Leopold, the Ecological Society of America, and the National Park Service had been among many parties to send letters commending the Conservation Commission for undertaking a well-conceived plan of prairie acquisition. Four years later, the Exploratory Committee of the USDA Bureau of Plant Industry cited Iowa as being among the leaders in prairie preservation.[56] There is no doubt that whatever accolades Iowa received were attributable chiefly to Ada Hayden and Louise Parker, her principal ally on the commission. Years later, Parker recalled the initial "lack of enthusiasm the Commission indicated about saving prairie areas." Bruce Stiles, ICC director from 1948 to 1959, apparently also was "luke warm" to the idea until he returned from a large national meeting and informed Parker that "the first question a number of delegates asked him was about [Iowa's] prairie areas."[57] Only gradually did the commission and the staff come to support prairie preservation, but there was little public interest in prairies. Moreover, shortly after Hayden's death, J. M. Aikman relinquished the reins of the IAS Conservation Committee, thereby loosening the ties these two academy members had made with the State Conservation Commission. As if to cap the initial prairie preservation effort, Ada Hayden was honored posthumously in 1950 when the Howard County tract was renamed Hayden Prairie.[58]

Even though Hayden and Parker finally convinced the commission to begin prairie preservation, there was confusion as to where these new lands fit into the system. For instance, the commission purchased Hayden Prairie with Fish and Game funds, Kalsow Prairie through the Lands and Waters budget. Additionally, biennial reports of the 1940s and 1950s never mention the prairie acquisition initiative, further indication that prairie management did not fit conveniently into the established organizational structure. When the commission reclassified its holdings in 1955, Kalsow Prairie officially became a state preserve under the management of the Lands and Waters Division.[59] Hayden Prairie, however, was not given the same status because, having been purchased with Fish and Game funds, it was technically under Fish and Game

Division management. This confusion continued for another decade until the State Preserves Act was passed in 1965.

One can always argue that more was possible, but that would disregard complex historical forces. The real question is why, given the recommendation in the *Twenty-five Year Conservation Plan*, did the commission slight prairies for so long? Part of the answer surely must lie in the tacit understanding that resource conservation activities would not vigorously challenge agricultural interests. Any attempt to acquire a large prairie tract would do just that. In fact, farmers in the vicinity of Kalsow Prairie were not long in lodging complaints that Canadian thistles were spreading from the preserve to nearby fields and that uncut prairie grasses were interfering with a new waterway that they had jointly financed.[60] The years since then have done little to reverse this widespread attitude. In his 1982 book, *Where the Sky Began*, John Madson recounts a chance encounter at Kalsow Prairie with an adjacent farmer who thought the "whole place was a damned silly waste of good land."[61] But opposition from farming interests accounts for only part of the commission's lassitude. Lack of public interest made it easy to ignore prairie conservation. Prairies offered little in the way of public recreation, and taxpayers often opposed land acquisitions that had no obvious recreational benefit to justify taking land out of tax-generating use. Consequently, the commission could evade prairie conservation until pressed from the outside, as the Iowa Academy of Science did in 1944.

Even so, Iowa once again proved to be a leader, sluggish perhaps, but nonetheless a pathbreaker. Wisconsin was the other state that began to preserve natural areas during the 1940s. In 1945, Aldo Leopold convinced the Wisconsin Conservation Commission to establish a Natural Areas Committee to acquire botanical areas for scientific study. The committee's work preceded 1951 legislation authorizing the State Board for the Preservation of Scientific Areas, which functioned within the administrative structure of the Wisconsin Conservation Department. Missouri became active in the 1950s, designating its first prairie preserve in 1957. In the mid-1950s, Ohio also began to restore a prairie area in the Killdeer Plains Wildlife Area. Illinois dedicated its first nature preserve, Illinois Beach in Lake County, in 1964. In 1969, Indiana set aside Beaver Lake Nature Preserve, a wet prairie in the Grand Marsh of the Kankakee, as a prairie chicken refuge. Nebraska entered natural areas conservation belatedly in the mid-1970s.[62]

In the larger scheme of things, Iowa's fledgling prairie program reflected the evolution of ecology as a science and a related concern among resource experts for the fate of the nation's "breadbasket," the great Middle West. In the United States, the 1915 founding of the Ecological Society of America marked the emergence of a particular school of ecological thought initially associated with Frederic Clements of the University of Nebraska and Henry C. Cowles of the University of Chicago. Clements and Cowles held that plant communities change and develop within limits determined by climate and landform, inexorably maturing toward a final "climax" stage. Mature plant communities, therefore, were distinctive of the region (or microclimate) in which they occurred, and vice versa. Although later scholars would modify the construct of climax theory, the history of ecology in the United States is inextricably linked with grasslands research, first by Clements but especially by his followers, John Weaver and Evan Flory, also of the University of Nebraska.[63]

Clements and Cowles were enormously influential figures in the scientific world, and Hayden's own research fell in line with the course of ecological thought they stirred; her doctoral dissertation demonstrates as much, and it is likely that she studied under Cowles during her brief period at the University of Chicago. It was precisely because prairie patches offered botanists the opportunity to study plant succession all the way to the mature, or climax, stage that scientists and conservationists began to value them. The historical record of regional biodiversity was obtainable in prairie remnants. Hayden's 1919 piece on "Conservation of the Prairie" and the *Twenty-five Year Conservation Plan* expressed this view in general terms. Then the monstrous dust storms of the "dirty thirties" brought new cogency to grasslands ecology and conservation. Each dust storm of the long drought eroded long-held notions about inexhaustible natural resources. Fifty years of Great Plains farming finally exposed the fallacy of Charles Dana Wilber's 1880s land-promoting slogan, "rain follows the plow."

Provoked by an environmental crisis with devastating economic consequences for agriculture, the federal government responded in the 1930s with a host of initiatives designed to tackle land-use problems on more than one front. The Soil Conservation Service, aided by the Civilian Conservation Corps, worked with individual farmers and local soil conservation districts to check wind and water erosion by modifying the landscape with gully dams, waterways, shelterbelts, and terraces. More ambitiously, the Roosevelt administration sought to establish regional land-use planning through such bodies as the Tennessee Valley Authority, the National Resources Planning Board, and the Great Plains Drought Area Committee. Ecologists suddenly found themselves sought as land-use advisers. The results of 1930s regional land-use planning efforts are debatable, but the Dust Bowl, as Donald Worster has noted, "laid the groundwork for a more scientifically fueled conservation movement in America, one that would pick up steam in the decades ahead."[64]

In Iowa, the conservation movement was in the beginning "scientifically fueled." But scientists had been unable to rouse interest in prairie preservation until the 1940s. The public awareness of ecology that came with the Dust Bowl, particularly through Paul Sears's 1935 book, *Deserts on the March*, no doubt created greater interest in Hayden's prairie research, and possibly caused her to redouble her own efforts. How else does one account for her many lantern-slide presentations during this period? She also became active in the Ecological Union and the Grassland Research Foundation. Her major contribution, however, was in systematically developing the scientific database from which the Conservation Commission could make informed decisions about land acquisition. In this respect, she continued the planning approach based on field study that was embodied in the 1933 state conservation plan. The goal in the 1940s was to preserve at least one each of the major types of prairie landscapes in Iowa, prairie museums, in a sense, where Iowa's distinctive natural history could be observed and studied by future generations. Two preserves hardly realized this goal. However, when the environmental movement got underway in the 1960s, Hayden's prairie studies were right there to guide a renewed prairie conservation effort.

More Lakes

In 1944, the commission outlined a $7 million budget to develop areas already under state ownership. Additionally, it proposed an ambitious $28 million *extension* budget to create fifty "multiple purpose" artificial lakes, purchase 500,000 acres of marginal land for forests, develop ten new state parks and preserves, and acquire 2,000 miles of public access to fishing streams. The timetable was left open, but the commission offered five-year, ten-year, and fifteen-year options, depending upon the level of annual appropriations. Considering that the commission's annual appropriation was approximately $250,000 in 1945, the price tag for completing the *Twenty-five Year Conservation Plan* suggests how far the state had to go and the degree of optimism that reigned at the end of the war. In retrospect, though, it was probably a conservative estimate of what it would take to reach the goal of "providing maximum opportunities for outdoor recreation" as part of a "comprehensive land and water use plan."[65] As events—and budgets—unfolded during the next decade, the dollars shrank and so did the goals. An "extension budget" became part of the financing picture, but it was never enough to achieve "comprehensiveness." The lion's share of limited funds went to developing artificial lakes. In the end, eight, not fifty, new lakes were created, but all were designed to expand recreational opportunities in state parks.

The earlier program in the 1930s to develop artificial lakes was, in part, a strategy for creating wildlife habitats and unpolluted fishing areas. Accordingly, Lake Wapello (Davis County), Lake Macbride (Johnson), Lake Keomah (Mahaska), and Beed's Lake (Franklin) projects were started in 1934 under the Fish and Game Commission's jurisdiction. That same year, Lake Ahquabi (Warren) and Lake of Three Fires (Taylor) got underway as cooperative projects with the Board of Conservation, and the two agencies jointly cooperated to create artificial lakes in some existing state parks. Additionally, the Fish and Game Commission handled restoration projects at several lakes technically under the jurisdiction of the Board of Conservation.[66] Even though several artificial lakes began as Fish and Game projects, most of them were classified as state parks, and they proved to be enormously popular for boating, water-skiing, and swimming, as well as fishing. While attendance at state parks systemwide rose a scant ten percent between 1935 and 1936, attendance at the new artificial lake parks nearly doubled. Combined attendance figures for Beed's Lake, Lake Ahquabi, Lake Keomah, Lake Macbride, Lake Wapello, and Lake of Three Fires (still known simply as the Bedford Tract) increased about ninety percent. Such a dramatic rise is partly attributable to their newness. Nonetheless, these areas continued to experience greater than average increases. During the next two years, attendance at the six lake parks increased twenty percent annually, while the systemwide increase held to a steady ten percent.[67]

The wartime suspension of new construction placed the artificial lake program on hold, but public interest did not wane. In fact, when park use began to return to normal, attendance figures continued to show greater increases at parks having artificial or natural lakes. Thus, even before the war was over, agitation for more lakes began to pick up. During the summer of 1944, the commission seriously discussed a written request from citizens in Shelby County for a recreational area in that vicinity. Such communications had been received with some regularity for at least two years, and various individuals had appeared before the commission to press their case in person. Now the commission decided that it was time to look beyond the war. In keeping with recommendations in the state conservation plan, the commission began to investigate possible sites.[68]

Studies and discussions concerning artificial lake development continued for the next few years. While investigations focused on southern Iowa sites, delegations from elsewhere in the state came to the commission seeking support for proposed lake restoration or artificial lake projects in their locales.

Louise Parker, in particular, favored a large recreational lake to serve the Des Moines metropolitan area population. In addition, some commissioners voiced the opinion that restoring natural lakes was as important as creating new lakes. As a result, the commission tried to balance resource allocations between dredging and new construction.[69] Complicating environmental factors also slowed the decision-making process. In late 1944, the commission's chief engineer reported that "satisfactory sites" in southwest Iowa were "almost non-existent . . . because of complex siltation and shore erosion problems."[70] In general, everyone now agreed that soil conservation in the surrounding watershed should be considered as part of any lake restoration or new lake construction project.

Rapid siltation in several artificial lakes constructed in the 1930s led the commission to reevaluate the whole program. As early as 1941, commission staff members met with representatives from the Soil Conservation Service to address the erosion and siltation problems in Backbone State Park. Studies showed that in less than eight years, the lake, constructed in 1934, had lost 6.6 percent of its storage capacity. The story was the same at other artificial lakes. Continued discussions with the Soil Conservation Service and the State Department of Agriculture led to a formal watershed protection program. The value of agricultural land, not to mention the value that Iowan's placed on farming, prohibited the commission from even thinking about purchasing entire watersheds. A feasible alternative was watershed improvement and watershed maintenance agreements with private landowners. Accordingly, in 1946, the state legislature passed a law enabling the commission to work with private landowners to improve the watersheds of state-owned lakes. A special $2.7 million appropriation in 1947, the first of several "extension budgets," allowed the effort to go forward. About half the appropriation was tagged for artificial lake development. Another $77,000 was slated for soil erosion work on watersheds.[71]

Watershed protection thus became a major activity in the late 1940s. At least fifteen artificial lakes, either existing or under construction, received some level of erosion control effort between 1947 and 1950. Natural lakes were not exactly neglected during this period. The shorelines of several natural lakes were riprapped to control bank erosion, and others were dredged to improve water quality. However, the primary reason for the watershed protection program was to make artificial lakes feasible. Even though it quickly became apparent that $77,000 was insufficient for the task, the commission began constructing spillways and silt dams, riprapping stream banks, and generally improving soil erosion control features on farms located above existing and planned artificial lakes. To implement the watershed protection program, the commission required cooperative agreements with affected soil conservation districts. In most places farmers willingly cooperated, although this was not always the case, and local chapters of the Izaak Walton League often lent their support.[72]

Under the special appropriation, which was augmented four times by subsequent legislative actions, eight new artificial lakes were approved, all in southern Iowa. By the end of the decade, four of the projects were nearing completion: a 400-acre lake in Washington County that would be christened Lake Darling; a 200-acre lake in Geode State Park, located in Henry and Des Moines counties; a small lake at Cold Springs Recreational Reserve in Cass County, and a seventy-acre lake in Nine Eagles State Park, located in Decatur County.[73] The commission chose to name the first of the lakes in honor of Jay N. Darling as the "instigator of the Twenty-Five Year Plan and the present commission form of conservation management."[74] It was one of many public honors Darling received during his lifetime for his dogged persistence in conservation efforts. Lake Darling was dedicated on September 17, 1950. Cold Spring, Nine Eagles, and Geode were dedicated the next year. In 1952, Rock Creek Lake in Jasper County was dedicated. Green Valley Lake near Creston, a cooperative project with the City of Creston and the Rural Electrification Administration, was underway, although engineering problems delayed completion until 1956.[75] Engineering and land acquisition issues also delayed construction of artificial lakes in Shelby and Montgomery counties (discussed in chapter 6).

A Greater Place for Wildlife

Aldo Leopold deserves much of the credit for the game management policy Iowa initiated during the 1930s, although it solidified in the 1940s. The wildlife biologist who eloquently stated the case for a biotic land ethic in *A Sand County Almanac* (1949) had a brief professional connection with conservationists in his native state in 1931–1932 when he conducted a game survey as part of the *Twenty-five Year Conservation Plan*. The Iowa survey, however, was part of a much larger study conducted for the Sporting Arms and Ammunition Manufacturers' Institute (SAAMI). Between 1928 and 1932, Leopold directed systematic field investigations in a block of eight North Central states including Iowa.[76] The game survey was the first of its kind ever attempted in the United States, and the results provided baseline data to justify the need for scientific game management.

Six months into the SAAMI survey, Leopold was selected by the American Game Conference to head a committee charged with drafting a national game policy. That policy, adopted by the conference in 1930, placed primary responsibility for game management with landholders; called for recognizing game management as a distinct profession with appropriate training programs; and emphasized the need for sport hunters, conservationists, and nature lovers in general to cooperate on matters of wildlife conservation. The policy statement drew praise from many quarters because it marked the emergence of a "coherent national strategy" aimed at directing "the previously disparate activities of sportsmen, administrators, researchers, and (its framers hoped) landowners."[77]

The game survey continued into 1932 largely because Iowa convinced SAAMI to fund a more thorough resurvey of Iowa as part of the state conservation plan. Leopold's major findings actually were contained in his 1931 *Report on a Game Survey of the North Central States*. Packed with statistics on game and habitats, the report secured Leopold a place of respect among conservationists nationwide. At the same time he was writing the report and directing the Iowa game survey, he also wrote a textbook, *Game Management*, published in 1933. This four-year odyssey crystallized Leopold's thinking about wildlife management, and the principles he developed during this time guided wildlife biologists for several decades. His philosophy was rooted in a realism that tried to reconcile the complexities of human nature and Western society—that is, the modern technological world—with the vulnerability of the natural world.[78]

The North American game survey confirmed what wildlife conservationists had long known: the old pattern of relying on a combination of artificial game propagation (game farms and fish hatcheries) and restrictive laws that set bag limits and regulated hunting seasons were simply not enough to stem the decline of wildlife populations. In pointed terms he asked, "Does anyone still believe that restrictive game laws alone will halt the wave of destruction which sweeps majestically across the continent, regardless of closed seasons, paper refuges, bird-books-for-school-children, game farms, Izaak Walton Leagues, Audubon Societies, or the other feeble palliatives which we protectionists and sportsmen, jointly or separately, have so far erected as barriers in its path?"[79]

Iowa was an early beneficiary of Leopold's influence. During early 1932, while he was in the state working on the game resurvey, Leopold met three times with the Fish and Game Commission. As a result of these discussions, the commission and the Board of Conservation jointly approved a wildlife refuge policy before the state plan was completed. The Iowa Conservation Plan Refuge Policy adopted in February 1932 established two classes of wildlife refuges: (1) protective refuges to provide sanctuary for rare species, such as prairie chickens, ruffed grouse, and wild turkeys, and (2) productive refuges "to produce an outflow" of game species.[80] Several state parks were declared wildlife refuges. Additionally, several dry lake beds were maintained as refuges. As articulated in 1933, the game conservation policy called for increasing the production of quail, pheasants, and other upland game on privately owned lands; acquiring more public land and restoring lakes and marshes for migratory waterfowl refuges; and regulating hunting on public shooting grounds.[81]

By the time the State Conservation Commission took over the reins in 1935, the distinctions among park, preserve, and refuge were consciously blurred. Since Iowa's parks had been intended from the beginning to function, in part, as wildlife sanctuaries, this was nothing completely new. More to the point, it was the increasing recreational use of parks, increasing concern for declining populations of game species, and dismay over increasing stream and lake pollution that led Iowa conservationists to focus on creating new places where wildlife could reproduce. In the 1930s, artificial lakes were seen as one strategy. These turned out to be heavily used recreational lakes, not quite the places conservationists had in mind as wildlife refuges, but they allowed the commission to establish beachheads from which to expand landholdings in order to restore wildlife habitats, particularly in southern and western counties.

The wildlife refuge system that took shape in the 1940s reflected the policies that Leopold helped to establish, the organizational peculiarities of the State Conservation Commission, and the 1937 Federal Aid in Wildlife Restoration Act. With regard to organizational peculiarities, the commission's fiscal arrangements reflected a delicately balanced bureaucracy. When the commission was set up in 1935, the method of financing each of its parent entities simply transferred to the new agency. This meant that hunting and fishing licenses financed operations of the Fish and Game Division; the Lands and Waters Division received appropriations from the general treasury. Thus, the commission did not suffer any fiscal loss in the merger, but segregated financing and the disparity between division income augured potential conflict. The Fish and Game Division received nearly three times as much in receipts (approximately $300,000 per year at the time) as the Lands and Waters Division received from the legislature (approximately $100,000 per year). Much of the Fish and Game budget was devoted to enforcement, but the division also began to acquire land for refuges.[82]

Federal aid and emergency state appropriations during the Great Depression skewed the financing picture in favor of parks for a few years during the 1930s, when additional money paid for state park development and forest acquisition. Some emergency aid, however, was earmarked for wildlife refuges and game management areas. In practical terms, it was often difficult to distinguish a state park from a wildlife refuge or a game management area, since these functions often overlapped, and both divisions were involved in land acquisition, management, and maintenance. Confusing the record-keeping picture even more, both divisions tended to report acreage figures according to function. Thus, if an area served more than one function (say, it was both a recreational lake and a wildlife refuge), the acreage showed up in more than one tally. Here was the multiple-use philosophy thoroughly bureaucratized.

By 1940, land acquired specifically for wildlife refuges totaled no more than a few thousand acres, but this tells only part of the story. Through land acquisitions, leases, and game management agreements with private landowners, the commission carried out an extensive game conservation policy that followed Leopold's recommendations. By the time the U.S. went to war, the commission managed ninety-one wildlife refuges ranging in size from a few acres to more than 5,000 acres. Over half, forty-eight, were also classified as state parks, lake reserves, or preserves. As a result, the commission could claim about 50,000 acres in wildlife refuges. Additionally, nearly 250,000 acres of private land were in the Game Management Program, established in 1937. Under this program, the commission furnished pheasant and quail chicks to farmers and sportsmen's groups, who reared them for release. Artificially propagated species later were expanded to include ruffed grouse, prairie chickens, partridges, and fur-bearing animals. The latter initiative proved to be highly successful, so much so that by 1942 the commission reported that "no other single activity . . . has served so well to cement the interests of sportsmen and farmers, teach them the fundamentals of wildlife conservation, and secure their active support and cooperation."[83]

A new source of federal aid led to a greater emphasis on wildlife refuges. In 1937, Congress passed the Federal Aid in Wildlife Restoration Act, commonly known as the Pittman-Robertson Act. Financed by excise taxes on sporting arms and ammunition, the act provided a new source of federal money to fund state-initiated wildlife projects, which typically meant land acquisition, game management programs, and research. The act is named for the two congressmen who sponsored the legislation, Senator Key Pittman of Nevada and Representative A. Willis Robertson of Virginia; however, Carl Shoemaker, secretary of the Senate Special Committee on Conservation of Wildlife Resources, and Jay Darling are generally acknowledged as the creators of a national wildlife agenda in the 1930s. Shoemaker drafted the Pittman-Robertson Act and shepherded it through Congress. Darling, during his brief tenure as chief of the Bureau of Biological Survey, established the Cooperative Wildlife Research Unit Program with land-grant universities. These schools trained many wildlife biologists who found professional staff positions with state and federal conservation agencies. Together, Darling and Shoemaker cofounded the National Wildlife Federation in 1936, a move designed to cohere fragmented and sometimes narrowly focused support among sport hunters into a broader, united constituency for wildlife protection.[84] This combination of federal funding, an increasing cadre of professional wildlife biologists, and grassroots support created a powerful force for which there was, at the time, no counterpart in historic preservation or prairie preservation.

Under Pittman-Robertson, twenty-nine states emphasized land acquisition from the beginning, as did Iowa, which entered the program in 1939. Iowa and Minnesota both adopted an early focus on wetlands acquisition. The term "wetlands" would not come into common usage until the mid-1950s, when the National Wildlife Federation began campaigning to "Save Our Wetlands," but the intent was the same: to replenish game supplies by protecting the breeding habitats of waterfowl and aquatic furbearing mammals.[85] As such, Pittman-Robertson fostered a more environmental approach to game management. However, it was not an approach that would be consistent with ecological principles more clearly articulated by Leopold and others in the 1940s. The value of wetlands for protecting nongame species, for protecting aquatic plants, and for flood control were not obvious considerations of the Pittman-Robertson program as it was conceived and implemented. Its goal was still to produce more ducks, geese, pheasants, quail, beaver, muskrats, and assorted other game species for hunters. The strategy, however, was in keeping with scientific game management principles set it motion in the 1930s: stocking game birds and animals in areas where the natural environment was the most advantageous for abundant reproduction. By the mid-1940s, this goal had not only spawned hundreds of narrowly focused scientific studies, it also led to a perspective that nearly equated wildlife conservation with farming. A quotation from ICC Director Bruce Stiles pretty much sums up this position: "Wildlife must be considered as a crop, and the surplus should be harvested each year in the same manner as a farm crop."[86]

Although the various wildlife initiatives of the 1940s advanced the recommendations contained in the *Twenty-five Year Conservation Plan*, in spirit there was a marked departure from the game management and wildlife refuge philosophy that Leopold initiated in the 1930s. Nonetheless, whatever its limitations in terms of environmental thought, the Pittman-Robertson program worked on a practical level. By 1950, federal allocations enabled Iowa to purchase more than 10,000 acres of sloughs, marshes, and small lakes. In addition to acquiring completely new areas, the commission bought land around lakes and marshes already in public ownership. Most of the land acquisition and development projects in the early years were marshes and sloughs concentrated in the northern counties, particularly the Iowa Great Lakes region: Dickinson, Emmet, Clay, and Palo Alto counties.[87] In this manner, the Pittman-Robertson program enhanced the recreational potential of some state parks, but park enhancement was incidental to the main goal of protecting and restoring the habitats of game species. By 1948, the Pittman-Robertson program had reached such proportions that the commission established a separate Federal Aid Section to handle land acquisition and all the other associated activities: surveys and investigations, wildlife habitat development and management, and ongoing maintenance. In operation, the former Game Management Program worked out with private landowners now fell within the purview of the Federal Aid Section, which tried to balance land acquisition with habitat development and game management.[88]

Increased allocations in the 1950s and a second federal program designed for fisheries development allowed the commission to accelerate land acquisition. By 1960, federally financed land purchases for wildlife totaled approximately 29,700 acres in Iowa. Whereas earlier land acquisitions were mostly located in the northern prairie pothole region, acquisitions in the 1950s were more evenly spread throughout the state. The focus remained, however, on small lakes, ponds, marshes, sloughs, and river access areas. Pittman-Robertson aid also financed the operation of five large "game management units" in Iowa, each with its own headquarters. Game management units were concentrations of land already under public ownership, land purchased through Pittman-Robertson, and privately owned land on which the Fish and Game Division developed and maintained wildlife habitats.[89]

Rice Lake (Winnebago and Worth counties) was a made-to-order Pittman-Robertson project, and the Rice Lake Game Management Unit became one of the largest in the system. At one time, the lake had covered as much as 800 acres, deep enough for steamboats and reportedly as beautiful as Clear Lake. An unsuccessful attempt to drain the lake in 1903, however, left about 200 acres of marsh and water. Conservationists lamented the loss of Rice Lake, a view shared more widely because the drainage effort had failed to yield good farmland. In 1924, the old Board of Conservation established a small state park, about fifty acres in size, adjacent to the remnant lake. Restoration would have been a priority had money been available then, but it was not. The opportunity to reclaim the area finally came in the mid-1930s, when the Conservation Commission used state emergency conservation funds to acquire a few hundred acres for migratory waterfowl habitat, and CCC crews improved the state park with a shelterhouse. These developments revived local dreams of returning the lake to its original size, even enlarging it into a major recreational spot.[90] In the early 1940s, the commission used Pittman-Robertson funds to purchase several hundred acres around Rice Lake for a migratory waterfowl refuge. Pressure from local citizens finally induced the commission to raise the water level during the late 1940s and early 1950s, expanding the lake to about 700 acres. At about the same time, Rice Lake became the nexus of a large game management unit comprising eight state-owned areas in four counties that together totaled about 5,000 acres.[91] Not all game management units established under Pittman-Robertson included state parks, as did the Rice Lake Unit, but many state parks still included, or were situated adjacent to, game management areas.

The Pittman-Robertson Act enabled all states to create more space for wildlife, and this one piece of legislation accounts for many of today's wildlife success stories: the millions of Canada geese that now migrate through flyways from arctic Canada to Mexico, thriving populations of wild turkeys, an overabundance of white-tailed deer, rebounding herds of pronghorn antelope on the Great Plains, and a limited prairie chicken comeback on certain ecological islands. Among North Central states, Iowa consistently emphasized land acquisition to the extent possible in a state where agricultural land has always commanded premium prices. Shifting the focus of game management from artificial propagation to wildlife habitat protection began the long process of reviving diminished wildlife populations and, in some cases, threatened species.

The Pittman-Robertson Act also contributed to a more balanced conservation agenda during the 1940s. After the war, the commission took some giant strides toward implementing the comprehensive agenda contained in the *Twenty-five Year Conservation Plan*. Federal aid benefited wildlife the most. However, Louise Parker saw to it that resource preservation was not neglected. She attempted a fledgling historic preservation program, and, in concert with Ada Hayden, succeeded in inaugurating prairie preservation. All these initiatives were revealed, in one way or another, in the changing landscape of state parks and preserves. This, however, was also the decade that the commission washed its hands of water pollution issues (discussed in chapter 4), so there were definite limits on what the State Conservation Commission could or would do. Moreover, federal flood control projects loomed on the horizon, some of them slated for construction near state parks; and the state still had no coherent water policy. The balance toward which the commission moved in the 1940s would prove to be fleeting.

6

The Center Does Not Hold

There aren't enough technically qualified conservationists in the state to form a good Conservation Commission. You will understand, of course, that in speaking of conservationists I am not talking about a shortage of sportsmen and their particular branch of wildlife conservation, nor of bird lovers nor wild flower fans. While they have done more than anyone else and paid all the bills up to date, few of them understand that you can't restock a barren lake or stream with fish until you have restored the balanced chemistry of the waters.

—JAY N. "DING" DARLING [1]

In 1955, as spring was warming to summer, the Iowa Outdoor Recreation Association invited fifty legislators to dinner, then peppered them with little speeches about "dirty beaches, insufficient parking areas, outmoded shelters and bathhouses and deteriorating picnic grounds." [2] A year later, the Iowa Ikes called upon the Conservation Commission to "take the initiative" and present "a realistic budget" that would fund a new survey of state recreational needs, pay for the repair of existing facilities, and allow the commission to complete half-finished projects. The *Iowa Waltonian* followed with a muckraking article depicting crumbling picnic grills, bottomless garbage cans, and rotting roofs.[3] These incidents capture the essential reality of Iowa's state parks during the 1950s and early 1960s. The state park system fell into disarray even though it did not stop growing. Appropriations for maintenance and operations lagged farther and farther behind an increasing number of visitors.

Economically, America was on top of the world. Americans enjoyed more leisure time than ever before, and consumer spending climbed steadily. Conservation, however, was now an idea that grated against the American psyche. After husbanding resources during the depression decade of the 1930s and years of enforced savings during World War II, Americans, it seemed, were in no mood to conserve anything. With more free time and more money to spend, families packed picnic baskets, loaded up tents, hitched up boats or campers, and set out in record numbers with their baby-boom children for parks near and far. No state park agency was well prepared to receive the press, and Iowa let its leadership position slip away.

Ironically, Iowa was spending considerably more money on parks during the 1950s, but millions of dollars in special appropriations were earmarked for specific capital improvements, notably artificial lakes, or emergency repair projects, both in an effort to keep up with increasing use. Routine maintenance and operations were beggared. In 1950, the Lands and Waters Division received an annual appropriation of about $400,000. Concession fees, cabin and camp rentals, dock and boat licenses, and miscellaneous other fees added another $70,000 to $80,000 a year in receipts. A decade later, the annual appropriation had increased to only $550,000, every dollar of which had to be stretched to care for more than ninety parks, preserves, lakes, and forests. Biennial reports chronicle progressively worsening conditions. In the 1952–1954 biennium, for instance, when park attendance approached five million per year, the commission pulled money from the capital improvement fund to expand picnic areas and parking lots. Water, sanitary, and road systems, however, were taxed beyond their limits, and there was not enough money even to maintain them properly. Expanding or improving these systems was impossible.[4]

The forces that led to decline were complex, and their effects reached far beyond state parks and preserves. State politics played a small role. During the late 1940s and early 1950s, commissioners were caught up in the politics of two gubernatorial administrations, those of Robert Blue and William S. Beardsley. At the same time, wage and salary issues generated considerable employee dissatisfaction internally. Additionally, an expanding federal presence complicated matters in contrasting ways. First, more money was available for wildlife and fisheries enhancement, which reflected both concern for continuing declines in wildlife populations and public demand for more hunting and fishing areas. Second, after the war, federal flood control projects were imposed on the state with little or no coordination. Iowa responded by creating a new state agency whose authority overlapped that of the State Conservation Commission, a step that led to interagency conflict.

Whatever strides the commission had made toward balanced resource policies up to the late 1940s, the momentum vanished in the 1950s. In the process, the park system, which had been the opening for establishing a resource conservation agency, lost status within the State Conservation Commission. A mixture of turmoil, confusion, and ennui marked the period from the late

1940s to the mid-1960s. For the park system, the most notable changes were completing the artificial-lake program and beginning to downsize through cooperative arrangements with county conservation boards.

The Gabrielson Reports

Two reports prepared for the commission by Ira Gabrielson of the Wildlife Management Institute, one in 1947 and another in 1954, suggest why parks were languishing. Gabrielson, an Iowa native, succeeded Jay Darling as head of the Bureau of Biological Survey, later the U.S. Fish and Wildlife Service. He left the agency in 1946 to head the WMI. By this time he was one of the country's foremost authorities on wildlife conservation, having authored two books on the subject.[5] He was, therefore, a logical choice when the commission sought outside expertise in matters relating to long-range conservation planning. In 1947, the commission engaged Gabrielson to appraise the commission's organizational structure, policies, and programs.[6]

Gabrielson may have been a logical choice, but politics prompted long-range planning at this particular time. The commission's action followed a series of complaints by Darling that Gov. Robert Blue's resource conservation policies were shortsighted. Until 1946, Darling's criticisms were mainly off the record. That began to change in 1945 when J. D. Lowe, an attorney from Algona, resigned from the commission before his term was up.[7] When Blue announced that Lowe's replacement was Ewald Trost, Webster County chairman of the Republican party, this was too much for Darling, even though he was a steadfast member of the Republican party himself. He charged that Trost had no background or expressed interest in conservation and ran a scathing cartoon depicting Blue and Republican State Chairman William B. York as criminals breaking into the "nonpolitical State Conservation Commission Safe." Precisely because Darling *was* a Republican, the cartoon raised a stir, and other newspapers also ran the story.[8] The flap over Lowe's replacement quickly died down, but the specter of political interference remained. Accusations of graft and corruption that never made the news concerned supposed "contributions" to the state Republican Party that either York or Governor Blue reportedly tried to coerce from State Conservation Commission employees, presumably in exchange for keeping their jobs.[9]

When Gabrielson's report appeared in late 1947, it actually said little about parks. This was not an oversight. Even though he and his team purportedly used the *Twenty-five Year Conservation Plan* "as a basis for measuring results," the report was considerably narrower in its sweep than the 1933 plan had been. There were several recommendations concerning water pollution and soil ero-

"Add Crime Wave." "Ding" Darling cartoon, *Des Moines Register*, 21 August 1946. *Courtesy Cowles Library, Drake University.*

sion control, stream bank and forestry acquisitions, and wildlife habitat protection, but the heart of the 1947 report concerned organizational and personnel matters. A note appended to the report stated simply that just because "such items as parks, the artificial lake program, the restoration of natural lakes and marshes and similar well advanced programs are not given prominence does not imply that these programs are not important. They should be continued along present lines." The park and preserve system was thus dismissed as operating smoothly. The report focused attention, instead, on low salaries as the "greatest single weakness" of the agency. Gabrielson's team was specifically requested to investigate alleged political interference, which it did and "failed to produce any evidence" that such was the case, a finding that ruffled a few feathers but was not publicly challenged. The report, however, did point to the "confusing financial set-up," meaning the methods by which the commission's two main divisions were financed, as a serious hindrance to effective operations.[10]

In 1954, Gabrielson prepared a follow-up report, this one considerably more critical. The second report noted that, in the intervening seven years, "few if any" of the legislative recommendations of 1947 had been adopted. Even more disturbing, the personnel situation had deteriorated. The reason cited was a 1950 amendment to the Iowa Code that gave the Office of the Comptroller, through the Executive Council, greater control over hiring practices, job classifications, and wage and salary rates for state employees. Prospective employees of the commission were now routinely asked to divulge their political party affiliation and voting habits. "It is obvious from this and other information received," the report stated, "that there is an effort being made to place political appointees in the conservation department."[11] Without naming names, Gabrielson charged that Governor William S. Beardsley's administration (1948–1954) had succeeded in reintroducing political patronage to the detriment of the commission and its employees.

Parks also received closer scrutiny in the 1954 report. Gabrielson acknowledged that the 1947 report had said little about the state park system, "which at that time seemed to be in relatively better shape." However, this was no longer the case. After visiting several parks, he concluded that "many of the buildings are in urgent need of repair and much of the equipment for maintaining roads and facilities is so old that an excessive amount of time is needed to keep it operating." While noting the success and popularity of the artificial lake development program, which will be discussed further on, Gabrielson nonetheless counseled that it should be continued only "after provision has been made for full development of the present park system."[12]

A prison labor program inaugurated in 1955 alleviated some maintenance burdens in the state parks. In that year, the legislature increased the Lands and Waters annual appropriation by $75,000 specifically to pay for prison labor projects; the labor, of course, came free. During the next several years, inmates from Anamosa and Fort Madison fabricated picnic tables, camp grills, portable latrines, boat dock sections, signs, buoys, and nature trail boxes; erected shower buildings, garages, and at least one beach house; laid new water and sewer lines; planted trees; and harvested firewood. By the early 1960s, the program was beginning to make a dent in the maintenance backlog, and it also provided any number of young men a chance to work on a better future. Parks in eastern Iowa benefited directly from the prison labor program, particularly Geode, Palisades-Kepler, Lake Macbride, and Wapsipinicon, where work camps were located. Inmates also spent thousands of hours working in Shimek State Forest. After 1960, students from the Boys Training School in

Eldora worked in Pine Lake State Park. Indirectly, *all* parks benefited from the many structures and items that were fabricated in carpenter shops and distributed throughout the state.[13]

If the prison labor program faintly echoed the Civilian Conservation Corps, it was still no substitute for routine custodial care and professional park management. Park road maintenance ate up much of the budget. Gravel roads, once the preferred surface because they were aesthetically compatible with the natural landscape, were inadequate to carry heavy traffic, and park users began to complain loudly about the dust. Aging equipment needed to be replaced. Roofs needed new shingles. Buildings needed paint. Shorelines needed more riprap protection. Above all else, commission staff and conservation officers in the field needed a decent wage. Employees were leaving in droves, and many positions remained unfilled for long periods of time. Morale was running low.

On this issue, the Izaak Walton League stepped up to lobby on behalf of the commission's 450 employees. In 1956, the Ikes conducted a nationwide survey to see where Iowa ranked. Their findings were embarrassing. Nationwide, the average salary of state conservation directors was $9,766; Iowa paid its director $6,000. Only one state, North Dakota, paid its director less. The situation was comparable for professional staff on down the line. In the field, junior officers averaged $3,280 nationwide. By comparison, new hires in Iowa fared a little better; they received a $3,310 starting salary, but they had to buy their own uniforms. Moreover, experience and training were of little benefit. Experienced conservation officers in Iowa topped out at $3,600, far below the nationwide average maximum salary of $4,216.[14] For a state that was average in population, wealth, and land area, the showing was poor. For a state that ranked among the highest in the nation in terms of park attendance, the situation was incomprehensible. The league's study was moderately successful in accomplishing its intended purpose. In 1957, the legislature authorized annual merit increases for conservation officers up to a maximum salary of $4,200, and the director's salary was increased to $7,500. Two years later, though, the Lands and Waters Division cut five caretaker positions and eliminated forestry research in order to maintain park vehicles.[15]

Gabrielson's 1954 report also charged that agency financing was still confused, a situation that now greatly disadvantaged the Lands and Waters Division. Federal aid for wildlife and fisheries programs, along with steadily increasing revenue from hunting and fishing licenses, allowed the Fish and Game Division to expand its technical staff and field force. The Lands and Waters Division, in contrast, did not have enough money to staff all of its existing positions.

Throughout the 1950s, the legislature repeatedly called for parks to be self-supporting, and the Budget and Financial Controls Committee was charged with finding ways to accomplish this end. It recommended three strategies: increasing license fees, a program of privately financed housing accommodations in parks, and requiring the Fish and Game Division to pay a more equitable share of maintenance costs for parks that included fishing areas.[16] None of these recommendations was implemented.

The State Conservation Commission also studied options to increase revenue for state park maintenance. Revenue bonds were considered, but such a plan would have led to increasing the number of rental cottages in state parks as the most feasible means of generating enough income to retire the debt. A survey conducted in 1949 indicated that a majority of visitors would be willing to pay a gate fee at parks, but the majority was not large. When other states were queried, the commission found that the cost of collecting admission fees was fairly high in relation to the return, and that charging gate fees tended to decrease attendance. A follow-up survey conducted in 1955 produced findings that also suggested caution with respect to imposing entrance fees. While the vast majority of Iowans, more than eighty percent, now wanted to see the state park system expanded or improved, more than half of the respondents felt that such projects should be financed through appropriations from the general revenue fund. Less than half were willing to pay an entrance fee, either paid at the gate or through the purchase of an annual park use stamp.[17]

The State Conservation Commission marked the twenty-fifth anniversary of the *Twenty-five Year Conservation Plan* with yet another study conducted by the Wildlife Management Institute. This one, the third in a decade, was supposed to update the survey data in the 1933 plan, ascertain how much of the original plan had been accomplished, and determine whether any unfinished projects were still feasible. But the 1958 study was not nearly as large an effort. The commission assigned staff members to assist WMI, and two outside consultants prepared the park and forestry sections. No effort was made to involve experts from the state's universities and colleges or to solicit information and advice from local organizations, community groups, and individuals. Additionally, Gabrielson decided that the new plan would project only ten years into the future.[18]

The ten-year plan appeared in late 1958 and quietly disappeared a few months later. It was not what the commission had anticipated, although precisely what commissioners *did* anticipate, individually or collectively, is not clear. A lengthy discussion in February 1959 produced agreement only that the plan was not one of "development," but one of internal "reorganization." Moreover, internal reorganization was not as straightforward a proposition as it might appear because Gabrielson's recommendations were predicated on substantial changes in state law that would give the commission greater organizational flexibility. Consequently, even though commissioners could further agree that it was a "good plan for reorganization," most of them were hesitant to take any action. Reorganization "would require a lot of study" and the commission "[did] not have sufficient time" to prepare recommendations for the legislature. Additionally, it was "a very controversial issue," and, indeed, an earlier recommendation by the Little Hoover Commission to consolidate several state agencies into a proposed department of conservation and natural resources had been vigorously opposed. Consequently, most of the plan was shelved. Out of more than seventy-five recommendations, the commission took action only (1) to add a planning section under the director's office, (2) to enlarge and strengthen the public relations section, and (3) to strengthen the operations of the forestry section.[19]

In dismissing reorganization as unfeasible, the commission threw the proverbial baby out with the bath water. Several recommendations, intriguing in retrospect, were never given serious consideration. For state parks, the WMI plan suggested that the commission foster a State Parks Citizens Committee to help create a "state park idealism . . . that has substance and meaning to the average taxpayer." This might have enabled the commission to obtain some sort of "state tax . . . dedicated to support state parks." In order to create a public image of the "ideal" state park, the report further recommended developing Ledges, Backbone, Lacey-Keosauqua, Waubonsie, and Stone state parks "as demonstration areas to show . . . what benefit can be attained through more liberal expenditures." Such recommendations were part of an overall thrust to reshape the state park system to meet the constant demand for outdoor recreation sites, particularly sites with access to water. At the same time, the ten-year plan recognized the pressing need for more coordination between the commission's two major divisions and the multiple sections of each. The terms "collateral values" and "common values" appear time and again, stressing the need for ongoing, integrated planning.[20]

In the end, nothing much changed. The legislature granted another small increase in the annual appropriations in 1960; and the commission slightly increased park concession, campground, and lodge rental fees. The Lands and Waters operating budget still totaled less than $1 million annually, while park attendance had escalated to more than seven million a year. The commission never acted on the recommendations for agency reorganization contained in the ten-year plan. However, when Glen Powers took over as director in 1959, he began a piecemeal approach to reorganizing the staff. It was not until about 1964 that the commission had a functioning planning section, and the impetus for taking this step came not so much from Gabrielson's reorganization plan as it did in response to federal mandates.[21]

The Ascendance of Fish and Game

Sometime in the early 1950s, the focus on fish and game began to overshadow other areas of commission responsibility. The shift came gradually, but the fiscal situation as of 1960 clarifies the trend. By then, fifty-three percent of the commission's income came from the sale of licenses, twenty-eight percent from federal aid and miscellaneous receipts, and nineteen percent from state appropriations. Federal dollars were funding more resource conservation than state dollars. The distribution of funds showed an even more pronounced disparity. The Lands and Waters Division, which handled parks, spent twenty percent of the budget, a percentage nearly equal to state appropriations. Sixteen percent of the total budget supported administration, public relations, and engineering. Fish and Game Division operations, including land acquisition, commanded the lion's share, sixty-four percent.[22] The funding formula established in the 1930s, combined with federal aid, now heavily favored the Fish and Game Division.

If there is a point where the balance began to shift, it would be the years 1951–1952. In 1951, Congress amended the Pittman-Robertson Act to prevent the executive branch from impounding funds. Thereafter, apportionments to the states were fully funded. A year later, funds from the Dingell-Johnson Act were first available to the states.[23] Iowa immediately began to increase its land acquisitions for wildlife and fisheries. In the eight-year period between July 1952 and June 1960, federal aid helped to pay for more than 17,000 acres, more than half the total land acquisitions made since 1939.[24]

A nationwide survey to determine the economic importance of hunting and fishing justified federal expenditures. Conducted in 1955–1956 under the auspices of the U.S. Fish and Wildlife Service, the survey found that the sale of hunting and fishing licenses, nationwide, had increased fifty percent in ten years. Iowa, moreover, ranked second in the nation in the percentage of hunters per total population. It also was part of a seven-state area in the Midwest containing the largest percentage of fishers. State-specific survey data indicated that an active sportsman (no doubt some were women) lived in fully half of all households in Iowa, and that nearly ninety-three percent of them did their hunting and fishing within the state. In 1955 alone, approximately 525,000 Iowans spent in excess of $42 million on hunting and fishing.[25] Federal programs for wildlife conservation and the surge in hunting and fishing as a popular form of outdoor recreation went hand in hand, especially because sportsmen traditionally were better organized political constituencies.

As the Fish and Game Division's budget, landholdings, and staff increased, parks received relatively less attention. Data compiled by the National Park Service between 1952 and 1960 reflect the declining status of Iowa's state parks within a nationwide framework. In 1952, the states reporting the highest number of visitors were, in rank order, New York, Illinois, California, Ohio, Pennsylvania, Oklahoma, Oregon, and Iowa. New York reported approximately eleven million, Iowa slightly more than four million.[26] Iowa ranked third in the nation in terms of money spent on capital improvements, including land acquisition, a total of $817,400. Only New York and Ohio spent more. Finally, even though the annual appropriation of $475,000 for the Division of Lands and Waters seemed small compared to the revenue that accrued to the Fish and Game Division, only eight state legislatures appropriated more money for state parks. However, since many states reported considerably higher revenues from operations than did Iowa, which took in about $73,000 that year, state appropriations are not a particularly revealing statistic. For instance, the Indiana state park program received only a $207,500 appropriation in 1952, but operating revenues brought in more than $865,000. The Missouri State Park Board, on the other hand, received a $475,000 appropriation, the same as Iowa's, but it had no additional sources of income either from operating revenues or special appropriations.[27]

Eight years later, Iowa still ranked high in attendance, tenth in the nation, although with 6.8 million visitors it was far behind the heavily urbanized or coastal states of New York, California, Michigan, Ohio, Pennsylvania, and Oregon, all of which reported attendance in excess of ten million. As measured by capital expenditures, Iowa had slipped from third to eleventh place. California and New York outspent other states by wide margins; but Pennsylvania, Illinois, Massachusetts, and New Jersey each spent one or two million dollars during the 1960 reporting year. By comparison, Iowa, which spent $577,500 on capital improvements and land acquisition for parks, ranked behind Ohio, Washington, Kansas, and Maryland. More revealing, Iowa ranked number twenty in expenditures for operation and maintenance.[28] One of the reasons that Iowa dropped so low in the rankings is that states in the southeast, particularly Tennessee, West Virginia, Kentucky, and Florida, were suddenly pouring more money into maintaining their state park systems. No doubt these states were attempting to meet increased public demand that came with demographic shifts favoring Sun Belt states. Still, the data clearly indicate that Iowa was falling behind in taking care of its park system.

Iowa was not alone in falling behind. Nationwide trends showed that from 1954 through 1957, attendance at state parks, collectively, increased at a much higher rate than attendance at national parks. During this period, the estimated number of visitors at all state parks rose from roughly 166 million to 216 million, a thirty percent increase. At the same time, national park use rose a more modest twenty percent, with attendance increasing, roughly, from fifty to sixty million visitors annually. Thus, not only was outdoor recreation booming, state parks were absorbing relatively more of the crowd. The states as a whole, however, were not acquiring new areas in sufficient numbers or acreage to keep pace with increasing use. Only California and New York were engaged in substantial land acquisition or capital improvement programs. Moreover, only a handful of states were increasing expenditures for supplies, equipment, and improvements; most states were showing decreases in these spending areas.[29] Generally speaking, few states were keeping up with demand.[30] Nonetheless, given Iowa's long tradition of commitment to its state park system, it was doing a much poorer job than might have been expected.

The funding arrangement institutionalized in the 1930s greatly disadvantaged state parks. Without a source of funding tied to park use, the State Conservation Commission had no built-in mechanism for increasing revenue when public demand escalated. To a certain degree, all figures reported for state park attendance and expenditures in Iowa are misleading, since the actual operations of the Lands and Waters Division and the Fish and Game Division blurred considerably. People who purchased fishing and hunting licenses were among those who used state parks. Likewise, money spent enhancing fisheries or wildlife areas benefited many state parks, although not necessarily in ways obvious to the typical user. Nonetheless, the commission had no way of reallocating funds internally in order to balance expenditures for the two divisions. By mid-century, the disparity of funding had rendered Lands and Waters "a relatively small division" within the State Conservation Commission, as Gabrielson noted in his 1954 report.[31]

Contributing to the decline were subtle shifts that seem to have taken place in the overall perception of state parks. Some of these shifts reflected national trends; others reflected changes within the commission. Throughout the 1950s and 1960s, the commission itself became much more closely identified with hunting and fishing interests. Gabrielson's 1954 report, which devoted more than twice as much space to Fish and Game matters than it did to Lands and Waters, suggests the relative weight the two divisions carried within the agency. Moreover, even a cursory review of the *Iowa Conservationist*, the commission's monthly publication, reveals a preponderance of articles on hunting and fishing. Wildlife conservation, forestry, parks, preserves, and conservation education certainly were not ignored topics during these decades, but sport hunters and fishers were far more likely to find feature articles of interest in every issue.

The dynamics of the State Conservation Commission also changed during the 1950s and 1960s. For the first time since the old Board of Conservation was organized in 1919, no one emerged as a forceful leader to establish a clear sense of priorities or direction. There was nothing much in the occupational backgrounds of commission members to suggest that change was taking place. The representation included attorneys, business people, farm owners, two dentists, a veterinarian, and a clergyman. A notable difference after 1961, however, was the lack of women on the commission. Up to this point, one seat always had been held by a woman, although after Louise Parker left the commission in 1949, none of the women commissioners commanded much of a presence. From 1961 to 1969, the commissioners were all men. The trait these men were most likely to share was a love of hunting and/or fishing. Many of the commissioners were members of the Izaak Walton League or some other sportsmen's club.[32]

It would be wrong to say that the men of the commission all thought alike, but many of them were primarily interested in conservation from the sportsman's point of view, that is, they wanted to increase game populations, wildlife habitats, hunting grounds, and fishing areas. In this sense, the commission both reflected and principally responded to the rekindled concern for wildlife conservation. Propelled by federal money through Pittman-Robertson and Dingell-Johnson, the weight of funding tipped to fish and game. Dollars, in turn, gave the commission greater control over wildlife management at a time when its authority in other areas, particularly water policy, was being challenged. As the funding scale tipped, the commission lost a balanced perspective on its mandate. A casual remark by commissioner Robert Beebe speaks volumes. During a lengthy discussion of hiring field supervisors, Beebe stated that he felt "the main function of the commission [was] to provide hunting and fishing from the moneys derived from license sales."[33]

Jay Darling was not so kind in his assessment of the commission during this period. When Norman Erbe was elected governor in 1960, Darling fired off one of his typical broadsides reminding the governor-elect that "it was the Republican Party, Republican Governor and Republican Legislature who in 1930 [sic] set up the original Conservation Commission and voted the 25-Year Program. Those folks aren't all dead yet and they remember, even if you don't. Three Governors in a row making such stupid appointments are enough to wreck the effectiveness of the Commission."[34]

Not only did the commission's vision narrow, but the balance power within the agency as a whole shifted to the director's office. To a certain degree, this was bound to happen as the bureaucracy grew; and, since the late 1930s, the

M. L. Hutton, ICC director,
1935–1941. *Courtesy Iowa Department of Natural Resources.*

G. L. Ziemer, ICC director, 1946–1948. *Courtesy Iowa Department of Natural Resources.*

Bruce Stiles, ICC director, 1948–1959. *Courtesy Iowa Department of Natural Resources.*

commission had gradually transferred more and more decision-making authority over routine matters to the director's office. However, beginning with Bruce Stiles, who assumed this position in 1948, the director came to set the tone and character of the organization. Importantly, the two most influential directors during this period, Stiles and Everett B. Speaker, came up through the ranks of the Fish and Game Division. In contrast, two of the three directors between 1935 and 1948 were engineers by training: M. L. Hutton (1935–1941) and G. L. Ziemer (1946–1948). Stiles and Speaker are remembered as among the most dedicated directors, and both certainly were well qualified for the post. Even so, they both had stronger personal and professional ties to the Fish and Game Division than they did to the Lands and Waters Division, and this bias accentuated the decline of parks.

Bruce Stiles began his career with the State Conservation Commission in

Everett B. Speaker, ICC director, 1963–1968. *Courtesy Iowa Department of Natural Resources.*

1938 as a conservation officer. Four years later he moved up to become chief of the Fish and Game Division; four years after that he became assistant director. In 1948 he rose to director, a position he held until shortly before his death in 1959.[35] His conservation interests were wholly focused on wildlife. At one time or another, Stiles served as an officer of several professional organizations that reflected his interests, including the International Association of Game, Fish, and Conservation Commissioners; the Midwest Association of Game, Fish, and Law Enforcement Officers; the Mississippi Migratory Waterfowl Flyway Council; the Iowa Ornithologists' Union; and the Iowa Academy of Science. Stiles was also the first director to engage extensively in public speaking and to publish. Among other things, he compiled a chronology of wildlife legislation in Iowa and coauthored *Wildlife Resources of Iowa.*[36]

Glen G. Powers filled the post of director in 1959, although he resigned four years later to become the first director of planning.[37] Everett B. Speaker succeeded Powers. Speaker, like Stiles, had a long history with the commission. His father, Dr. Everett E. Speaker, a well-known park and conservation advocate from Lake View, had served on the first State Conservation Commission for a short time, 1935–1936, before ill health forced him to resign.[38] After studying forestry at the University of Iowa and Iowa State University, Everett B. went to work for the old Fish and Game Department in 1931 as an assistant to the state fish pathologist. By 1935, when the State Conservation Commission was established, he had advanced to superintendent of fisheries, a post he continued to hold in the new organization. In 1948, he became superintendent of biology, and moved from there to the director's position in 1963. Speaker, too, resigned as director, in 1968, to become special projects coordinator. After resigning, he served on the State Preserves Advisory Board and the Iowa Water Pollution Control Commission.

Shortly before Speaker died in 1971, he was elected to honorary membership in the American Fisheries Society and inducted into the Iowa Conservation Hall of Fame. Like Stiles, he had been very active in professional organizations, including the American Fisheries Society, the Midwest Fish and Game Commissioners Association, the American Institute of Fisheries Research Biologists, and the Iowa Academy of Science. Throughout most of his career with the commission, he also maintained an alliance with conservation officers in the field as a U.S. deputy game warden. Speaker also published many scientific and popular articles on fish and fishing and coauthored a book by that title, *Iowa Fish and Fishing*. With Stiles and others, he coauthored *Wildlife Resources of Iowa*.[39]

Notes that Speaker typed before a 1968 interview reveal much about his concept of parks in relation to conservation. On the topic of "future potential for conservation and outdoor recreation," Speaker framed his thoughts mostly in terms of the miles of streams, the acres of lake waters, and the number of public shooting areas available for hunting and fishing. He also addressed the need to work with farmers to protect wildlife habitats and to purchase wetlands to build up game populations. The topic of "parks, preserves, and recreation areas" appears almost as an afterthought in his outline. "Other forms of recreation can, if need be, be man-made to some degree" he jotted. "Man-made" was explained to mean "artificial lakes, picnic and camp grounds, swimming beaches, trails, boat-launching facilities and marinas, large reservoir impoundments, and other facilities."[40] For Speaker, and Stiles before him, parks clearly were associated with forms of outdoor recreation other than fishing and hunting. Parks were no longer the center of a system of

conservation lands and waters. Indeed, parks were almost divorced, concep-
tually, from the resource conservation functions of the commission.

Water Policy and the Natural Resources Council

The construction of large-scale flood control projects in the post–World
War II era, which opened the potential for expanding outdoor recreation
areas, also reshaped the way people thought about state parks. Once again,
water policy, or the lack thereof, was a critical issue. Gabrielson's 1954 report
pointed out that the Waters Section of the Lands and Waters Division was
nothing more than a "paper organization." It did little more than inspect
boats, conduct water safety and first-aid training sessions, monitor lake concessions, oversee sand and gravel extraction from state waters, and inspect
construction projects in state waters. The only substantive conservation program assigned to the Waters Section was erosion control on the watersheds of
lake parks. Gabrielson thought this section should be doing much more. Because of "the complexity of [the nation's] water problem," he recommended
that the commission develop a "sound water management program" based in
the Waters Section.[41]

What Gabrielson failed to grasp was that Iowa had already handed off the
responsibility for developing water policy to another agency, the Natural Resources Council, established in 1949. The reason the Waters Section was a
"paper organization" is that the State Conservation Commission no longer
had a clear sense of its role with respect to the state's water resources, even in
state parks. Decades of fragmented authority over water pollution, wetlands
drainage, hydroelectric power, river navigation, and lake protection had left
the state vulnerable when the federal government, through the Army Corps of
Engineers, began to impose large-scale flood control projects. The legislature
took steps to remedy this situation in 1949 by creating a new agency, authorized "to establish and enforce an appropriate comprehensive state-wide plan
for the control of water and the protection of the surface and underground
water resources of the state."[42] This law placed the Natural Resources Council,
that new agency, squarely in conflict with some of the duties and responsibilities of the State Conservation Commission.

On one level, state legislation establishing the Natural Resources Council
can be viewed simply as a response to a 1944 federal law that directed the
Corps of Engineers to consult with affected states during the course of its
investigations and planning for flood control projects.[43] At the time, such consultation in Iowa would have involved no fewer than four separate state entities: the State Department of Health, the State Soil Conservation Committee,

the Iowa Geological Survey, and the State Conservation Commission. Of the four entities, the State Conservation Commission had a long history of opposing power dams on meandered rivers. When the issue of high dams for flood control purposes first surfaced in the late 1930s, the commission also went on record in opposition. For instance, when the COE proposed in 1939 to dam the Des Moines River near Madrid, the Commission opposed the project because it had the potential to flood Ledges State Park.[44] Likewise, when the War Department announced a few months later that it would build a flood control dam on the Iowa River near Iowa City, the commission requested that the COE's Rock Island District provide information concerning the relationship between the proposed dam and Lake Macbride State Park. After Col. C. P. Gross apologized for being unaware that the state *should* have been consulted in the matter, the commission adopted a resolution requesting that the federal government abandon the project.[45]

There was no question that the issue of flood control was vitally linked with the future of state parks located on or adjacent to lakes and rivers. In the end, the Corps built four large flood control dams in Iowa: Coralville, Saylorville, Red Rock, and Rathbun. The first two affected existing state parks. The relationship between Saylorville Dam and Ledges State Park, in particular, took on greater significance after the National Environmental Policy Act was passed in 1969, a topic that will be detailed in the next chapter. However, during the late 1940s the more immediate issue was which state agency would formulate state water policy. The timing of federal flood control projects complicated the issue because it coincided with the State Conservation Commission's own policy of constructing artificial lakes to meet public demand for outdoor recreation. Increasingly, the public associated outdoor recreation with lakes, and Iowans wanted more of them. This placed the commission in an awkward position when local communities or special interest groups championed flood control projects because the resulting reservoirs could be used as recreational lakes.

Still, the State Conservation Commission was the most logical state agency to take responsibility for developing a coordinated state water policy; indeed, it had agitated for coherent water policies in the past. Why, then, did the legislature set up a competing agency charged with this function? The answer is undoubtedly complex, and it is beyond the scope of this study to address state water politics fully. Nonetheless, it is useful to highlight certain aspects of the controversial Natural Resources Council as it affected state parks.

Just as Jay Darling had been the force behind the *Twenty-five Year Conservation Plan*, he also was behind creation of the Natural Resources Council. Dissatisfied with progress on the 1933 state conservation plan, he seems to have felt compelled to make another attempt to establish ongoing, coordinated conservation planning. As early as 1943, he met with Governor Bourke Hickenlooper to outline a plan for coordinating under the State Conservation Commission all existing state agencies that had any jurisdiction over natural resources. Darling later recapped the meeting in a letter to commissioner Fred Poyneer, stating that his plan "wasn't a very clearly outlined program and perhaps not the right one but the Governor accepted it as a fruitful thought for consideration." [46] The Corps of Engineers's proposed flood control projects in Iowa, first announced in 1939, also rekindled Darling's quest for coordinated resource planning and management. Although the Corps had set aside plans for these projects during the war, but it was only a matter of time before those plans resurfaced. Thus, in late 1943 he also met with the governor and members of the State Conservation Commission in order to press for a "better comprehensive program for conservation . . . than would likely be found in any projects designed by Army engineers in Washington, D.C." Darling saw this as "the only practical way of defeating any Federal program which might have adverse effect on our Iowa resources." [47]

A year later, Darling and state geologist H. Garland Hershey promoted a similar idea through the Postwar Rehabilitation Commission, established during Robert Blue's tenure as governor (1944–1948). Both men served as advisory members to the commission's conservation committee, which recommended a natural resources council composed of representatives from state agencies having jurisdiction over resource conservation. This time it was more clear that the greatest concern was pending federal projects. The committee warned that "the State of Iowa is without an effective mouthpiece when such proposals as the Missouri Valley Development, the Red Rock Dam, the Coralville Reservoir, and similar projects are under consideration." [48] Hershey also worked through the Iowa Academy of Science to mobilize further support for a natural resources council that would act as a clearinghouse for coordinating state agencies involved in natural resource problems. [49]

Blue's administration demonstrated little sincere interest in conservation, but Darling never let politicians sleep in peace. "Here's that man again," he wrote to the governor in July 1946, and went on to campaign for a state entity authorized to collect scientific data on Iowa's water resources. [50] Prods such as these took on added force after Darling's cartoon of the "safe-cracking governor" hit the papers a month later. Blue publicly responded that he was now "intensely interested in a study and revision of [state] law with reference to water use" and offered to set up a special natural resources committee to study the issue. Darling, with help from *Register and Tribune* columnist Ries Tuttle, immediately drafted a lengthy proposal for a research agency tentatively titled the "Iowa Natural Resources Council." The proposal was backed by a score of sportsmen's clubs, principally Izaak Walton League affiliates.[51]

Major floods in the summer of 1947 prompted the state legislature to appoint an Interim Flood Control Committee to study interrelated aspects of state water policy. It was out of this committee that finally came a recommendation for a new state entity, which emerged in 1949 as the Iowa Natural Resources Council. Specifically, the interim committee called for a "state water control and resources council" with authority to coordinate activities of the federal, state, and local governments for flood control, soil conservation, forestry, drainage, and water supplies.[52]

Darling, once again, had been the force that set events in motion. It is therefore worth comparing earlier proposals for a natural resources council with the resulting permanent body bearing that name. Such a comparison does not fully explain why the Natural Resources Council was constituted as it was or why its authorized functions overlapped those of the State Conservation Commission. What it does elucidate is the tension that existed between the two agencies, particularly in the 1950s.

The initial proposals called for a fact-finding body authorized to conduct scientific research, to assemble data, and to prepare reports and recommendations at the request of state departments charged with managing natural resources. The entity Darling and Hershey had in mind was to be advisory only, with no planning, policy-making, or administrative functions. Its proposed members were five experts drawn from biology, geology, agriculture, hydraulic engineering, and industry. Darling admitted that some State Conservation Commission members initially objected to a *new* agency, even if only advisory in nature, with the potential to assume its functions. However, "in the minds of most people," the conservation commission was "naturally prejudiced in favor of fish and game," and its opinions were therefore "disqualified . . . by industry, agriculture, flood control and hydroelectric interests." For this reason, the Interim Flood Control Committee and its advisors apparently agreed that the new council "should be wholly devoid of any taint or preconceived prejudice or suspicion of prejudice."[53]

Governor Blue also made it fairly clear that he would block any attempt to place authority for flood control with the State Conservation Commission or any existing state agency that had a well-developed constituency. "Through my own experience with the Conservation Commission," he announced in a 1948 speech to the Interim Flood Control Committee, "I am of the opinion that it would be a mistake to place the full authority with any one group. . . . If you put it in the Conservation you will get fish and game and bring a terrific pressure on that department to build their flood control with that in mind."[54]

Contrary to the initial proposals, however, the Natural Resources Council as established in 1949 was not a fact-finding body; and its seven members, appointed by the governor, were not required to have any professional expertise useful for making water policy. It was given sole authority "to establish and enforce" a statewide program of flood control and, in addition, to formulate a comprehensive plan that addressed "the conservation, development and use" of the state's surface and underground water resources. To carry out its functions, the council was given jurisdiction over the public *and* private waters of the state. Not only was the council to be wholly independent of other state agencies, but the State Conservation Commission was specifically required to seek council approval for any project related to flood control.[55]

The legislature established, deliberately or unintentionally, an entity that was bound to create conflict with the State Conservation Commission at a time when a united front was needed.[56] Darling called the Natural Resources Council "the hand of Esau but the voice of Jacob," not at all the research body he saw as essential. "No layman's Resources Council," he wrote, "can provide the factual data and analysis" he felt Iowa needed to place the state in the best possible position to cope with the federal government's flood control initiative.[57] Darling was not an uncritical supporter of the State Conservation Commission, but he could see no logic whatsoever in the legislature's action. Whatever merit there may have been in creating a separate agency to handle flood control matters, it was clear from the beginning that the State Conservation Commission would not be terribly cooperative. Aside from the pending conflict inherent in the legislative mandates of each agency, organized sportsmen had been on the losing side.

The long view helps one to see why 1949 marked a point of divergence. Up to that point, state parks had been considered part and parcel of resources conservation and management, in theory at least if not always in fact. But large-scale, federal flood control works, which were coming ready or not, forced states to adjust their thinking about natural resources policy, particularly water policy. It would appear, in retrospect, that by the mid-1940s the public image of the State Conservation Commission already was cast as an agency that primarily protected fish and game interests, despite Louise Parker's initiatives in prairie preservation and historic preservation. Moreover, it had demonstrated time and again its opposition to high dams regardless of their purpose. Rightly or wrongly, the commission seems to have been perceived by many as too narrow in its focus to be entrusted with formulating state water policies that would affect industry and agriculture as well as wildlife and outdoor recreation. The Natural Resources Council not only assumed control over flood control and water resource planning, in the decades to come it also became the state agency more closely identified with environmental policy in general.

The 1950s thus became a period of sorting out just what administrative control the State Conservation Commission would retain over state waters. Key personalities initially made the adjustment easier, but relations were nonetheless uneasy. Garland Hershey, a long-time supporter of the State Conservation Commission, was appointed chair of the first Natural Resources Council. He served with Louise Parker, who had gained considerable respect among sportsmen during her twelve years on the commission. Additionally, G. L. Ziemer, former director of the commission, became the council's first director. Shortly after the council was formed, Hershey and Ziemer appeared before the commission to discuss joint policies in cases where the authority of the two agencies overlapped. Hershey took a conciliatory stance, explaining that the council did not wish to interfere with established procedures of any state department and assuring the commission that the council would "pass on certain plans presented by the Army Engineers in which they believe the Conservation Commission would also be interested." [58]

The commission's role in federal flood control projects was reduced to commenting on the effect that such projects would have on hunting, fishing, and outdoor recreation. Its first official statement in this new role seemed to reflect an underlying resentment. Commenting on an early COE proposal for siting the Red Rock Dam, the commission noted that it had confined its remarks to the recreational aspects of the project, but then went on to state that it "had completely disregarded any gain or loss to the people or the nation which might be social, economic, agricultural or industrial." [59] As time went on, the commission's role in COE large dam projects became clearer, and a little bit stronger. It had no power to stop such projects, but it did exercise some control over mitigative measures when COE reservoirs threatened existing state parks. The commission also dealt directly with the Corps to develop recreational areas and facilities in conjunction with flood-control reservoirs.

Interagency relations were further strained when the Natural Resources Council began to expand its activities beyond coordinating federal flood control projects. The catalyzing issue this time was farmland irrigation, prompted by intermittent years of below-average rainfall in the early 1950s. Although Iowa law generally followed the principle of riparian water rights for nonconsumptive uses, there were many gray areas. In response to increasing demands on rivers and lakes for irrigation, the legislature increased the NRC's jurisdiction over the state's water supplies. Amendments passed in 1957 declared that water occurring in *any* basin or watercourse or other natural water body was *public* water, and vested the Natural Resources Council with authority to control the use and development of *all* water resources.[60] Further conflict with the commission, therefore, was inevitable.

In its entirety, the 1957 Water Resources Conservation Act, as the amendments are known, was a far-sighted piece of legislation. It established a doctrine of water rights based on public benefit (even though vaguely defined) rather than prior appropriation or riparian rights. In part, the law was enacted to curb the indiscriminate taking of water for farmland irrigation, although some critics charged that certain farm groups were really behind its passage. That question is not of immediate importance here; however, shortly after the law took effect the state water commissioner, a position created under the NRC, was soon inundated with applications from farmers who wanted to draw water from lakes and streams for irrigation.[61] Before the Natural Resources Council could act on these applications, the State Conservation Commission registered its opposition to issuing any permits, claiming that drawing down the water level of lakes and streams would be detrimental to fishing. The Iowa Division of the Izaak Walton League weighed in right behind, urging the NRC to deny all applications for irrigation.[62]

The commission's immediate opposition to using state waters for irrigation probably could have been anticipated, and it did prompt the Natural Resources Council to appear for a lengthy conference in early December 1957. Hershey again acted as spokesperson, opening the discussion by pointing out that in passing the Water Resources Conservation Law the legislature intended to allow some irrigation. In the interests of satisfying farmers as well as fishers, he suggested that perhaps the two agencies could agree upon establishing a low-flow minimum with irrigation permissible during high flows. To this suggestion, the commission responded that "permitting the use of water from our rivers and streams for irrigation is a reversal of our policy whereby we are purchasing ponds, sloughs, and access areas for the purpose of retaining water."[63]

Although the commission and the council agreed to continue discussions on the matter, this did not happen. Two weeks later the Natural Resources Council approved three applications to use water from the Maple River in Ida County for irrigation. The commission immediately appealed the decisions on the grounds that the Maple River barely had enough water to support fish life and outdoor recreation. The council defended its action in a lengthy statement insisting that it had not ignored the possible effects on fish and wildlife and insisting that it was not the council's "policy or intent" to dry up Iowa's streams. The commission was not swayed. Realizing that there were principles at stake if water rights for fishing and recreational use were to be protected adequately, the commission went on record as favoring a policy that "at no time should anyone be allowed to take an excessive amount of water from a stream for irrigation purposes after . . . July first in any one calendar year."[64]

The Iowa Ikes wholly supported the commission. Judge John Tobin of Vinton, a past national president of the IWLA, noted that the conflict was not really whether a few farmers on the Maple River should have irrigation permits. The issue was whether farmers would be granted irrigation rights before there were adequate scientific data to establish average or minimum flow levels, thereby setting a precedent that would be difficult to reverse. A related issue was whether the term "beneficial use," now codified in law, included irrigation, which wasted most of the water through evaporation. Tobin called for redefining "beneficial use" to recognize the benefits of recreational use, pointing out that the "rights of nonconsumptive recreational use of the water by the public [was] a vested right, established by usage long before the [1957] law was enacted."[65]

When the NRC went ahead and approved the irrigation permits, the commission sought to get a compromise policy in writing. The situation escalated when the commissioners learned through a newspaper notice that the state water commissioner had given the State Highway Commission permission to take water from Red Haw Lake for road construction. The Conservation Commission had never been notified, let alone consulted, even though the activity was slated to take place entirely within a state park. At this point, the commission filed suit against the Natural Resources Council, joined by the Wildlife Management Institute, the National Wildlife Federation, and the Sport Fishing Institute. Before proceeding, however, it offered to suspend legal action if the NRC would agree to refrain from approving any more irrigation permits until the two agencies could draft another amendment that would place some limits on the taking of water for irrigation purposes. The NRC rejected the commission's offer, and the matter dragged on for a year and a half before the two agencies finally agreed to work together to establish base flow levels in order to protect streams and lakes from depletion through excessive withdrawals for consumptive uses.[66]

Flood control projects and the use of state waters for irrigation were the two issues that erupted in open conflict, but the commission also was forced to adjust its administrative procedures in other areas. NRC approval was required, for instance, when construction on lakes might change the flow characteristics of an inlet or outlet, or when dock placement would obstruct more than five percent of the flow of a river. After the NRC made an issue out of the commission's long-standing prerogative to grant permits for removing sand and gravel from river beds, an administrative responsibility it had held since the early days of the Board of Conservation, the commission finally agreed to send the NRC a copy of all future removal agreements. County conservation boards proposing to construct dams or otherwise modify state waters had to seek approval from both agencies.[67] While the Natural Resources Council may have placed Iowa in a stronger position to deal with the federal government on flood control matters, it also had the effect of layering the state bureaucracy and increasing red tape. It also meant that the State Conservation Commission spent precious time and energy dealing with interagency conflict, especially during the 1950s, when the commission also was trying to cope with inadequate salaries, increasing park use, and static appropriations.

Rough Jewels: More Artificial Lakes

"God, the legislature and the conservation commission have combined to give us the rough jewel." So declared a local speaker at the 1950 "gate setting" ceremony of the dam at Lake Darling. "Now it must be polished."[68] Even the bright spots amid park problems and politics, however, were not without their own blemishes. Some minds may have seen the hand of a deity fashioning new state park and recreation areas, but the legislature and local boosters clearly had a lot more clout than the State Conservation Commission. The ceremony, in fact, was filled with irony. Ewald Trost, the commission appointee whom Jay Darling had denounced four years earlier as little more than a political patron, now patronized "the honored guest, whose vision and leadership resulted in the *Iowa Twenty-five Year Conservation Plan* and the resultant commission form of conservation administration." Darling managed to work a barbed retort into his own brief comments. "If there ever should be a return of conservation to politics," he reportedly said to those assembled, "I'll turn over in my grave if you don't get up and fight."[69] As the artificial lake program expanded into a multifaceted large-lake program, the commission also acted more and more like a broker agency, negotiating the best deals for creating new lake parks that were environmentally compatible with existing natural features and that, at the same time, served the greatest number of people. Conservation politics went hand in hand with large lakes and flood control reservoirs.

Flood control issues may have usurped time and energy, but they did not stand in the way of launching a new round of artificial lake construction. Lake Darling was the first of the "new" artificial lakes, that is, the first of six artificial lake projects authorized by the legislature in 1947.[70] When it was dedicated in 1950, Iowa had about twenty artificial recreation lakes. By the mid-1960s, there were several more. Three small lakes were completed in 1951: one in Geode State Park, another in Nine Eagles State Park, and a third in Cold Springs State

Park.[71] Rock Creek Lake between Newton and Grinnell was the next to fill. When it was dedicated in 1952, Rock Creek was hailed as "the largest lake in the United States constructed solely for recreational purposes."[72] Located within easy reach by 150,000 people, Rock Creek Lake offered anglers, picnickers, and boaters 640 acres of water and 1,000 acres of park space behind fifteen miles of shoreline.

Two more artificial lake projects in southwestern Iowa completed the lands and waters development initiative begun in the 1940s: Viking Lake (first known as Stanton Lake) in Montgomery County and Prairie Rose in Shelby County. Feasibility studies were done in 1946, but both projects were delayed for a number of reasons. Problems finding suitable sites from engineering and cost standpoints took several years.[73] In the interim, ominous cracks appeared in the dams at Geode, Rock Creek, Nine Eagles, and Creston. Consulting engineer A. L. Alin, hired by the commission to study the situation, leveled harsh criticism at the various design-engineering firms involved. A minor furor erupted when the study was released to the press before the engineering firms had a chance to review it. The *Des Moines Register* chose to run its story as headline news in the Sunday edition. A long article cited extensively from the Alin report, focusing in particular on obsolete design features and an overall tendency to build cheaply rather than for durability. As a result of the Alin study and the ensuing controversy, the commission decided to subject future designs for large dams to review by "competent outside authorities." In addition, remedial work on the defective dams increased the overall cost of the artificial lake program.[74]

Work finally got underway in 1956 on the Montgomery County lake. After the *Stanton Viking* sponsored a naming contest, the commission selected from the entries, choosing Viking Lake State Park to commemorate the Scandinavian heritage of most Stanton-area residents (table 7). The 150-acre lake and surrounding park were dedicated in October 1957 with festivities involving fifteen satisfied communities.[75] The Shelby County lake, which had first priority in the mid-1940s, was the last to be completed. No fewer than sixteen possible sites were considered before engineers, in 1952, settled on a spot in the vicinity of a long-gone pioneer village known as Prairie Rose. As things turned out, the proposed lake park had a name well in advance of its creation. No sooner had the commission settled on the Prairie Rose site than a small group of farmers began to fight it. The chosen site required acquisition of one complete farm and parts of ten others. Landowners were split over the Prairie Rose decision. Some were willing to sell, others vowed not to sell at any price. Seemingly at issue was the high productivity of agricultural land in this particular area, although the opposition actually was slightly more complex than that. One absentee landowner was upset at the thought of losing her birth-place, which the planned lake would inundate. Another couple and their two sons operated a multigenerational family farm. They were upset because the lake project not only would take half the farm but would split up the family partnership as well.[76] Lengthy condemnation proceedings delayed the project for several years. The contract for lake construction was not awarded until 1959, and Prairie Rose State Park was quietly dedicated three years later.[77]

A comparison of artificial lake construction during the New Deal and post–World War II eras reveals that the lakes constructed during the 1930s, while more numerous, tended to be small bodies of water, mainly suitable for fishing and beach activities (table 6). Those constructed between 1948 and 1965 tended to be larger, reflecting the need to serve larger crowds of people, the growing popularity of motorboating, and the influence of federal flood-control projects.

Table 6.
Artificial Lakes Constructed in State Parks, 1933–1942[78]

Location	Biennium Completed	Lake Acreage
Backbone SP, Delaware Co.	1934–36	125
Beaver Meadow SP, Butler Co.	1936–38	30
Beeds Lake SP, Franklin Co.	1936–38	130
Echo Valley SP, Fayette Co.	1934–36	12
Heery Woods SP, Butler Co.	1936–38	50
Lacey-Keosauqua SP, Van Buren Co.	1936–38	30
Lake Ahquabi SP, Warren Co.	1934–36	130
Lake Keomah SP, Mahaska Co.	1934–36	82
Lake Macbride SP, Johnson Co.	1934–36	138
Lake of Three Fires SP, Taylor Co.	1936–38	125
Lake Wapello SP, Davis Co.	1934–36	287
Mill Creek SP, O'Brien Co.	1936–38	25
Palisades-Kepler SP, Linn Co.	1936–38	125
Pine Lake SP (Lower Pine Lake), Hardin Co.	1936–38	63
Pine Lake SP (Upper Pine Lake), Hardin Co.	1934–36	70
Red Haw Hill SP, Lucas Co.	1936–38	72
Springbrook SP, Guthrie Co.	1934–36	27
Swan Lake SP, Carroll Co.	1934–36	130
Walnut Woods SP, Polk Co.	1936–38	12

Table 7.

Artificial Lake Construction, 1948–1968 (Lands and Waters Division only)

Lake Name	Location	Biennium Completed	Lake Acreage
Darling	Lake Darling SP, Washington Co.	1948–50	302
Cold Springs	Cold Springs Recreation Reserve, Cass Co.	1948–50	16
Geode	Geode SP, Des Moines Co./Henry Co.	1950–52	205
Nine Eagles	Nine Eagles SP, Decatur Co.	1950–52	56
Rock Creek	Rock Creek SP, Jasper Co.	1950–52	640
Green Valley	Green Valley SP, Union Co.	1954–56	390
Viking	Viking Lake SP, Montgomery Co.	1956–58	150
Macbride +	Lake Macbride SP, Johnson Co.	1956–58	950
Prairie Rose	Prairie Rose SP, Shelby Co.	1960–62	218
Upper Pine +	Pine Lake SP, Hardin Co.	1960–62	101
Lake Anita	Lake Anita SP, Cass Co.	1966–68	182
+ enlarged			

Lake Ahquabi, late 1930s. *Courtesy Iowa Department of Natural Resources.*

Viking Lake under construction, 1956. *Courtesy Iowa Department of Natural Resources.*

Although the state committed far more money to park development in the late 1940s and 1950s than it ever had before, it still was not enough to keep pace with increasing park use. Visitors inundated parks everywhere in the postwar era. Attendance at Iowa parks rose from 1.5 million in 1946 to 3.5 million in 1951 and showed no signs of leveling off. In 1953 and 1954 attendance figures moved steadily closer to the five million mark, then soared to nearly six million in 1956 and 1957. By the end of the decade, annual park visitation stood at nearly seven million.[79] Lake parks consistently drew the largest crowds. A visitor survey showed that at least sixty percent of park users traveled no more than fifty miles from home, validating the park development approach set in motion with the *Twenty-five Year Conservation Plan*. However, although commissioners often talked about the need to serve large urban concentrations, statistics showed that townspeople were more likely to use state parks than city residents. Parks were even more popular among farm dwellers than among city folk.[80]

County Parks Revisited

An old idea, county parks, found new adherents in the 1950s. This proved to be a winning strategy for relieving the pressure of human use on state parks as well as for shifting some of the cost burden for maintaining state parks. Actually, the idea of small parks in every county, if not in every township as Thomas Macbride had once envisioned, never really went away. But in 1955, Iowa gave new meaning to the concept through legislation authorizing county conservation boards. The genesis of that law can be traced as far back as 1936 or 1937, when the State Conservation Commission began promoting county conservation committees to aid the state in its work. The need for a way to distribute game birds raised at the Boone County hatchery initially led the commission to begin working with local conservation and sportsmen's clubs. Before long, this initiative had grown into a program designed to enlist a wide variety of local organizations—women's clubs, sportsmen's clubs, civic groups, youth organizations, schools, and farm groups—in soil and water conservation efforts, wildlife programs, outdoor recreation, and conservation education. The response was encouraging. By June 1938, the commission had approved thirty-nine organizations and several more were organizing.[81]

It so happened, though, that the newly formed National Wildlife Federation started canvassing the country for affiliates at about the same time. In late 1938, the NWF approached the commission and suggested that the organized county committees, now nearing sixty in number, should be affiliated with the federation because the federation was having difficulty setting up local NWF units in counties already organized by the commission. This possibility cast a new light on the commission's initiative. What started out rather innocently as a strategy for nurturing a wide range of conservation activities at the grassroots level now began to look more like a strategy for organizing a political constituency. Consequently, the commission began to back away from the movement it had set in motion. In 1940, commissioners agreed it would be prudent to maintain "a line of cleavage" with the county committees.[82]

The commission did not actually retreat; rather, it reconceived a good idea. In early 1936, the Lands and Waters staff drafted a bill that would have enabled counties to establish park commissions, a proposal that seems to have been prompted by recommendations from the Iowa State Planning Board. In its 1935 report, the planning board called upon the state legislature to pass enabling legislation permitting counties to establish planning boards, or park and planning boards, with functions similar to those of a proposed permanent state planning board.[83] In the end, though, the legislature failed to pass bills authorizing either a permanent state planning board or county park and planning boards. As the New Deal lost its luster, Iowa, like most states, rejected the idea of centralized land-use planning.

Nonetheless, the idea of county-owned and managed parks began to take hold. Over the decades, scores of small tracts had assumed quasi-public status by virtue of constant use as picnic areas, walking trails, and fishing spots, but few of these tracts had been legally dedicated as public parks. According to survey data gathered by the Iowa State Planning Board, only eighteen counties had actually set aside land for parks as of 1935.[84] Greene County was among them, and it was the first to take advantage of New Deal relief programs to develop its county park. More than the State Planning Board's grand plan for perpetuity, the development of Squirrel Hollow, southeast of Jefferson, may have demonstrated the viability of a state-county park partnership under the wing of the State Conservation Commission. The timing of events suggests as much. When Greene County embarked on its development program in 1934, the commission indirectly supported the effort by providing assistance from the Central Design Office and by authorizing stream improvements to the Raccoon River, adjacent to the park site. Additionally, in 1936, the commission approved Squirrel Hollow as a state game refuge.[85]

Squirrel Hollow exemplifies the early pattern of county park formation in Iowa.[86] The idea for a rural park in Greene County reportedly began in the mid-1920s, when a few local citizens, including a county supervisor and the county engineer, put their heads together. Local talk was soon perking. In about 1928, the idea had progressed to the point where the county hired a college student to produce a topographic map of the area under discussion. Squirrel Hollow, as the site already was known, had long been used by county residents as a picnic ground. In addition, a scenic old county road followed the Raccoon River through the hollow, but it had been closed for several years because of flood damage. The planning was indeed preliminary because the county did not actually own any land until 1933, when the Jefferson Chautauqua Association, then in the process of disbanding, voted to use money remaining in its treasury to purchase twenty acres of timbered land along the Raccoon River and then deed it to the county. This move prompted citizens of Jefferson and Rippey to raise enough money to purchase an additional forty acres in 1934.

To develop the area into a park, the county first hoped to include Squirrel Hollow in a federally funded reforestation project, but, as events transpired, park development appears to have been piggybacked onto civil works projects that were officially authorized to rebuild the old county road. Local newspaper accounts report that CWA, PWA, and WPA funds all advanced the project to completion. Regardless of the channels through which federal aid flowed, the final product is not in doubt. In May 1935, approximately 500 people gathered to dedicate Squirrel Hollow County Park, which had been improved with a new county road, a massive rock retaining wall along the Raccoon River to contain overflow waters, park roads, a stone-and-timber shelter house, picnic grills and tables, trail steps, and latrines. In essence, it was a small-scale "state park" with all the attributes of 1930s park design aesthetics. Squirrel Hollow thus became something of a model for the type of county park development that the State Conservation Commission hoped to see take hold in the state.

By 1940, the idea of enlisting the aid of local organizations to promote resource conservation and the idea of a legislatively sanctioned county park system began to merge into a broader concept of county conservation boards. There is no question that this shift in thinking was linked with another commission goal: divesting the state park system of smaller units. Various commissioners expressed a desire to trim the number of state-owned areas as one body or another appeared to request a new park for their community. Typically, unsolicited proposals were turned down, at least initially. The only petitions the commission took seriously were those that proposed parks in areas recommended in the *Twenty-five Year Conservation Plan* and those in which there was promise of substantial community participation in terms of land acquisition—and, as time went on, those projects that obtained political backing from legislators. The commission also began, cautiously, to dispose of a few parks. In 1940, Flint Hills State Park, which included a public golf course, was conveyed to the City of Burlington. Several years later, in 1949, the commission transferred ownership of Farmington State Park, also known as Lotus Lake, to the Town of Farmington.[87]

The commission drafted the first bill authorizing counties to establish and maintain small recreation areas and parks in 1941. It is said to have been based, in part, on the 1910 Wisconsin County Park Law and the 1910 Illinois Forest Preserve District Law.[88] This initial bill failed to pass the legislature, and, for more than a decade, successive bills met the same fate. The sticking point, as usual, was financing. Provisions in the various county conservation bills that would allow counties to increase property taxes in order to acquire, develop, and maintain small parks continually met with resistance.[89] Meanwhile, maintenance costs for existing state parks increased as park attendance rose. The commission sought to resolve this dilemma by reshaping the park system: creating new lake parks that would accommodate greater numbers of people, and, at the same time, transferring small areas to other governmental entities. The legislature proved willing to appropriate money for expanding the state park system, yet state appropriations for maintenance remained static.

After the Izaak Walton League threw its weight behind several measures designed to improve state park funding in the mid-1950s, the state legislature finally passed the county conservation bill. The new law, adopted in 1955, specified procedures that allowed voters to decide whether to establish a county conservation board. Upon voter approval, counties then were empowered by the state to levy taxes for park funds and to create county conservation boards with authority to acquire, control, and manage several types of public areas, including parks, preserves, playgrounds, recreation centers, forests, and wildlife areas. Within a few months after the law took effect in 1956, twenty-one counties had proposals ready for the November ballot. Voters in sixteen counties approved county conservation boards. By the early 1960s, approximately two-thirds of Iowa's ninety-eight counties had established conservation boards.[90]

Importantly, the 1955 act required county conservation boards to obtain project approval from the State Conservation Commission, thereby linking their actions to the commission's overall policies and procedures. Consequently, from 1956 on, the commission was inundated with requests to approve land acquisitions or plans to construct dams, buildings, roads, or other improvements in the myriad areas that CCBs began to develop as parks, hunting grounds, wildlife refuges, and fishing areas. Proposed projects involving state waters, of course, also had to be approved by the Natural Resources Council. The process often delayed projects, or stopped them if the intended use was not compatible with other aims. Because of the volume, the commission streamlined procedures by delegating technical decisions to the staff, but county conservation matters nonetheless occupied much of the agenda during the late 1950s. Ultimately, the staff assumed the entire burden of screening proposals, developing viable projects in consultation with county boards, and then preparing recommendations for the commission to approve. This work was handled through a new Department of County Conservation Activities, established by the State Conservation Commission in 1960.[91]

Within a short time, the commission also began to transfer maintenance of smaller state parks to county boards, typically through long-term agreements rather than transfer of ownership. The process began in 1958, when the boards in Cass and Carroll counties entered into twenty-five-year agreements to maintain, respectively, Cold Springs and Swan Lake state parks. Shortly afterward, the Lands and Waters Division was instructed to reclassify state parks, ranking them according to their desirability for maintenance by the state as state parks and recommending areas that, from a maintenance standpoint, were good candidates for transfer to counties.[92] Many of these were small parks, such as Frank A. Gotch State Park in Humboldt County, which had been maintained by local communities or organizations for some years.[93]

Considering the immediate success of the county conservation program, it is difficult, in retrospect, to fathom why it took so long to pass authorizing legislation. World War II certainly slowed the effort, but that accounts for only a few years. A more fundamental reason seems to be that Iowa was breaking new ground. In this regard, it is worth noting that Indiana introduced similar legislation in the 1940s, patterned after the 1941 Iowa bill. Likewise, the Indiana legislature took just as long to pass a bill; its county conservation law also took effect in 1955.[94] At a time when the nationwide trend was toward more and more layers of governmental activity, both states pursued an innovative mechanism for decentralizing their parks, and the fiscal responsibility for maintaining them, by establishing a two-tiered system.

Thomas Macbride's vision of county parks finally materialized sixty years after he first broached the idea in 1895. Even back then, he acknowledged that there was nothing particularly new in his call for county parks, that is, "open grounds available for public use in rural districts."[95] Yet it took more than a half-century to establish a framework to encourage county park formation. The county conservation initiative succeeded in creating rural parks and outdoor recreation areas easily accessible to nearly every Iowan. It placed the financial responsibility for management and maintenance at the local level, where the majority of users resided. At the same time, it tied activities at the county level to statewide conservation policies and goals through the authority of the State Conservation Commission.

The county conservation program in Iowa merits its own history, but there is space here only to sketch the relationship of county conservation boards to the state park system, and the magnitude of their importance in fulfilling the conservation and parks mission of the state. Under the provisions of the 1955 County Conservation Act, the state has entrusted wildlife areas, river access points, trails, lakes, recreational areas, parks, preserves, woodlands, campgrounds, and open space areas to the care and management of county conservation boards. Dozens of state landholdings, including state parks and preserves, have been placed under long-term management agreements. In a few cases, the state has deeded former state parks to counties or municipalities (see appendix).

As of 1962, with sixty-two boards fully functioning, more than 14,000 acres were under county management. Most of the landholdings were managed either as parks, river access points, or multiple-use areas. Land use at the remainder ranged, in descending order, from game management areas and roadside parks (a couple dozen each) to historical areas, cooperative school projects, forests, and preserves. The total budget of ccbs approached $2 million annually.[96] By 1986, ninety-six county conservation boards owned and/or managed 1,156 areas totaling more than 82,000 acres, and collectively, the county conservation budget exceeded $17 million. The most recent figures show that the county system continues to expand steadily. As of October 1994, county conservation boards managed nearly 124,000 acres in 1,398 areas, including 478 park and recreation areas; 558 wildlife and forest areas; 222 water access areas; and 140 environmental education facilities, historic sites, or preserves.[97]

Although county conservation boards were, and continue to be, important agents in reshaping the state park system, state areas represent only a portion of total county conservation landholdings. Locally initiated land acquisition and development efforts benefited immeasurably from federal aid programs that were extended to counties through the State Conservation Commission. The Land and Water Conservation Fund Act of 1964, in particular, opened up a vast sum of federal aid for county conservation projects. More will be said about the LWCF later on, but it boosted land acquisition and development of recreational areas with funding, on a fifty-fifty match basis, channeled through a succession of Department of Interior agencies. The State Conservation Commission chose to reallocate half of Iowa's annual apportionment to local entities, a decision that significantly advanced the county conservation program.[98] County conservation boards also benefited from the Marine Fuel Tax grant program, a state fund available to develop boating facilities, and from snowmobile fees earmarked for snowmobiling-related projects.

Post–World War II forces to maximize public access to outdoor areas, combined with inadequate funds to maintain them, subtly reshaped the state park system. A reclassification scheme adopted in 1963 acknowledged the primacy of recreation. Based on new standards set by the National Conference on State Parks, Iowa reclassified its holdings into state parks, state recreation areas, and preserves.[99] On paper, *state parks* were now defined as "relatively large areas of outstanding scenic or wilderness qualities." Recreational activities in parks were not to "interfere" with resource preservation. In keeping with this definition, the State Conservation Commission would focus its efforts on maintaining the larger parks; counties would handle smaller areas that attracted mainly local users. By 1970, county conservation boards had assumed the management of thirty-nine state areas, nine of them still technically classified as state parks or preserves.[100]

State recreation areas were defined as "areas selected and developed primarily to provide outdoor recreation for more than local needs, but also having scenic qualities." The artificial lake program begun in the 1930s did not come close to satisfying public demand, which further increased when federal flood control projects made large bodies of water available for recreational use. Lake parks, particularly artificial lake parks, were now looked upon primarily as outdoor recreation spots. *Preserves* were defined as small areas "established primarily to preserve objects of historic and scientific interest and places commemorating important persons or historic events." This continued a long-standing tradition of recognizing that certain areas deserved greater care, although the commission still had steps to take toward adopting management techniques and plans that would effectively protect such areas. As events transpired, the commission soon inherited a new entity, the State Preserves Advisory Board, to oversee preserves management, a topic that will be discussed in the next chapter.

The 1963 classification scheme was much less cumbersome than the one adopted in 1941. In most respects, however, the distinctions were largely conceptual. In practical terms, the commission listed only two categories, parks and preserves, in biennial reports; and visitors were largely unaware of even these distinctions. Nonetheless, the new classification system reflected an evolving concept of state parks as larger recreational areas bureaucratically divorced from fish and game areas. By the mid-1960s, park management was geared to accommodate mass outdoor recreation, preferably tied to water—boating, swimming, and lakeside camping. Yet despite the fact that water played an ever-larger role in park use and park management, the State Conservation Commission had lost much of its authority over water policy to the Natural Resources Council.

7

The Lightning Rod Decades

Inasmuch as this Commission has, somehow, become the Lightning Rod of the State of Iowa, I hereby place on [permanent] loan to the Iowa Conservation Commission this genuine antique lightning rod. . . . Let it be displayed at each and every Commission meeting, as a symbol of our determination to quickly ground every lightning bolt that strikes within our jurisdiction.

—JOHN FIELD [1]

So opened the March 1980 meeting of the State Conservation Commission. Newspaperman John Field of Hamburg, a greenhorn on the commission, thought that stormy times called for a little levity.

During the confrontational 1970s, as social activism reached crescendo in the United States, the commission was periodically besieged by activists of one stripe or another. The most persistent of them was Bob Leonard, a farmer from Lamoni and the person to whom John Field aimed his lightning rod speech. There were many others, groups as well as individuals, but Leonard achieved legendary status. For several years he routinely attended commission meetings to demand stricter enforcement of hunting laws, protest the acquisition of more public land, or complain about the activities of specific commission employees. As of late 1979, the "commission gadfly" had filed no fewer than thirty complaints against the State Conservation Commission, most of which involved trespass by hunters. Other activists challenged the commission's policies and lack of leadership with some justification, but Leonard was not among them. He was simply the most voluble, and, as so often is the case, he received an inordinate amount of attention, not only from the commission itself, which was a captive audience, but from state politicians seeking ready-made publicity. When newspaperman John Field of Hamburg joined the commission in 1979, the activist era was in full swing.[2]

The lightning rod was a fitting symbol for the era. From the mid-1960s to the mid-1980s the commission was touched by strife, much of which came from competing impulses in the post–World War II environmental era. From one corner came a renewed interest in preservation—of endangered species, of history and culture, of wilderness. From another came even more pressure to expand publicly supported outdoor recreation. As evidence of burgeoning public demand for outdoor recreation areas and facilities, Congress established the Outdoor Recreation Resources Review Commission in 1958, during the Eisenhower administration. The ORRRC did not issue its report until four years later, however, during the Kennedy administration. Recommendations in the report influenced considerable federal legislation during yet another presidential administration. The 1964 Wilderness Act, the 1964 Land and Water Conservation Fund Act, the 1968 National Wild and Scenic Rivers Act, and the 1968 National Trails Act, all signed by President Lyndon Johnson, are traceable, in part, to recommendations contained in the 1962 ORRRC report. Parks, ever the turf where preservation confronted public access, once again became battle prizes. In some respects, Leonard represented those who opposed the expansion of government that was inherent in the expansion of public lands and public recreation. These were the crosswinds that buffeted the commission during this period.

After the Outdoor Recreation Resources Review Commission issued its report, Governor Harold Hughes appointed a twenty-four-member Advisory Committee on Conservation of Outdoor Resources to prepare a similar assessment for Iowa. The committee was composed entirely of men, three of whom were either current or future members of the State Conservation Commission. By one count, eighteen of the twenty-four men were members of the Izaak Walton League, including the chairman, Maynard Reece, the vice chairman, Arnold O. Haugen of Iowa State University, and all the subcommittee chairs.[3] This is not to suggest that the Iowa Ikes controlled the agenda and findings of the governor's committee, but it is offered as further evidence that by the 1960s environmental affairs and environmental politics were completely dominated by a male perspective. If Iowa reflected the nation, it suggests the degree to which the post–World War II outdoor recreation boom was gender driven.

In any event, the committee's *Report on Iowa's Outdoor Resources*, released in 1964, laid out an ambitious, $50 million recreation development plan that was wide-ranging in its scope.[4] On balance, though, the plan reflected an increasingly bifurcated view of natural and cultural resources and society's role in both using and caring for them. As a reflection of the times, therefore, it is an important document. Parts of it echoed new environmental concerns for preserving "wild" lands and other relatively pristine natural areas. For instance, in addition to recommending a system of state preserves, the plan also called for designating extensive sections of certain rivers as "wild rivers" and developing a subsidy program to encourage private landowners to maintain or restore wetlands. At the other end of the spectrum was a proposal to develop at least one 5,000-acre "master recreation-outdoor-education area." This concept was called "Hawkeye Naturama,"and the name was appropriate. Presented as conservation education for virtually everyone, Hawkeye Naturama was the flip side of the preserves system. It was a Disney-like notion that stretched the definition of outdoor recreation into the realm of entertainment. If traditional hiking or canoeing or fishing did not suit one's tastes, proposed options ranged from strolling through the Hall of Prehistoric Animals to quiet time in a meditation garden or—even further—to antique shopping.[5] As a working document, though, the 1964 plan was not of the same caliber as the 1933 *Twenty-five Year Conservation Plan*, and it was quickly superseded by the specific planning requirements of another federal program, the Land and Water Conservation Fund. The main legacy of the *Report on Iowa's Outdoor Resources* is the State Preserves Act of 1965.

Preserves: A New Paradigm

Watersheds in nature are among the more subtle features on the landscape. Watersheds on the political landscape, by contrast, almost never are subtle. This makes the 1965 State Preserves Act all the more intriguing, that a little bill of such importance could pass into law without so much as a whisper of dissent. But a watershed it was. Looking to the past, the State Preserves Act codified in the clearest language possible the kernel of intent at the center of the original State Park Act. In language reminiscent of the 1917 law, which authorized state parks to preserve areas of scientific interest, historical association, or natural scenic beauty, the 1965 State Preserves Act authorized a system of preserves to provide the highest form of protection to areas of "unusual flora, fauna, geological, archaeological, scenic or historic features of scientific or educational value."[6] Looking to the future, the State Preserves Act signaled a new era of environmental activism. Just as academy-based natural scientists had been the driving force behind the 1917 law, so too were botanists, biologists, and ecologists once again the driving force behind the 1965 act.

While state park facilities were literally falling apart for lack of maintenance, while the State Conservation Commission and half a dozen other state agencies were coping with the ramifications of new water resources issues, and while the public was clamoring for more outdoor recreation spots, a few conservationists, who now thought of themselves more as ecologists, were organizing in support of nature. As others have noted, the initial phase of post–World War II environmentalism evolved from widespread local and state concerns about and interest in outdoor recreation, open space, and natural environments.[7] These were not new concerns. In most respects, they were *renewed* concerns, but the disquiet was more widespread, or became so. *Silent Spring*, Rachel Carson's treatise exposing the devastating effect of chemical pesticides on wildlife, is often cited as the beginning of the "environmental age." Ironically, the same year it was published, 1962, Jay Darling died. Although postwar environmentalism has been portrayed as a movement without a history, it was not as if environmental awareness suddenly dawned and ushered in a new age. Rather, new voices found new, and greater, audiences.

Existing conservation organizations, such as the Sierra Club and the National Wildlife Federation (another Darling brainchild), witnessed rapid growth during the 1960s. Iowa was no exception in this trend, and the Nature Conservancy proved to be a key link in the chain of events that led to Iowa's State Preserves Act. Galen Smith, Roger Landers, and J. M. Aikman of the Iowa State University department of botany spearheaded the effort to organize a chapter of the Nature Conservancy in the early 1960s. Although local events played a part in galvanizing support for this move, the organizational meeting took place in conjunction with the 1963 annual meeting of the Iowa Academy of Science.[8] Thus, to a certain extent, the Iowa Chapter also emerged from long-standing concerns and activities of the IAS conservation committee.

Aikman, Ada Hayden's colleague in prairie preservation during the 1940s, had been instrumental in reorganizing the IAS conservation committee in 1941. Prairie preservation was the catalyst, but the committee quickly expanded its focus to include water policy, wildlife policy, soil conservation, antiquities, forestry, and state parks. Aikman, in particular, saw the committee as influencing the policies of the State Conservation Commission.[9] However, this did not really happen. Although E. B. Speaker, director of the State Conservation Commission from 1963 to 1968, was a member of the IAS conservation committee, the IAS as a whole was, once again, less interested in state environmental issues than the handful of members who gravitated to the committee. In addition, the IAS had less direct influence with commissioners than did the Izaak Walton League. Prairie preservation languished during the 1950s, and the scope of conservation issues was so broad that one small committee could not possibly do more than monitor what was happening. Nonetheless, the committee played an important role as a meeting ground for several individuals whose names either had been, were, or would be associated with resource policies in the state. In addition to Speaker and Aikman, the committee included geologist H. G. Hershey, chair of the Natural Resources

Council, archaeologist Charles R. Keyes, forester G. B. MacDonald, and soils expert F. H. Mendell. Beyond this committee, though, academic scientists were not actively engaged in any organized effort to protect natural areas. The Iowa Chapter not only attracted academicians, but enabled them to work with a broader spectrum of concerned citizens in order to acquire natural areas for preservation. It was, in this sense, the Iowa Park and Forestry Association reinvented for a new mission.

Although the IAS conservation committee had little success working with the State Conservation Commission during the 1950s, the concept of dedicated preserves as sanctuaries nonetheless was slowly evolving within the agency. Since 1928, when the commission declared Woodman Hollow a "preserve" in a move to stop the construction of a power dam on the Des Moines River, certain areas of the state park system as a whole had been accorded special status (see appendix). In 1941, when the commission first classified its park holdings, it created two preservelike categories that recognized this: "geologic-biologic" areas and "historic-archaeologic" areas. There were four of each at the time.[10] Then in 1955, the commission reclassified the areas under Lands and Waters management, reducing the categories from eight to three. This time, fifteen areas were designated as "state preserves," an increase that reflected, in large part, the acquisition of prairies and historic sites that took place in the 1940s.[11] Wildlife was accorded protection in all but three of the fifteen "state preserves" when they were designated "wildlife refuges" in 1962 under new provisions in the Iowa Code, a move consistent with the fish and game focus of the commission at that time.[12] But there was no state law establishing preserves themselves as inviolate areas, and no evidence that the commission at any time considered drafting such legislation.

The governor's Advisory Committee on Conservation of Outdoor Resources stepped forth to press for a preserves law. More precisely, since preservation of natural features and cultural sites were just two of many related aspects on a long agenda, one of its subcommittees took the lead. The Natural Features and Scientific Preserves Subcommittee, chaired by Galen Smith, produced the most detailed section in the entire committee report, which set forth specific recommendations for a preserves system. Smith, who was frustrated with the lack of organized support for protecting natural areas in Iowa, seized the opportunity to produce something of long-lasting benefit.[13] His subcommittee justified the need for a preserves system by noting that state and county park systems in Iowa had been "designed in part to preserve outstanding natural features." However, "with primary emphasis on public recreation . . . it has become apparent that a natural feature is not assured of preservation simply by being placed in public ownership in a park."[14] While most of the advisory committee's work focused on ways to expand Iowa's outdoor recreation areas and amenities, the preserves subcommittee addressed the need to protect natural and cultural resources from relentlessly increasing public use and from amorphous forces of "progress."

For assistance, Smith turned to George Fell of Rockford, Illinois. Fell, the first executive director of the Nature Conservancy, drafted and lobbied to passage the 1963 law authorizing the Illinois Nature Preserves System. Smith also sought out botanist Edward Cawley of Loras College and vice chair of the Iowa Chapter of the Nature Conservancy, to help draft a legislative proposal for a preserves system in Iowa. As a graduate student at the University of Wisconsin, Cawley had worked under John T. Curtis, who, along with Aldo Leopold, instigated the Wisconsin Board for the Preservation of Scientific Areas, established in 1951.[15] Cawley took up the task by reviewing legislation other states had adopted. The list was not long. At the time, only a handful of states had established systems for preserving natural areas, but these state initiatives represented the beginnings of an important focus within the environmental movement as a whole. In addition to Wisconsin and Illinois, Michigan, New Jersey, Virginia, and North Carolina had established fledgling preserve systems by the mid-1960s. In the end, the Iowa bill most closely conformed to the 1963 Illinois preserves law. However, to pass the bill more easily, it was rewritten to admit cultural areas—that is, archaeological and historic sites—as well as natural areas, to preserve status.[16] The Iowa State Preserves Act thus represented a significant departure in concept, but it was wholly consistent with the way conservationists in Iowa had traditionally thought about preserves.

State Senator Kenneth Benda of Hartwick and Representative John Kibbie of Emmetsburg, the legislative members of the governor's advisory committee, introduced the preserves bill in 1965, and it passed both houses without discussion. Cawley attributes the lack of debate to the language of the bill, which stipulated that preserves be created principally from existing state-owned areas or from donations. Therefore, the proposal did not entail major land-acquisition efforts; it simply set up an administrative mechanism for managing certain areas differently. As a result, the bill went through the legislature before any questions or concerns about it could arise from other state agencies or from outside organizations.[17]

By law, the State Preserves Advisory Board was placed under State Conservation Commission administration, which was to provide staff and a budget line from the commission's annual appropriation. Other provisions named the commission's director as a member of the advisory board and directed the State Conservation Commission, the conservation committee of the Iowa

Academy of Science, and the State Historical Society to submit names of possible appointments to the governor. From this list, the governor promptly appointed the six additional members, and the advisory board organized in 1965. The commission also provided a staff ecologist to help survey areas and design management guidelines. However, since the commission had not sought this entity, it was not immediately prepared to fund it. Consequently, the advisory board spent the first two years working to establish a budget, even fighting for its very existence while the legislature debated several plans to reorganize the State Conservation Commission and various other state agencies into a department of natural resources.[18] Thus, the advisory board did not actually dedicate any preserves until 1968.

The Preserves Advisory Board's principal functions were, and still are, to recommend areas for dedication as preserves either under the jurisdiction of the State Conservation Commission (now the Department of Natural Resources) or under the ownership of other public agencies, private groups, or individuals; to promote research and investigations of existing and proposed preserves; to draw up management agreements for preserves; and to oversee preserve management. For many years, however, the advisory board functioned in the shadow of the agency, not of it and not really understood or accepted by the staff. Internal suspicion and animosity finally began to dissolve in the mid-1980s. In recent years, the Preserves Advisory Board has been drawn into the organizational structure under the Division of Parks, Recreation, and Preserves. Functionally, however, it is fluid because only about half of the preserves in the state system are actually owned by the state; and of those that are, not all are managed through the parks division. In operation, the advisory board negotiates management agreements with private parties and other public agencies; it also coordinates preserves management with the DNR's Fish and Wildlife Division and its Forestry Division.

Appropriately, Hayden Prairie holds the distinction of being the first area dedicated as a preserve under the 1965 act; State Conservation Commission approval came in May 1968. Three months later, the commission approved eight more preserves, all state-owned areas: Sheeder Prairie in Guthrie County, Kalsow Prairie in Pocahontas County, the Turkey River Mounds Preserve in Clayton County, Fort Atkinson State Preserve in Winneshiek County, the Fish Farm Mounds Preserve in Allamakee County, Wittrock Indian Village Preserve in O'Brien County, Pilot Knob State Park in Hancock County, and White Pine Hollow in Dubuque County.[19] A decade after the State Preserves Act took effect, thirty preserves had been dedicated, and they represented diversity bordering on eclecticism among the state's distinctive features. As described by the board ecologist at that time, dedicated preserves included "six native prairies, a native White Pine stand, the state's only Sphagnum bog, a Balsam Fir stand, some of the oldest exposed rock outcrops in the world, an ancient fort, a fen, several Indian mound groups and a historical cemetery."[20]

From an uncertain launch, the state preserves system is now well established and increasingly recognized as one of the state's environmental assets. The system now includes nearly ninety sites, variously owned by the state, by county conservation boards, or by some other public entity. A few are owned by private individuals, and some, like Berry Woods in Warren County, are owned by the Nature Conservancy. The total area in preserves is still very modest, approximately 9,000 acres, but the system is among the most diverse in the country.[21] The act, as implemented, admits four types of resources—archaeological, biological, geological, and historical—into one preserves system.

In this manner, the State Preserves Act accomplished what the 1917 State Park Act intended but failed to fulfill: protect significant natural and cultural features in perpetuity. The 1965 act thus codified into law a management philosophy that the State Conservation Commission had long acknowledged but never fully implemented because providing for intensive recreational use had so dominated the commission's management style since the 1930s. This legal step involved more than just a conceptual segregation of preserves within the park system; it was based on a new paradigm. Preserves, once dedicated, are subject to the highest standards of land stewardship. They represent vestiges of the past that cannot be replaced or reproduced elsewhere; therefore, preserves are to be held in trust as public assets, protected by each generation for the next.

As the preserves system has grown, however, cultural properties have come to represent a smaller percentage of the whole. Undoubtedly the system would contain a greater number of historic and prehistoric areas today except that the responsibility for historic preservation, in general, was transferred from the State Conservation Commission to the State Historical Society in 1974. The circumstances behind this division of administrative responsibilities, however, were fairly complex. Although the commission continued to pursue acquisition of cultural sites during the 1950s and 1960s, notably the Julian Dubuque gravesite and the Walter House at Quasqueton (Cedar Rock), designed by architect Frank Lloyd Wright, there was less and less agreement that such areas belonged under the commission's jurisdiction. As early as 1956, the commission rejected a suggestion that it create a position of state archaeologist, even though protecting archaeological sites clearly fell within its purview.[22] Three years later, the commission discussed setting up a historical sites division, even going so far as to instruct the staff to study the matter and make recommendations. Further discussions took place at a conference with the Iowa Society for the Preservation of Historical Sites, the State Historical Society, Iowa State University, and the National Trust for Historic Preservation, but no action resulted.[23] At best, historic and prehistoric sites had been treated with benign neglect from the time Louise Parker left the commission in 1949 until the preserves system was organized. Symbolic of the commission's waning interest in cultural resources, it turned over management of the Abbie Gardner Sharp Cabin to the State Historical Society in 1960.[24]

Passage of the National Historic Preservation Act in 1966 came on the heels of the 1965 State Preserves Act, an interesting accident of history, but the two pieces of legislation emerged from different constituencies. As a result, the care and management of Iowa's state-owned cultural properties became hopelessly confused during the late 1960s and early 1970s. Under the National Historic Preservation Act, states were eligible for federal money to survey and protect historic and prehistoric places. Governor Ray designated William Petersen, director of the State Historical Society, to administer Iowa's program. The society's board, however, had no interest in accepting federal money for historic preservation, so nothing happened until 1971, when Adrian Anderson, then assistant state archaeologist, replaced Petersen as the state historic preservation officer. With this change, administration of the program, such as it was, moved from the State Historical Society to the Office of State Archaeologist and then, abruptly, to the State Conservation Commission.[25]

The administrative tie with the State Conservation Commission gave Anderson an opportunity to work with the Preserves Advisory Board, which he did, but he was unsuccessful in his attempts to introduce historic preservation planning into the larger structure of commission operations. By this time, the commission staff had virtually ceased to think of historic places as being part of "their system." In Anderson's words, "They didn't admit that they had any historic sites. They had parks, and they had recreational facilities. I tried to point out that the ccc [Civilian Conservation Corps] buildings and structures were indeed historic, but they didn't believe that."[26] Thus, when an opportunity came to transfer back to the State Historical Society, Anderson worked with the new director, Peter Harstad, to do just that.

In 1974, the Office of Historic Preservation finally found a permanent berth. At this point, historic preservation functions became officially associated with the State Historical Society. The mandate of the Preserves Advisory Board remained unchanged, though. It continued to manage historic and prehistoric sites designated as preserves, and the State Conservation Commission continued to acquire cultural properties. It was not until after 1986 that the two agencies began to coordinate their responsibilities in the area of historic preservation.

The Land and Water Conservation Fund

A system of dedicated preserves was the most important outgrowth of state-initiated planning in the 1960s. Federal initiatives, however, once again drove the public recreation side of state park systems. The Land and Water Conservation Fund Act of 1964 stimulated both national and state park expansion and development at a level that had not been seen since the 1930s. In stark contrast to the 1930s, the socioeconomics of outdoor recreation now received greater attention. Policy analysts pointed out, for instance, the need to reconcile the demand for outdoor recreation with the ability of government to provide for it at low, or no, cost to actual users; to decide how much of the natural resource base under public management should be allocated to recreation; and to determine how far government should go in making outdoor recreation conveniently available to economically and socially disadvantaged groups.[27] These new concerns, coupled with a rising class of professional outdoor recreation planners, gave the Land and Water Conservation Fund program a much different thrust.

The LWCF came on the heels of recommendations contained in the 1962 report of the Outdoor Recreation Resources Review Commission, which, among other things, called for strengthening the states' role in meeting public demand for outdoor recreation. Under the 1964 act, Congress authorized the creation of a new Bureau of Outdoor Recreation within the Department of Interior, charged with developing a national outdoor recreation plan and administering a twenty-five-year grants-in-aid program to states.[28] Despite severe cutbacks in the early 1980s, the LWCF was, by almost any report, the most important source of funds for acquiring and developing state and local parklands. Between 1965 and 1987, two years before the LWCF was set to expire, federal grants to states totaled approximately $3.2 billion, an amount that funded more than 30,000 projects nationwide. Overall, the LWCF enriched the pattern of state, local, and regional parks throughout the country.[29]

Under the provisions of the federal law, funds could be used for three purposes: comprehensive recreation planning, land acquisition, and development of outdoor recreation facilities. Grants could not be used to maintain existing facilities. Federal money would pay half the cost of an approved project, the other half to be paid by the recipient. Early on, the State Conservation Commission decided to reallocate half of Iowa's annual apportionment to local agencies, in large part because this was the only source of government support available to county conservation boards at the time.[30] To qualify for federal grants, states were required to prepare comprehensive, statewide plans and update them at regular intervals. This process eventually forced the State Conservation Commission to refocus on the state lands system as a whole in relation to public demand for a wide variety of outdoor recreation activities. However, the degree to which the planning process fostered serious or sustained reexamination of the state's role in providing outdoor recreation is debatable.

Iowa submitted its first SCORP, or statewide comprehensive outdoor recreation plan, in early 1966. It set out an ambitious five-year program designed, in the main, to expand water-based recreation areas, which was clearly the nationwide trend in outdoor recreation at the time. The plan proposed to add more than 40,000 acres to state lands, with an estimated price tag of nearly $20 million. At the top of the list were projects to acquire and develop three large-lake recreation areas, to acquire and develop lakes on two P.L. 566 watershed areas in conjunction with the Soil Conservation Service, to develop new parks near two federal flood control reservoirs, and to develop recreational facilities in four state forests. Further down the list were proposals to acquire and develop more than eighty boat-access areas on the Mississippi, Missouri, and inland rivers, and on the flood control reservoirs; to begin land acquisition along the Upper Iowa River with an eye toward preserving it as a "wild" river; to expand and increase waterfowl habitats, public hunting grounds, and stream fishing accesses; to expand the teachers' conservation education camp at Springbrook State Park; and, at the end of the list, to develop Hawkeye Naturama at the Saylorville Reservoir.[31]

Unlike federal aid programs of the 1930s, the LWCF did not produce a signature architectural style or significantly increase the *number* of state parks and recreation areas, so the tangible legacy is much less obvious. Moreover, because Iowa, among all the states, had the smallest percentage of land in state or federal ownership, the commission directed considerable LWCF spending toward land acquisition; construction and development were given lower priority. During the first year of operation, 1966, the commission authorized land acquisitions to increase the size of Clear Lake, Lacey-Keosauqua, and Lake Macbride state parks, and Shimek and Stephens state forests. It also approved state projects to acquire more land and upgrade facilities at Beeds Lake, Bellevue, Lake Anita, Lake of Three Fires, Pikes Point, and Waubonsie state parks. Nonacquisition projects approved by the commission included repairing spillways at Lake Ahquabi and Lake Keomah; overhauling the water systems at Backbone, Pine Lake, Rock Creek, and Springbrook; constructing new bathhouses at Green Valley Lake and Lake Darling; and a variety of minor improvements, such as parking lots at Prairie Rose and Spring Lake state parks.[32]

Table 8.

State Park Projects Receiving More Than $500,000 LWCF, 1966–1986 [33]

State Park	LWCF Grants 1966–1986	Purpose
Lake Manawa/Pottawattami	$2,343,774	expansion and development
Big Creek/Polk	2,316,324	acquisition and development
Pleasant Creek/Linn	2,241,357	acquisition and development
Volga River/Fayette	1,709,504	acquisition and development
Mines of Spain/Dubuque	1,494,741	acquisition
Brushy Creek/Webster	1,104,282	acquisition
George Wyth/Black Hawk	1,033,613	development
Ledges/Boone	772,541	redevelopment

These projects typified the ways in which the State Conservation Commission used LWCF funds. Overall, the emphasis was on improving the size and quality of existing areas. However, the LWCF helped to underwrite the cost of five new recreation areas—Big Creek, Volga River, Pleasant Creek, Mines of Spain, and Brushy Creek. Table 8 provides a summary of the major park expansion or acquisition and development projects funded in part with LWCF grants from 1966 through 1986.

Initially, the LWCF was intended to run until 1989 and to function much like a trust fund. Neither happened, and annual appropriations fluctuated widely. The beginning of the end came in 1981, when the Reagan administration cut funding drastically, eliminated the administering agency, and transferred a much-reduced program to the National Park Service. By the mid-1980s, the program had all but disappeared as a source of funds for state park programs.[34] Nonetheless, from 1966 through 1986, the Land and Water Conservation Fund injected more than $43 million into park and recreational development in Iowa. Of this amount, approximately half was passed on to county conservation boards and municipalities. Most of the spending took place between 1972 and 1984.

The legacy of the Land and Water Conservation Fund might have been greater, but even in its reduced scope, the matching-funds requirement stimulated greater state spending for capital improvements. For the period 1965–1968, the state appropriated nearly $8 million for capital improvements to the state park system, a figure eclipsed by $11 million in special appropriations during the last half of the 1970s (table 9). Combined federal and state funds

Table 9.

Land and Water Conservation Fund Expenditures, 1966–1995 [35]

Year	Annual	Cumulative
1966	$ 718,027	$ 718,027
1967	1,162,423	1,880,450
1968	1,884,633	3,765,083
1969	1,041,446	4,806,529
1970	815,843	5,622,372
1971	204,784	5,827,156
1972	2,951,899	8,779,055
1973	1,937,929	10,716,984
1974	5,462,179	16,179,163
1975	821,023	17,000,186
1976	1,655,683	18,655,869
1977	7,136,076	25,791,945
1978	3,678,237	29,470,182
1979	3,527,671	32,997,853
1980	4,584,204	37,582,057
1981	1,346,457	38,928,514
1982	793,676	39,722,190
1983	1,461,167	41,183,357
1984	723,805	41,907,162
1985	957,947	42,865,109
1986	769,384	43,634,493
1987	423,539	44,058,032
1988	402,010	44,460,042
1989	933,363	45,393,405
1990	139,133	45,532,538
1991	165,552	45,598,090
1992	162,071	45,860,161
1993	152,417	46,012,578
1994	193,428	46,206,006
1995	476,933	46,682,939

allowed the State Conservation Commission to increase the size of the state park system, as measured in acres, by seventy-three percent. As of June 1966, before LWCF projects began, Iowa tallied ninety-one parks and preserves totaling more than 30,000 acres of public lands, excluding lakes. Eight of these areas were managed by county conservation boards or other local entities. Annual visitation stood at about nine million.[36] Two decades later, even though the system had been vastly reordered, the size had increased to nearly 52,000 acres, again excluding lakes. Park attendance now exceeded thirteen million annually.[37]

For the first time in decades, state park expansion kept pace with increasing park use, although there was still a mounting backlog of deferred maintenance. During the 1970s, the commission tapped other sources in an effort to maintain parks. Iowa's Green Thumb program, inaugurated in 1975, provided additional state funds to employ senior citizens as part-time help in state parks.[38] The Young Adult Conservation Corps, a jobs program aimed at the other end of the age spectrum, had a much shorter life. Federal legislation enacted in 1977 provided funds to put youth to work in parks and forests. Using YACC funds, the commission established resident youth camps at Red Haw State Park and Yellow River State Forest. Although many people hailed the YACC as a long-overdue revival of the Civilian Conservation Corps idea, federal budget cuts during the Reagan administration forced the program to shut down in 1982.[39] At about the same time, the commission began to encourage volunteer efforts in state parks and to accept private donations of goods and services.[40] In an effort to focus attention on the ongoing neglect, park officers even solicited volunteer services and donations from individuals and local organizations. In routine gestures of gratitude, the commission acknowledged private donors who favored the parks with everything from snow removal to birdhouses, with the value of goods and services sometimes reaching as much as $30,000 a month.[41]

Cutbacks in LWCF funding prompted many states to seek new sources of revenue, including general obligation and revenue bonds, user fees, and special taxes. In 1981, the State Conservation Commission held its first serious discussions of adopting a park user fee in order to renovate existing park facilities, especially the aging CCC structures. These discussions were based on a thorough study of approximately thirty other states that, by this time, were relying on a variety of revenue sources in order to stabilize financing.[42] Missouri voters, for instance, approved $60 million in bonds to finance capital improvements to state parks *and* a modest sales tax increase for ongoing state park operations and soil and water conservation efforts.[43] After four years of rejection and continued debate, the Iowa legislature finally approved a park user fee in 1985, which took effect on January 1, 1986. Contrary to the indications of survey data, however, the fee proved to be highly unpopular. During the second year, the fee structure was modified; free permits were eliminated for the disabled and for welfare recipients, and the general price was lowered.[44]

The Large Lakes Program: Volga River and Brushy Creek

The tangible legacy of the LWCF in Iowa is, for the most part, bound up in the new parks and recreation areas that were established, particularly through what was known as the "large lakes program." Importantly, the Governor's Committee on Conservation of Outdoor Resources revived long-dormant hopes for an artificial lake near Fort Dodge. When the *Des Moines Register* announced, in June 1964, that the long-range plan included artificial lakes near several major urban areas, one of which was *not* Fort Dodge, former commissioner Ewald Trost (of Fort Dodge) tapped out a letter to ICC director Everett Speaker.[45] A few weeks later, Trost appeared before the commission, accompanied by a local delegation selected for maximum influence. The group included representatives from the Chamber of Commerce and the Izaak Walton League, a member of the Natural Resources Council, a state senator, and a state representative. Trost reminded the commissioners that eighteen years earlier, when he was a member of that body, an area along Lizard Creek near Fort Dodge had been surveyed and found favorable for artificial lake construction. At that time, though, sites in the southern part of the state had top priority, and, in addition, "it seemed to be considered 'bad taste' to expect development in your own community." Now Trost was calling in the chits. Having been on the receiving end of such missives during his twelve years on the commission (1945–1957), he knew how to achieve an effect. "We are here to firmly and respectfully plead our cause to establish an artificial lake in accordance with the previous investigation. . . . After eighteen years we say with the Psalmist (Psalm 6, verse 3), 'My soul is sore vexed, but thou, O Lord, how long?'"[46]

By November of that year, the vexing question was answered. The commission's chief engineer came up for a quick look around, and a few weeks later Lizard Creek was on the list of artificial lake sites under consideration.[47] During the next year, the engineering section investigated about twenty different sites in the Des Moines, Sioux City, Fort Dodge, and Waterloo areas. When the first SCORP was submitted to the Bureau of Recreation for approval in 1966, the list of feasible sites had been pared to three: Elliott Creek near Sioux City, the Volga River in Fayette County, and Brushy Creek near Fort Dodge.[48] During the next year, the legislature effectively eliminated the Sioux City site by choosing to fund Big Creek reservoir near Des Moines, discussed later in this chapter in connection with the Saylorville Dam. In June 1967, the legislature approved all three projects—Big Creek, Volga River, and Brushy Creek—and appropriated an initial $1 million for each. These artificial lake appropriations were easily the most controversial elements of an $8 million outdoor recreation package bill that was heavily criticized as pork-barrel spending.

Such criticism was justified with the Volga River project, which the powerful Speaker of the House, Don Avenson, pushed through the legislature. Brushy Creek also reflected the power that a determined group of local supporters could wield in the legislature. Funding the Big Creek project, however, was largely inevitable given the Corps of Engineers' plans to construct flood control works near Des Moines. In any case, a few months later the commission authorized land acquisition for the Brushy Creek and Volga River sites and applied for LWCF funds to cover half the estimated cost.[49] The third project, Big Creek, was acquired and developed as a cooperative project with the Corps of Engineers.

Land acquisition for the Volga River project proceeded smoothly, but the lake aspect was problematic from the start. As early as 1964, State Geologist Garland Hershey had pointed out that the underlying shale stratum was fractured and, as a result, potential sites in the area probably would not hold water. The project was approved anyway, largely for political reasons. Local support for it was mixed, but House Speaker Don Avenson, who represented Fayette County, wanted the project; and the legislature gave him what he wanted.[50] After land acquisition was completed in 1971, there was some discussion of developing the area without a lake, since further engineering studies found that impounding the Volga River was not feasible. Avenson, however, was adamant that a lake would be constructed, so engineers searched for ways to design one that would hold water. Meanwhile, the estimated cost kept rising.[51]

Eventually, the legislature relented and authorized three small lakes rather than one large lake, but this did not really solve the engineering problem. Outside consultants confirmed that it would be necessary to compact a "blanket" of clay at every conceivable site in order to keep water from seeping through rock fissures. In addition, soil conservation practices in the watershed area were not at acceptable levels. With this evidence in hand, the commission finally voted, in 1977, to abandon the lake plan unless directed otherwise by the legislature. The legislature directed otherwise. Avenson and his supporters in Fayette County were constant in their demand for a lake, and the legislature consequently forced it. Eventually, the project was negotiated down to one lake of about 135 acres, which was constructed in 1978–1979. Ultimately, the

Volga River project consumed $1.7 million in LWCF grants, one of the largest commitments of federal funds to any state park project. Even so, it attracted relatively little attention from the media or environmental groups. Political pressure overrode all other concerns, which, in any case, were raised mainly within the confines of the State Conservation Commission. In the end, the state acquired more than 5,400 acres to develop the recreation area with only one small lake, yet only forty-five percent of the watershed was within acceptable soil loss limits, far below the seventy-five percent level the commission desired.[52]

The Brushy Creek project progressed very differently, and much more slowly. Water pollution surfaced as a potential problem early in the project. Land purchase negotiations had just begun in 1967, when the commission learned that the water and sewer systems from two towns, Duncombe and Lehigh, as well as an industrial plant, drained into the proposed lake area. This information came in addition to the usual concerns over agricultural runoff and soil erosion from nearby farms. Realizing that pollution control in the Brushy Creek watershed could be a serious matter, the commission tried to deal with the situation before lake construction was even considered. Among other things, it began to put pressure on local proponents to agitate for county zoning. If local residents wanted the lake and recreation area, they would have to help assure that the project met some environmental standards. Commissioner Keith McNurlen of Ames made it clear that he would not support spending state money just to build a "sewage lagoon." The commission also alerted the new state Water Pollution Control Commission to potential problems, although it was not clear what this new agency could do.[53]

Ewald Trost played a key role in facilitating zoning discussions among Webster County supervisors and the Fort Dodge business community. The Fort Dodge Chamber of Commerce, one of the project's principal backers, supported county zoning, and the county supervisors seemed receptive to the idea, at least at first. But they also wanted something in return. From their perspective, which was mainly economic, the supervisors and businessmen wanted some private development near the lake in order to recover the loss of revenue when land was taken out of the property tax base. The commission had hoped to avoid private cottages and commercial development near the lake, but it did not immediately reject the idea. By early 1968, the commission seemed to be heading toward a compromise with business interests in Fort Dodge that would allow limited private development within the park in return for a county zoning ordinance.[54] Trost began lining up support from other civic organizations, and the local Izaak Walton League chapter agreed to study county zoning as it might apply to Webster County.[55]

Trost and company pushed steadily to keep the project moving forward, and in the process everyone seemed to forget that real people would have to sell their farms and homes in order to amass the three or four thousand acres needed for Brushy Creek. As it turned out, many of the people living in the proposed "take" were unwilling to sell at the prices the state offered. By August 1968, the main issue was no longer county zoning, but how long it would take to settle condemnation lawsuits. Landowners charged that commission employees mistreated them and offered below market prices. As a result, approximately half of the thirty-one affected landowners eventually opted for condemnation proceedings. Brushy Creek divided the commissioners as no other project had. In a move that was virtually unprecedented, two commissioners made public statements critical of the project as a whole, a rare exposure of internal disagreement. William Noble of Oelwein stated, for the record, that "if the project is typical, it was picked by a group of local businessmen who sold it to the legislature." His colleague, Edward Weinheimer of Greenfield, noted that "the ones pushing it were not the ones who live out there."[56]

Land acquisition came to a temporary halt in 1969, and condemnation proceedings dragged on for six years. The last tract of land was finally acquired in late 1975, by which time the state had purchased about 4,200 acres. During this period, the commission opened the area for public use, although no development took place. Basically, Brushy Creek was managed (minimally) as open space, although reportedly used quite heavily—by "several hundred thousand" people a year—for horseback riding, camping, picnicking, hunting, and snowmobiling.[57] At the same time, the Fort Dodge Chamber of Commerce kept pressure on state legislators and on the State Conservation Commission to keep moving the project forward. And, until he died in 1972, Ewald Trost functioned as the primary liaison.[58]

With land acquisition finally complete in 1975, the commission turned its attention to development. By this time, it was standard practice to solicit public comment during the preparation of master plans, and Brushy Creek was no exception. However, there is no indication that the commissioners themselves, who were removed from the planning process, were at all prepared for the opposition that erupted.[59] If anything, the commission anticipated having to negotiate with business interests who wanted the proposed lake opened to high speed motorboats and resort development along the shore. Instead, it found itself at odds with newly organized groups who wanted the commission to abandon the lake project altogether. Two things had happened between 1969 and 1975 that changed the course of Brushy Creek's development. First, local users had established recreational land-use patterns that differed from the 1960s development concept driving the project. Second, environmentalists had organized for political action.

By the time the master plan was completed in 1977, public opinion on the Brushy Creek project was on the way to becoming polarized. The Committee for the Preservation of Brushy Creek Valley and the Iowa Public Interest Research Group appeared before the commission in October 1977 and asked for reconsideration of the project. Specifically, both groups requested that the commission prepare a full environmental impact statement, in accordance with the 1969 National Environmental Policy Act, before undertaking any development. More important, and this was their real intent, they asked that the proposed project "be modified to a multi-purpose low development recreation area and that no dam be built at Brushy Creek." Commissioner John Link of Burlington was noticeably peeved at this turn of events, querying the groups as to how many people they really represented and suggesting that proponents of the lake be invited to the next meeting for an equal-time hearing.[60]

Much of the November 1977 commission meeting was given over to a public hearing on developing Brushy Creek. An ad hoc organization known as the Concerned Iowans for Brushy Creek marshaled a cross-section of supporters who included State Senator Joseph Coleman of Clare; the editor of the *Fort Dodge Messenger*; the mayor of Badger; and representatives from the Fort Dodge Parks and Recreation Department, the Webster County Board of Supervisors, the Fort Dodge and Eagle Grove chambers of commerce, the Central Webster Community School District, the Fort Dodge Betterment Foundation, and the Lions Club. After rebuttal from opponents—the Committee for the Preservation of Brushy Creek and the Iowa Public Interest Research Group—the commission then voted unanimously to reaffirm the master plan approved the previous year.[61] The commission, in effect, dismissed the opposition and proceeded with the project as planned, even though some of the commissioners themselves were lukewarm about a large lake.[62]

The following year, the legislature appropriated $1.7 million to begin construction of the dam and spillway, and the commission applied for a matching LWCF grant to cover half the estimated cost. Applying for federal aid accomplished what a handful of opponents could not. The Heritage Conservation and Recreation Service, successor to the Bureau of Outdoor Recreation, ordered an environmental impact statement. The size of the proposed impoundment raised environmental questions, but it was really the public controversy that caused HCRS to stall. As a result, the $1.7 million state appropriation was transferred to other projects, and the commission hired an outside consultant to prepare an EIS.[63]

In large part, the EIS was a referendum on public opinion. Through public meetings and with the aid of a citizens' committee, the consulting firm, Brice, Petrides & Associates of Waterloo, confirmed that public opinion was polarized, but the EIS also clarified the nature of that polarity. The majority of opponents were "recent users" who resided all over the state, and horseback riding was their "predominant" activity. They had little in common with the environmentalists who opposed the lake principally because it would inundate some picturesque wooded hillsides. Those who favored a large lake and extensive development were mainly community leaders, organizations representing business interests, and residents of Fort Dodge, Lehigh, and Duncombe. To figure out what the real "silent majority" wanted, the consultants surveyed residents in a defined "project service area" and found a rather wide diversity of opinion. On balance, though, the majority of potential users favored only moderate development for a variety of uses, some water-based and others land-based. Respondents also tended to favor maintaining the area's scenic qualities, namely its wooded hills.[64]

When the draft EIS was ready, the commission reviewed it at another meeting crowded with supporters and opponents. The purpose of the meeting was to select an official "proposed action" from among six options that ranged from a 980-acre lake with extensive development to no lake and limited development. Nineteen individuals made comments that revealed no discernible shift in public opinion: business interests and community leaders still favored a large-lake project, while the Pony Express Riders of Iowa, the Iowa American Saddlebred Association, and the Committee for the Preservation of Brushy Creek Valley called for no lake with either no development or moderate development. The Iowa Chapter of the Sierra Club also opposed the lake plan, but it opposed equestrian use as well. The commission's decision, however, did not turn on public opinion as much as it did on certain environmental factors brought to light by the EIS. Among its salient findings, the EIS identified several remnants of native prairie and upland forest, about a dozen rare or endangered animal species, and more than sixty prehistoric American Indian sites, six of which were deemed eligible for the National Register of Historic Places. After considering the environmental qualities of Brushy Creek, only two commissioners still supported a large-lake development plan. The rest voted to abandon the large lake in favor of a 470-acre lake with moderate development.[65]

Depending upon one's perspective, the environmental review process had either worked to produce a better land-use decision or provided the commission with a document that justified a compromise development project, something the commission was desperately seeking in 1981. By then, however,

LWCF funding for state projects had been cut substantially. Development of Brushy Creek thus was placed on hold, where it languished for the next four years. Michael Carrier, who had the unenviable task of resolving the issues of Brushy Creek when he joined the agency as head of the Lands and Waters Division in 1985, succinctly sums up the situation he inherited this way: "They completed an environmental impact statement, they completed a master plan, they spent a lot of money, but they basically wallowed in indecision. Their hearts weren't in it because of the element of public controversy." [66] While the environmental review mandates of the 1969 National Environmental Policy Act provided a much needed mechanism for involving the public in environmental affairs, that same mechanism also served to produce political gridlock, which is precisely what Brushy Creek symbolized in 1986.

Dam the Des Moines, But Save Ledges

After 1936, the year Congress assigned to the U.S. Army Corps of Engineers responsibility for eliminating flood hazards nationwide, the Corps became a major player in state environmental affairs. The role of the Corps in reshaping Iowa's landscape deserves its own study, but one project in particular, the Saylorville Dam, is pertinent to the reshaping of Iowa's park system during the 1960s and 1970s. In general, federal flood control projects skewed state park development even more toward outdoor recreation. Such projects were alluring because they created big reservoirs of water that registered as "lakes" in most people's minds. Even better, their creation came mostly at federal expense. From the beginning, there was a clear difference in the way the U.S. Army Corps of Engineers conceived of flood control projects and the way they were perceived by local communities facing the prospect of having one of these projects constructed nearby. Corps engineers talked in terms of conservation pools, flood pool elevations, and release rates. Local boosters talked in terms of multiple-purpose reservoirs. Large-dam flood control was promoted as a win-win proposition. Business, industry, and property owners downstream would be protected from the ravages of floodwaters, while boaters and fishers would get thousands of acres of "lake" waters.

Many conservationists, of course, saw flood control quite differently. During the latter 1930s and the 1940s, the National Resources Planning Board, the Department of Agriculture, and the Department of Interior, among others, advocated flood control as part of multipurpose drainage basin planning, a policy that recognized the value of controlling floodwaters at their source.[67] But effective flood control at the source depended upon too many variables, including massive voluntary participation by private landowners, and the idea never found enough adherents in Congress to mount an effective challenge to the Corps of Engineers, which focused its engineering prowess entirely on large downstream dams. Controlling floodwaters at their source also held no potential to create large lakes.

Four large flood control dams were constructed in Iowa between the late 1950s and 1977: Coralville on the Iowa River, Rathbun on the Chariton River, and Red Rock and Saylorville on the Des Moines River. Collectively, they increased the area available for water-based recreation by 30,000 surface acres. Additionally, two new state parks were added to the system: Honey Creek, overlooking the Rathbun Reservoir, and Big Creek, overlooking the 900-acre Big Creek Reservoir, a companion reservoir to Saylorville. In addition, Lake Macbride State Park was expanded, and a 15,000-acre game management area was established around Red Rock Reservoir.

The story of federal flood control is no different in Iowa than it was anywhere else. The State Conservation Commission was nearly powerless against the Corps of Engineers and local proponents who focused on the short-term benefits. In part this was because, as of 1949, the Natural Resources Council became the official state agency to deal with flood control and the Corps. At best, the commission could negotiate tradeoffs that enhanced outdoor recreation. Only when proposed flood control projects affected existing state parks, which happened twice, did the State Conservation Commission have any noticeable influence on the outcome.

The commission first tangled with the Corps in the late 1940s over the Coralville project, which threatened to inundate Lake Macbride. However, since Lake Macbride was an artificial lake, the commission came to a compromise solution more readily. In this instance, the commission was more interested in protecting the investment the state had made in the 1930s to develop the area for outdoor recreation. Without much fuss, the Corps agreed to acquire more land to expand Lake Macbride from 140 acres to 950 acres of water and to build a barrier dam between Coralville Reservoir and the lake. The work was carried out during the late 1950s while Coralville Dam was under construction, and the federal government paid for most of it. Except for a minor disagreement concerning where to place the sewage disposal system and the inconvenience of park reconstruction, everyone was satisfied.[68]

The Saylorville project elicited a more contentious response because it threatened to overflow Pea's Creek in the very heart of Ledges State Park. Saylorville Dam is important because it represents both sides of the federal flood control controversy and because it reflects the continuing duality of conservationism as it merged with postwar environmentalism. On the one hand, the Saylorville project provided a long-sought opportunity to create a recreation area close to the state's largest city, Des Moines. On the other hand,

Lake Macbride, 1956, before it was enlarged from 150 acres to 950 acres when the Coralville Dam was constructed. *Courtesy Iowa Department of Natural Resources.*

it placed at risk one of the most highly prized scenic areas in all of Iowa's state parks, the small sandstone-ledged canyon cut by Pea's Creek as it flows into the Des Moines River. Ledges State Park was, and still is, one of the most popular state parks, partly because it is easily reached from Des Moines, but also because it is among the most scenic. Once again, park defenders rallied. The second fight to save Ledges was a key battle in the environmental movement as it was played out in Iowa during the 1970s.

The Saylorville saga opened in 1939, and from start to finish, the project took about forty years. Proceeding under authority of the Flood Control Act of 1938, the Army Corps of Engineers began investigating possible dam sites along the Des Moines River. When the State Conservation Commission got wind that the Corps was proposing a dam at Madrid, it immediately opposed the plan because the dam would permanently flood part of Ledges. The commission's opposition provoked its own flood of indignant responses from local boosters. A delegation from Boone, Madrid, and Des Moines appeared before the commission proclaiming that *all* the proponents were conservationists and asserting that the question was not *whether* a dam would be constructed but where the Army engineers would decide to put it. Lloyd Waddell, representing the Des Moines Defense Industrial Committee (later known as the Des Moines River Valley Project), wanted to know what business the commission had making a statement in opposition to the dam anyway, since it had not been asked to go on record for or against the project. He claimed that thirty-four private engineers had been consulted, none of whom raised any objections (as if engineering questions were the only ones that mattered). After the delegation stormed out, Louise Parker, a persistent advocate herself for a state park near Des Moines, called for the question. The commission unanimously reaffirmed its opposition.[69]

World War II interrupted the flood control program, but after heavy rains in 1947 caused another round of flooding in Iowa, the Corps geared up to proceed with the Des Moines River projects. By this time, the Corps had settled on two Des Moines River sites: Saylorville and Red Rock. Support for the dam projects kept growing, but the commission remained steadfast in its opposition. Essentially, the commission allied itself with those who challenged the wisdom of holding floodwaters behind large downstream dams without a comprehensive program of soil conservation and upstream watershed protection to check floodwaters at their source. At a public hearing in October 1948, the commission issued a formal statement declaring that "downstream flood control on the Upper Mississippi watershed concerning itself with high dams and levees proposes to treat the symptoms and ignore the cause. . . . No flood control program is justified that concentrates its efforts on measures that are both limited in general benefits and doomed to short life in effectiveness."[70] By now, though, the state legislature was debating the merits of what eventually became the Natural Resources Council, so the commission shied away from taking a direct stand against the Saylorville project. In the end, it hardly mattered because the next year the legislature muzzled the commission by placing flood control matters in the hands of the Natural Resources Council.

Thus removed from the front line of defense against large-dam flood control projects, the commission eventually dropped its *complete* opposition to the Saylorville dam and adopted a new stance: compensation for damages. In retrospect, this was probably the strongest position the commission could take at the time. After a spate of damaging spring floods in 1951, 1952, and 1954, there was little hope of stopping the project. The Natural Resources Council came out in favor of it, calling the "impairment" to Ledges "unavoidable." Seeking to minimize the consequences to the park, the NRC issued an optimistic understatement to the effect that depending on how the rains fell, Pea's Creek Canyon might not suffer *any* impairment for years.[71]

The canyon had always been subject to routine flooding for short periods of time, but the 890-foot flood pool elevation contemplated by the Corps meant that during extreme flooding, water would remain in the park long enough to kill trees and deposit a heavy load of silt on the canyon floor. It was a matter of time, and the prognostications on that score ranged as high as 150 years. Realizing that the lower reaches of the park eventually would suffer damage of unknown proportions, the commission sought to make its new position as clear as possible. Long before the National Environmental Policy Act was passed, mandating the environmental review process, and years before a new wave of environmental activism began to swell, the commission insisted "that the state be fully compensated for any remedial work at Ledges State Park, and that full consideration be given to the aesthetic loss in the flooding of Pease Creek canyon." (The name is spelled both ways.) It further demanded that the Corps enter into an agreement with the commission before any remedial work began.[72]

Judging from those who took the time to write letters or to speak at public hearings held by the Corps and by the Natural Resources Council, proponents of the dam were at least three times as vocal as opponents. At the very least, they were, initially, much better organized. Municipal agencies, business interests, and labor unions overwhelmingly supported the dam for the obvious reasons: communities might lose industries if floods were not controlled, and dam construction would provide lots of jobs. Organized opposition came principally from the Izaak Walton League in the 1950s, although a few county conservation groups and the Boone County Board of Supervisors also went on record against the dam. Both the Iowa Ikes and the national IWLA opposed Saylorville because excessive flooding of Ledges was inevitable, eventually. The Boone County supervisors likewise pointed to the submergence of "our beloved Ledges State Park." The State Soil Conservation Committee at first declined to take a position, then joined the opposition, urging the Corps to allow upland water control and proper soil conservation practices to be given a fair trial before constructing large and expensive dams. Elmer Peterson, who had just made a name for himself with *Big Dam Foolishness*, also made an appearance in order to trade barbs with the Army engineers.[73]

The opposition never had much of a chance. They were heard, and that was all. Congress authorized the Saylorville project as part of the Flood Control Act of 1958. Still, nothing much happened until funds were appropriated. Construction finally began in 1965, but the fate of Ledges did not resurface as an issue until several years later. In the meantime, the Governor's Commission on Conservation of Outdoor Resources issued its 1964 report recommending, among other things, construction of Hawkeye Naturama somewhere near a body of water (either natural or impounded), near a high-density population, and near an interstate highway. No location was suggested, but it was clear that Des Moines was a favored venue.

Hawkeye Naturama was not a concept that came from the State Conservation Commission, but the commission inherited it, at least in name. In 1967, state politicians, not the commission, decided that Hawkeye Naturama would be constructed near Polk City, north of Des Moines. The decision was designed, in part, to mollify Polk City residents who were upset because the Saylorville reservoir would isolate their community.[74] But it also was a convenient excuse to give Des Moines something it had wanted for decades: a lake park near the capital city. Thus, in 1967 the legislature appropriated $1 million to begin acquisition of land near the proposed Big Creek Reservoir, which was a Corps project. It was an appropriation initially unsought by the commission, and the single most controversial item in the $8 million outdoor recreation appropriation bill passed that year. With an earmarked appropriation, though, the commission had no choice but to follow through. Land acquisition was aided with a federal Land and Water Conservation Fund grant of nearly $500,000 in 1969.[75]

Somewhere between land acquisition and development, the "naturama" concept quietly devolved into another state park. After Big Creek reservoir was filled in 1972, the adjacent land was designated as Big Creek State Park and Recreation Area, not Hawkeye Naturama, and the commission decided that it would be developed much like other lake parks. A somewhat confused delegation from the Governor's Commission on Outdoor Resources appeared before the commission in 1974 inquiring where the 1967 appropriation for Hawkeye Naturama had gone. When they realized that the money had been spent to acquire land for what was now a designated state park, though still undeveloped, they tried to persuade the commission to adopt the naturama concept in the development plan. The delegation was politely thanked, but that was the effective end of Hawkeye Naturama.[76]

Hawkeye Naturama would be only an amusing footnote except that it demonstrated where the State Conservation Commission was positioned with respect to environmentalism in the 1960s. At one extreme were those who wanted nature served up Disneyland style, or so it seemed at the time.[77] At the other extreme were those who wanted to carve preserves out of the state park system in order to protect them from nature lovers, and, implicitly, from the intensive recreational uses that the State Conservation Commission sanctioned. It was hard to tell where the Iowa Ikes stood anymore, since they supported both proposals, and they were curiously absent from the furor over

Ledges in the 1970s.[78] The commission was caught somewhere in between. Individual commissioners may have been more interested in land acquisition and development for outdoor recreation than in maintaining existing parks or protecting unique scenic resources. But it is also true that the commission's function had been reduced too many times to negotiating the best "deal" for natural resources in the face of political pressure from legislators bent on appeasing special, or local, interests. Moreover, as of 1949, the commission had been effectively excluded from serious debate over water policy, especially when it came to federal water projects.

While land acquisition and construction of Big Creek reservoir were taking place, the commission also began to consider a redevelopment plan for Ledges so that it could negotiate a mitigation agreement with the Corps. Discussion concerning the redevelopment plan, adopted in 1972, revealed considerable disagreement among commissioners over whether the park should be expanded on the periphery to offset eventual losses in the canyon area and the degree to which more recreational use of the park should be facilitated.[79] More important, though, the timing coincided with the rising tide of environmental activism. In preparing the redevelopment plan, Gerry Schnepf, head of the commission's planning staff, sought assistance from ISU's landscape architecture department. The resulting ISU report is notable chiefly because it exposed benign neglect and uncontrolled visitor use as additional and perhaps more serious threats to Pea's Canyon. Researchers cited vandalism, trampled vegetation in overused picnic areas, and serious erosion on the sandstone cliffs caused by hikers.[80]

The commission's plans for redeveloping Ledges also stirred local interest. In late 1972, the Iowa Citizens' Alliance to Save Ledges State Park organized. Led by Hans Goeppinger, a young farmer from Boone, the Citizens' Alliance attracted, as the local paper reported, "housewives, professors, farmers, conservationists and students" whose age difference "varied by an easy 30 years."[81] It most surely reflected the character of liberal activism in the 1960s and 1970s: generations united by zeal in a common, and usually single-issue, cause. Organizers included a number of ISU professors.[82] Other environmental groups represented were the Sierra Club, the Iowa Confederation of Environmental Organizations, the Ames Conservation Council, and the Ames Audubon Society. Farmers prominent in the alliance, besides Goeppinger, were Elmer Dobson of Glidden, who had earlier organized the Citizens United to Save the Valley in order to challenge a Corps's plan to dam the Raccoon River in Greene County, and Lester Sturtz of Boone, who initially stirred up the controversy locally by suggesting that the Corps build a barrier dam across the mouth of Pea's Creek to keep floodwaters from flowing back into the canyon. John "Jack" Nystrom, State Senator from Boone, was the group's political voice.

The Citizens' Alliance lost no time in showing the Corps that it was prepared to fight. When the Corps casually dismissed the barrier dam as "too costly" and announced that it would award contracts for the final phase of construction on the Saylorville dam and spillway, the alliance went to court. Charging that the Corps had not prepared an environmental impact statement, as required by the National Environmental Policy Act, the alliance, backed by the Sierra Club and by Iowa Citizens for Environmental Quality, won a temporary restraining order preventing the Corps from awarding further construction contracts. A few weeks later, the plaintiffs were granted a preliminary injunction stipulating that the Corps must prepare an EIS and, in good faith, evaluate alternatives that would minimize injury to Ledges State Park.[83] Goeppinger maintained that the environmental coalition was not trying to stop construction of Saylorville Dam, and, in fact, work proceeded while the environmental study was in preparation. The alliance, however, advocated a barrier dam as the only remedial option that would truly "save" the park, a position that the State Conservation Commission refused to endorse.[84] Therein lay the rub.

By the time the draft EIS appeared in July 1973, the commission and the environmental groups were at loggerheads. Instead of opting for the barrier dam as the preferred alternative, the Corps offered to pay the commission a cash settlement up front for the loss of aesthetic values along Pea's Creek. Goeppinger immediately went on the attack and brashly accused the State Conservation Commission of downplaying the damage that would occur: "They love to quote the statistic that only once in 76 years will the river ever back up to the maximum flood pool situation." (How frequently severe flooding would occur was still anybody's guess, but 76 years was the number then in vogue.) Noting that the Corps had agreed to construct a $3 million barrier dam near Polk City in order to protect a cemetery, he implied that the commission had sold out. "If 75 acres of Ledges is not of comparable value [to the cemetery], then I guess my values are mixed up," he told reporters.[85]

The great irony, of course, was that twenty years earlier, the commission would have walked arm-in-arm with such determined opponents. Now it was being taken to task for holding to the only compromise position that seemed viable in the mid-1950s. But environmental activists were not dealing with the same commission that opposed flood control dams in the late 1940s and early 1950s. The commission of the early 1970s was more inclined to promote outdoor recreation than to preserve unique natural features. Broad policy goals adopted in 1970 focused the commission on encouraging private investment in the development of park facilities and programs and providing better access to park areas and facilities for the disabled, the underprivileged, and the elderly. The overall goal was to "assure a quality outdoor recreational experience" for as many Iowans as possible.[86]

In truth, though, the commission also was pressing for more than cash compensation in the Ledges settlement. It supported an alternative that would require the Corps to lower the operating level of Saylorville reservoir, in return for which the commission would accept some flooding in the park. To prepare for the latter eventuality, the commission would take measures to reduce the number of people using the canyon, which also would address problems caused by overuse, and plant more water-tolerant trees, shrubs, and grasses.[87]

In the months between the draft and final environmental reports, tensions ran high. By the time the Corps scheduled its required public hearing, another environmental group had entered the fray. The Iowa Student Public Interest Research Group (ISPIRG) at ISU, part of a nationwide student activist group, had organized college and high-school students to "Save the Ledges." Students packed into Veterans Memorial Auditorium on September 26, 1973, adding a new chorus to the voices of dissent. There was the anticipated litany of opposition from the Citizens' Alliance and allied environmental groups, but students also went tit-for-tat with various city officials from Des Moines and communities located below the dam. The contest squared off two camps. Des Moines and communities downstream wanted maximum reservoir capacity and water retention in order to minimize flood damage and, at the same time, maximize the outdoor recreation area. Upstream park defenders wanted the opposite.[88]

Tempers flared even higher after the Corps settled on a three-part solution, explained in its 1974 EIS. First, the Corps would increase the release rate from 8,000 cfs to 12,000 cfs in order to reduce the duration of flooding in Ledges. Because this would increase the amount of water flowing downstream, the Corps additionally proposed to acquire a nine-mile "environmental corridor" (read floodplain) below the dam and develop it for "compatible" recreational uses, such as walking and bicycle riding. To offset the loss of scenic and recreational values in Ledges, the Corps would relocate facilities to higher ground, purchase approximately 380 acres of land to expand the park, and pay for a vegetative management plan in the canyon area.[89]

Importantly, the Corps rejected building a barrier dam to protect Pea's Creek Canyon. The Citizens' Alliance was outraged. The Corps, however, had studied four barrier dam options, each of which entailed constructing sizable earthen dikes in an attempt to seal off the little canyon from floodwaters, thereby creating one or more *visual* barriers between the park and the Des Moines River. The barrier dam was a complex solution with environmental problems of its own; nonetheless, for the alliance it had become the only acceptable solution. Even before he had read the final report, Goeppinger told reporters that the Corps was "throwing some crumbs" to park users while selling them down the river. He called the vegetative management plan "a high-sounding euphemism for a hairbrain[ed] scheme." (Indeed, everyone acknowledged that it was an experimental plan.) Admitting that the nine-mile greenbelt below the dam and the planned addition to Ledges would be "wonderful" benefits for central Iowans, Goeppinger insisted that they were no substitutes for eventual loss of the canyon—"*that* is the portion of the park carried in people's memories and is the portion which gives the park its essence and unique flavor."[90] Many park users and environmentalists shared his sentiments even as they winced at his rhetorical posturing.[91]

Similarly, the Corps rejected options that would lower the floodpool below 890 feet. In a lengthy defense of its position, the Corps concluded that even though a lower floodpool would reduce the flood potential in Ledges to near-normal conditions, the benefit was not great enough "to justify the attendant loss of flood control potential" downstream.[92] In other words, the Corps would not give up any of the floodpool, which would entail costly modifications to the dam and spillway then under construction. Even the *Des Moines Register*, which might have been expected to weigh in on the side of maximum flood control benefits for the capital city, raised an editorial eyebrow over that logic.[93] All things considered, the Corps demonstrated that it was willing to bend only a little under the weight of public pressure. Shortly after the final environmental report was released, the Corps went back to court and won a motion to dismiss the lawsuit filed in 1972 on the basis that it had fulfilled its legal obligation to prepare an EIS.[94]

Having exhausted legal remedies, or so it seemed because an appeal was beyond their financial resources, environmental groups turned to politicians for help. The Citizens' Alliance, the Iowa Student Public Interest Research Group, and another group, the Save the Ledges Coalition, met with congressional representatives Neal Smith and William Scherle as well as with Governor Robert Ray. Additional public meetings were held in Ames and Boone. Smith, who was instrumental in getting Congress to appropriate funds for the Saylorville project in the first place, straddled the political line. Governor Ray promised to study both the Corps's final environmental report as well as a critique of the EIS prepared by environmentalists, which he did.[95] In turn, the governor asked the Iowa Inter-Agency Resources Council, an ad hoc group that included the directors of the Natural Resources Council and the State Conservation Commission, to study a nine-point proposal prepared for his office by the ISU Water Resources Research Institute. The governor's proposal was an attempt to find acceptable middle ground. Accordingly, it called for a give-and-take solution that involved variable rates of water release depending upon seasonal fluctuations, a lower floodpool elevation, a lower permanent conservation pool (which would reduce the "lake" available for recreational uses), an environmental corridor below the dam, better floodplain management at the local and state levels, and further study of the vegetative management plan for Ledges.[96]

The governor also rejected the barrier dam as a workable solution on the basis of engineering unknowns, cost, and aesthetics. The Citizen's Alliance and ISPIRG remained equally firm in their conviction that a barrier dam was the *only* feasible option that would "save" Ledges. In an act of desperation, the two groups decided to confront the State Conservation Commission hoping to revive the barrier dam option. By this time, the outcome was predictable. Two years of tension erupted in a bitter exchange of rhetoric at a September 1974 meeting. Goeppinger opened his remarks by accusing the commission staff of acting with "irresponsibility and disregard for public interest." John Liepa, representing ISPIRG, criticized the commission's master plan for redeveloping Ledges because it did not rate the sandstone ledges as the park's major attraction, presumably as justification for shifting park use to the bluffs and thus freeing the canyon for flooding. In a statement that he may have regretted later, Liepa asserted "that he would rather see it [the canyon] destroyed by overuse than by floodwaters from the Saylorville reservoir." Commissioner Thomas Bates of Bellevue wanted to know why the group had waited a year after the master plan was completed to raise any objections. Could it be, he wondered, "that by failing to pursue an audience before the Commission their group evidently made the decision that it had to be a political issue?" Chairman Jim Bixler of Council Bluffs cautioned that pressure tactics would not be successful with the commission. He dismissed Goeppinger and Liepa by stating that "when both groups became concerned they should have contacted the State Conservation Commission," adding, to emphasize his point, "the agency which owns the Ledges in the name of the people of Iowa."[97]

The Citizen's Alliance also failed in its bid to enlist the aid of the President's Council on Environmental Quality, the environmental watchdog in the executive branch.[98] State Senator Jack Nystrom made a last-ditch effort (unsuccessful) to enact legislation that would prevent the State Conservation Commission from entering into a mitigation agreement with the Corps unless the agreement stipulated that Ledges would not be flooded above the 870-foot elevation level. U.S. Representative (later Senator) Tom Harkin intervened on behalf of the Citizen's Alliance and asked the Corps to prepare a feasibility study of the barrier dam option.[99] Colonel Walter Johnson, head of the Rock Island District, maintained to Harkin that the Corps was neutral on the controversy and would do whatever the people of Iowa wanted. This was a reassuring thought, good for political use, but the reality was slightly different. Construction of the dam and spillway was nearing completion, and no major changes had been made in the original design. Moreover, slightly before the Inter-Agency Resource's Council released its own report to the governor, the Corps submitted a proposed modifications plan to Governor Ray in June 1975 and gave the state only two months to respond.[100]

By this time, the point of contention between the Corps and the Inter-Agency Resources Council had boiled down to whether the Corps would proceed with a release rate of 12,000 cfs, as it proposed in the modifications plan, or whether it would adopt a 16,000 cfs release rate during the snowmelt season, as proposed by the state. With time running out, the Citizen's Alliance and ISPIRG announced at a statehouse press conference that they were dropping their campaign for a barrier dam and would support the governor's proposal for a release rate of 16,000 cfs, as recommended in the Inter-Agency Resources Council's report. By conceding the barrier dam, the environmentalists made it much easier for the state to present a united front when the Corps's request for a final appropriation went to Congress.

Governor Ray praised the Citizen's Alliance and ISPIRG for their "diligent efforts to focus attention on Ledges State Park." His remarks undoubtedly were intended to soothe bruised egos, but they also contained an element of truth. "They [the Citizens' Alliance and ISPIRG] have made the public aware of environmental concerns, secured the submission of an environmental impact statement, and, through a public hearing, have provided a forum for the public's interest to be voiced."[101] Their methods might have been counterproductive at times, but grassroots activists had made a place for themselves in environmental politics.

The post–World War II environmental movement has been characterized as a time of increasing conflict between the "environmental public" and "environmental managers," a conflict that pitted newer ecological values against older conservation values, notably the multiple-use concept.[102] It is true that the government agencies that house most of the "environmental managers" typically were the targets of attack, on the state as well as federal levels. As the second battle for Ledges demonstrates, though, "the environmental public" did not necessarily exhibit a high degree of environmental awareness. The barrier dam that the Citizens' Alliance advocated was no more sensitive to the natural environment than the solutions reached through the environmental review process and settlement negotiations between the Corps and the State Conservation Commission.

If anything, the battle over Ledges demonstrates that environmentalists joined a long line of special interest groups who had been using the political process to shape environmental policies for a very long time. This was nothing new, but the NEPA-mandated environmental review process provided a mechanism for raising the process to the level of public debate, as opposed to "working the system." In this sense, it is not at all clear that the contest over Ledges demonstrated a conflict between the values of environmental managers and those of the environmental public. Ledges primarily reflected continuing conflict among values held by competing special interests. This is not to imply that the State Conservation Commission was a neutral party in the matter, but it also was not the villain that the Citizens' Alliance and ISPIRG perceived it to be.

One can argue that the environmental review process as a whole mainly served to move environmental politics into a more open forum, although the intent was to create a process of study and evaluation that would lead to more environmentally sensitive decisions concerning land use, management, and development. The Brushy Creek controversy buttresses this view. The EIS prepared by Bride, Petrides & Associates certainly addressed environmental issues, and this information ultimately influenced a compromise decision from the commission, but the controversy that provoked the EIS in the first place was defined by competing groups who wanted their own land-use preferences served. By the same token, "working the system" sometimes produced greater rewards, as passage of the 1965 State Preserves Act demonstrated.

In any case, after nearly three years of acrimony, the bickering over Ledges stopped. The Corps endorsed the governor's proposal, and congressional approval followed. In 1976, the State Conservation Commission entered into a mitigation agreement that required the federal government to purchase an additional 385 acres of land to expand the park on the northeast, to pay the commission $403,000 for relocating park facilities above the 890-foot flood-pool elevation, to pay an additional $387,000 for implementing the first phase of a vegetative management program, and to pay a negotiated sum for the completion of the vegetation plan based on the first ten years of reservoir operation. For its part, the State Conservation Commission agreed to abandon and relocate park facilities above the 890-foot elevation, implement a vegetative management plan that would gradually convert the lower park area up to the 870-foot elevation into a grassy open space, convey 128 acres of land on the west side of the Des Moines River to the federal government, and grant a flowage easement over 136 acres on the east side of the river, including Pea's Creek Canyon.[103]

The floodgates on Saylorville Dam were closed in April 1977, thirty-eight years after the Corps first announced its plans to construct flood control dams along the Des Moines River.[104] At about the same time, the State Conservation Commission began to update the park redevelopment plan approved in 1972. The intervening years had also given the commission time to rethink its approach to the planning process for park development. After the ISU planning report detailed the effects of park overuse, the commission closed the canyon road to vehicular traffic. This caused an uproar, with hundreds of people weighing in on both sides of the issue, all of which added fuel to the flood-waters controversy. The legislature intervened in 1974 to settle the matter with a political compromise, mandating that the road must be open to vehicles on a limited basis, particularly during the autumn leaf season. Thus, in an effort to avoid further controversy that would surely end in yet another political settlement, the commission began what would be a four-year planning process that was based, in part, on special environmental and interpretive studies and, most important, that solicited public comment through open meetings.[105]

The revised master plan, adopted in 1981, even had a theme: "Rediscovering Ledges State Park." It opened with a quotation from The Wooing of Earth by Rene Dubos: "Humanized environments give us confidence because nature has been reduced to the human scale, but the wildness in whatever form almost compels us to measure ourselves against the cosmos." Even if the soaring rhetoric was borrowed, it still hinted that a new ethic, or, more precisely, a refreshed old ethic, had crept into the commission and its staff. "Whether it be a discovery or a rediscovery," the master plan continued, "it is important to experience nature, to be awed by its wonders, and to feel a reverence for the natural laws which link human kind to the rest of creation."[106] These were words that Thomas Macbride would have understood. "Rediscovering Ledges State Park" echoed the past and promised better stewardship in the future. In addition to implementing the mitigation agreement, the commission continued to limit traffic in the canyon, banned rappelling down the canyon walls, eliminated the caged-animal exhibit known as the "Ledges Zoo," and initiated an interpretive program designed to increase public awareness of the park's natural and cultural features.

Transformations and Transitions

It is tempting to see the second battle for Ledges as a bookend to the first battle for Ledges in the 1920s, but the compromise decision on Brushy Creek, which came at the same time "Rediscovering Ledges" was adopted, gave evidence that the commission was still on uneven footing. Nonetheless, the character of the commission did change, once again, during the 1970s. The official record gives little indication of when the shift began or what triggered it, the change was so subtle, yet by 1980 there was a new attitude. Gerry Schnepf, who was on the commission's planning staff during the 1970s, generally attributes the change to Governor Ray's appointments during his fourteen-year tenure.[107]

One of those appointees, Marian Pike of Whiting, is more reflective in her assessment. When she joined the commission in 1975, she found a group dominated by "good-old-boy politics." "Most everybody on the commission in those days had an ax to grind or [particular interests to serve] in one section of the state or another. . . . I remember the first appointment I had with Ray. He said to me, just as I was leaving, 'I suppose you have a lake you want to dredge out there in western Iowa.' I laughed and said, 'Yeah, Blue Lake.'"[108] When she left the commission twelve years later, she felt there was less *quid pro quo* politicking among the commissioners. She attributes the change to two factors. First, Governor Ray reestablished the practice of appointing women to the commission. Among his first appointees in 1969 was Joan Geisler of Dubuque, who resigned before completing her term, citing health reasons. In 1973, he appointed Carolyn Lumbard (later Wolter) of Des Moines, and two years later Marian Pike joined the group.[109] Second, the environmental critique of the 1970s began to make a difference.

Marian Pike allowed her name to be submitted as a possible appointee when the chair of the Monona County Republican Central Committee learned that Governor Ray wanted to appoint more women to boards and commissions. She is fond of telling people that her husband, Herb, urged her to do it because he thought it was the *conservation* commission and she would be good at that. In fact, though, she and her husband had long been active in community affairs and farm organizations. "This sounds awfully conceited," she says, "but I lay some of the shift on the two women who were on the commission [herself and Carolyn Wolter]. We both had good sense, and we didn't have any irons in the fire or anybody that we were beholden to. We did what we thought was right. We didn't always do it unified, and sometimes Carolyn and I would be on opposite sides." But, she adds, "it was not just the women. There were some men of the same philosophy, and there were some on the commission who were environmentalists from the very beginning."[110]

In retrospect, Pike also acknowledges that the environmentalism of the 1970s influenced change, although it is difficult to find any direct cause-and-effect relationship. The confrontational tactics of groups such as the Iowa

Iowa Conservation Commissioners, 1979. *Left to right*: John D. Field, Hamburg; Richard W. Kemler, Marshalltown; Marian Pike (vice-chair), Whiting; John C. Brophy (chair), Lansing; Thomas A. Bates, Bellevue; Carolyn T. Lumbard, Des Moines; Donald E. Knudson, Eagle Grove. *Courtesy Iowa Department of Natural Resources.*

Student Public Interest Research Group provoked disgust, even sarcasm and occasional anger, and none of the commissioners had much patience for the "one issue and the hell with everything else" approach of single-issue activists. But, on a different level, the broad public interest in environmental affairs that arose during the 1970s did have an effect on the commission's overall point of view. "I think we began getting more and more environmentally oriented. Some [of the commissioners who were set in their opinions] backed off and saw the whole picture. We didn't really have many [commissioners] who were just absolutely rigid. We didn't have that. They might be rigid on one thing, and we [the majority] would either go around them or educate them." [111]

Perhaps the best indication that the State Conservation Commission was attempting to reestablish its claim as the state's lead resources agency occurred when it hired a new director in 1981. From 1968 to 1979, Fred Priewert had served as director of the State Conservation Commission; and although those who worked with him uniformly characterize him as an able administrator, his views were increasingly at odds with some of the commissioners. His at-

Fred Priewert, ICC director, 1968–1979.
Courtesy Iowa Department of Natural Resources.

tention was focused chiefly on fish and game matters, and, as Marion Pike reflects on his relationship with the commission, he was a man "who just didn't understand that women had a place in the world." Others recall a sign that hung in his office as a distinct clue to his personality: "Lead, Follow, or Get Out of the Way." By the late 1970s, the commission was increasingly under attack from a variety of sources. At the same time, commissioners were consciously attempting to create a more open atmosphere in its monthly meetings by dedicating time for public comment, and they were searching for middle ground on a range of issues. An unbending administrator at the top made it much more difficult to establish trust and understanding. Priewert thus was eased out of his position.[112]

Two different directors rotated through the office in as many years. Then, in an unprecedented step, the commission undertook a nationwide search for a new director. Realizing that there was no one within the agency who was capable of administering it in a way that was acceptable to all the commissioners, they hired a search firm to find candidates from outside Iowa. Each of the commissioners had his or her own ideas about the ideal qualities of a good administrator, but above all, they were looking for someone with experience, energy, and a fresh face. Former commissioner John Field recalls that "most of the candidates were at the peak of their careers looking for a nice soft berth out here to coast into oblivion." The one candidate who didn't fit this mold was Larry Wilson. "He was fighting for the top. He didn't have all the experience we wanted, but we thought we'd rather have somebody going up than somebody coming down. . . . He was young, he was fresh, he was eager." [113] It was a good fit. Having spent seventeen years as a fisheries biologist and then regional supervisor with the Utah Fish and Game Department, Wilson was looking for an opportunity to prove himself in a top administrative position. [114] After the formalities of hiring were completed, he took over as director in January 1981.

Within a few months, the commission had a mission statement. Whereas the 1970 "broad goals and objectives" had stressed "quality outdoor recreational experience," equal access to programs and services, and private investment, the 1981 mission statement harkened back to the founding purposes the state park system. It stressed the importance of assuring the "availability" of a "resource heritage" of "lasting integrity," of "preserving critical and

Larry Wilson, ICC director, 1980–1986;
DNR director, 1986–present. *Courtesy
Iowa Department of Natural Resources.*

unique resource features and protecting or enhancing natural resource systems." Signaling a definite break with past practices, the 1981 statement asserted that "public need does not necessarily equate to public demand. This differentiation may seem minor, but [it] is a critical factor in determining proper and balanced use, management, and long-term protection of the resource heritage."[115]

In response to sizable reductions in federal grants through the Land and Water Conservation Fund and belt-tightening at the state level, Wilson redirected efforts away from expansion and refocused the agency on multiple-use management with maximum regard for the resource base. Under his direction, the commission staff developed a protected waters program and assumed full responsibility for a natural areas inventory and database program initiated by the Nature Conservancy.[116] In addition, staff grappled with the ill-fated park user fee initiative. A new gubernatorial administration also brought changes. Shortly after Terry Branstad was elected in 1982, the governor's office requested a study of Iowa's recreation and tourism needs. With an initial legislative appropriation of $85,000 in 1984, the Recreation, Tourism, and Leisure Committee was launched. During the next few years, this committee—which included legislators from both houses and representatives from the governor's office, the Iowa Development Commission, the Iowa Natural Heritage Foundation, and the State Conservation Commission, plus an assortment of representatives from various special interests—worked to devise a plan that would coordinate the various state entities providing recreation, tourism, and cultural programs.[117] The RTL committee did not report until 1987, after the end date of this study, but its findings and recommendations led to some important initiatives, notably the 1989 Resource Enhancement and Protection Act (REAP) and new avenues of cooperation between state agencies, which are touched on briefly in the epilogue.

The Branstad administration also succeeded in effecting a complete overhaul of state administration. For the State Conservation Commission, reorganization took effect in June 1986, combining the agency with the state's other major environmental entity, the Department of Water, Air, and Waste Management, into a new Department of Natural Resources.[118] Abolishing the State Conservation Commission had been in the wind off and on since 1950, when the Governmental Reorganization Commission, better known as the Little Hoover Commission, recommended that the commission, the Soil Conservation Committee, the Geological Survey, and the Natural Resources Council (then only a year old) should be consolidated into a proposed department of conservation and natural resources.[119] No changes were made at that time, but the idea resurfaced in the latter part of the 1960s when legislation was introduced to consolidate several state agencies into a proposed department of natural resource management.[120]

Both times reorganization was opposed on the basis that it would shift the reins of power from a noncompensated citizen board to a governor-appointed director, and therefore open the commission to political interference. Part of Jay Darling's legacy is a persistent belief, held to this day, that the State Conservation Commission, now the "old" commission, was free of politics. This was true only in the most narrow sense of the term. By law, only four of the seven members of the commission could be from the political party in power. The commission, therefore, was *bipartisan* in its composition, and commissioners' terms were staggered, so no governor could manipulate it to serve partisan interests very effectively. But the commission always was at the mercy of the legislature for money to operate state parks, and purse strings *were* used to control park development, as the Volga River and Big Creek projects amply demonstrated—only two instances among many that could be cited over the decades. Local communities and groups routinely were advised to petition their legislators if they wanted specific projects funded; and if they were successful, such projects were adopted by legislative mandate. This was (and to a degree still is) the way the system operated, and everyone knew it. As long as no underserved interests squawked too loudly, the system functioned smoothly. But by 1970, the political climate was changing; newly organized or reinvigorated environmental groups were demanding to be heard. It was no longer business as usual.

If the commission opposed reorganization, it also was the first to acknowledge that administrative fragmentation often worked against resource protection and, in particular, placed the state at a disadvantage when dealing with powerful federal agencies such as the Corps of Engineers. Thus, in 1972, the State Conservation Commission took the lead in establishing an Inter-Agency Resource Council composed of representatives from the commission, the Natural Resources Council, the Department of Agriculture, the Soil Conservation Department, the Iowa Geological Survey, and the State Archaeologist's Office.[121] The council had no legislative authority, but it did provide a forum for debating matters of common interest. In one instance, at least—negotiations with the Corps of Engineers over the operating agreement for Saylorville Dam—the governor used the council to help forge a united front among state agencies.

The ensuing decade brought even more administrative fragmentation. In 1972, the legislature established the Department of Environmental Quality at a time when many states were creating such agencies in response to the

National Environmental Policy Act. Four new commissions operated within Iowa's DEQ: the Air Quality Commission, the Water Quality Commission, the Chemical Technology Commission, and the Solid Waste Disposal Commission, each exercising policy over its respective area of concern.[122] Following yet another reorganization study completed by the Governor's Economy Committee in 1979, during Robert Ray's tenure, the legislature took steps to reduce some of the overlapping jurisdiction by creating a new Department and Commission of Water, Air and Waste Management. This act, passed in 1982, collapsed the four DEQ commissions into one and abolished the Natural Resources Council. In addition, it absorbed the powers and duties of the State Department of Health as they related to private water systems, water wells, and private sewage disposal systems.[123] The 1982 reshuffling left the State Conservation Commission intact, but at a time when the new director was reasserting the agency's prominence in protecting natural resources. When Governor Terry Branstad assumed office in 1983, he made government reorganization a top priority. Practically every agency was affected in one way or another as he sought to streamline the state's administrative apparatus.

The State Conservation Commission "quietly fought reorganization tooth and toenail" for the same reasons reorganization had been opposed before: "We thought it put too much power in any one governor's hands."[124] That "quiet fight" ended in a compromise when it was clear that the legislature was finally going to follow the governor. The 1986 reorganization merged the State Conservation Commission and the Water, Air, and Waste Management Commission into the Department of Natural Resources. Both existing commissions, however, were continued, as was the Preserves Advisory Board. The Natural Resources Commission, successor to the State Conservation Commission, and the Environmental Protection Commission, successor to the Water, Air, and Waste Management Commission, now set policy under one umbrella agency served by one director.

Although it was a compromise of their own making, the sitting commissioners still found it hard to accept. Those who believe the old commission form of administration is the best antidote to political interference have remained critical of the 1986 reorganization because it removed the departmental director from under commission control and established the position as a gubernatorial appointment, subject to state senate confirmation.[125] It is a point of no small significance that Larry Wilson, director of the State Conservation Commission, underwent a bruising, highly partisan confirmation process before the state senate finally approved his appointment as director of the superseding Department of Natural Resources.[126] Senate confirmation ultimately rested on his demonstrated leadership and administrative abilities, and no doubt the legacy of the State Conservation Commission as well, but it also tended to validate fears of political control.

If the supposed political insularity of the commission and its director vanished in 1986, the present-day Department of Natural Resources nonetheless echoes, perhaps faintly, the wisdom of earlier generations who understood the complex linkages of resource protection and management. In this sense, resource conservation came full circle in 1986. During the 1910s and 1920s, conservationists in Iowa had looked to parks as a means for preserving unique, unusual, and interesting natural and cultural features, and as an instrument for fostering the conservation of natural resources in general. State parks were to be the building blocks of resource conservation. Today, state parks and preserves function as part of a more inclusive resources agency. Six divisions— Parks, Recreation, and Preserves; Forests and Forestry; Fish and Wildlife; Energy and Geological Resources; Waste Management Assistance; and Environmental Protection—administer a sweeping range of environmental programs and regulations that affect all Iowans. That sweep, however, still does not include soil conservation and other agricultural programs that affect environmental quality, which are still handled through the Department of Agriculture and Land Stewardship.

Some would also point out that the inclusiveness is more apparent than real. The Natural Resource Commission is more or less the old State Conservation Commission renamed, and it functions with essentially the same staff and field personnel. The same is true of the Environmental Protection Commission. The agency is perceived both inside and out as two distinct entities: the "resource conservation" side and the "environmental" side. The *real* difference is that both commissions now advise one director, and the director has considerably more power than was true before 1986. Also, the director's power flows from the governor's office. This means that strong leadership and coordination *must* come from the director. In the past, leadership could, and often did, emerge from among the many people who served on the commission over the decades, which is why these personalities are so integral to any history of Iowa's resources. Within limits, policy shifted as the personalities changed. Today, the director's office holds greater control over policy directions.

Criticisms duly noted, the 1986 reorganization nonetheless brought into being an old Progressive Era concept of comprehensive and coordinated resource management. It is hard to imagine Thomas Macbride or Louis Pammel or Jay N. Darling opposing the move to meld the State Conservation Commission into a much broader Department of Natural Resources, although Darling would surely protest the political structure. Nonetheless, here, in principle at least, was the "Iowa Commission for the Conservation of Natural Resources" that Macbride had envisioned in 1909, an agency that could defend the "environmental rights" of citizens, on the one hand, and, on the other, teach "the highest sort of earth-culture." [127] Politics had finally caught up to foresight.

During the 1920s, the Iowa state park system provided a point of entry into resource conservation and management. Over the decades that role changed more than once. Federal work-relief programs of the 1930s helped to promote a vision of parks, both state and national, as leisure outdoor recreation areas. Enticed by the availability of federal funds and the need to relieve widespread economic distress, most states eagerly developed their parks in accordance with the National Park Service's aesthetic guidelines. Indeed, landscape architects in Iowa contributed to the park rustic-design movement that flowered in the 1930s. Thus, it is clear that Iowa's park system was moving in this direction already, although the degree of facilities development would have been far, far less without federal assistance. Because of the 1933 *Twenty-five Year Conservation Plan*, park development and management might have been integrated more completely with forestry and wildlife management had New Deal programs not existed. Nonetheless, the plan provided an expansive multiple-use rationale quite in keeping with various federal initiatives to restore the national domain and manage natural as well as cultural resources in a more scientific, or at least professional, manner. The statewide plan also laid the groundwork for creating a comprehensive resources agency.

Through the 1940s the State Conservation Commission maintained its commitment to the goals articulated in the statewide plan. It strengthened the resource protection side of its mission, augmenting the park system with historic sites and prairie preserves, searching to balance the range of resources under its jurisdiction. However, new forces—notably, federal flood control projects, increasing public demand for outdoor recreation, and political interference—sidetracked this effort and eventually swamped it. Federal aid programs, in addition, favored certain types of land acquisition and development. While programs such as Pittman-Robertson and the Land and Water Conservation Fund were of enormous benefit, federal aid also made it easier for states to ignore their financial responsibilities. At the same time, federal aid diverted attention from the kind of comprehensive resource management approach envisioned in the 1930s. By 1958, the twenty-fifth anniversary of the *Twenty-five Year Conservation Plan*, the state had drifted far off the course charted in 1933, and the State Conservation Commission responded mostly to its fish and game constituents. The environmentalism of the 1970s brought a certain cacophony as new interest groups became politically active, but it also called into question just how well the state was managing public resources, and for whom. By then, state parks were only a small part of the public lands under state management, but their prominence within state landholdings made them focal points of debate.

On many levels, parks have been "the commons" that Thomas Macbride called for in 1895. They can be seen simply, as he suggested, "suitably warded" public lands "devoted to public enjoyment." [128] Vague terms such as "suitably warded" and "public enjoyment" are, of course, subject to broad interpretation. Through the decades, "the public" has claimed rights to ever more intensive uses in order to enjoy the great common outdoors. Thus, parks and preserves have had to function both as recreation areas and as resource banks. Garrett Hardin, in another context, exposed this as the "tragedy of the commons," arguing that unregulated access to common resources ultimately destroys the resource base as well as the life it sustains. [129] The very dualism of parks, as a concept, is reflected in the ways they have been managed. That arcane Progressive Era idea of "wise use," which meant balancing resource use with resource protection in order to achieve orderly development, gradually evolved into "multiple use," a concept that opened the way for maximizing the recreational potential of public lands, sometimes with little regard for resource carrying capacity. An enduring vision of parks as special places, however, has made them the common focus of debate between those who would enjoy them to ruin and those who want to cradle them into the future.

Ultimately, then, park management philosophies reflect the duality of human nature. Therein lies, one hopes, a path to common ground and, ultimately, toward a sustainable society. Undeniably, we have the knowledge and the skills to balance resource use with resource protection in our parks—indeed, throughout our entire resource base. Whether we have the collective will to adopt an ethic of sustainability is still very much an open question. The answer, however, is not a foregone conclusion. As Kenneth Boulding has observed, community is a learned concept. The process of unlearning concepts that are pointing us toward failure and building a new concept of community based on the goal of mutual sustainability may be "a process of appalling complexity," as he also points out, but it is not a process that is beyond human creativity. [130] In the face of massive environmental problems that challenge us, parks are but specks on the landscape. Nonetheless, they are the very places of quiet beauty we need in order to find answers. If we can succeed in caring for our parks, perhaps we will succeed in caring for the complex environment that sustains, and inspires, us.

Epilogue

What did the 1986 reorganization mean for state parks and preserves? At the administrative level, it meant that the visibility of parks was further reduced within the overall organization. Specifically, the Lands and Waters Division was split into two divisions, which separated forestry from parks and preserves. Michael Carrier moved from his position as chief of the Lands and Waters Division to take the helm of the Parks, Recreation, and Preserves Division. While the administrative title clarifies management lines of authority on the formal organizational chart, in operation there is still considerable management flexibility in the field. Fish and wildlife biologists, foresters, and park rangers often are involved in managing the same resource complexes cooperatively on a day-to-day basis. Moreover, administrative titles and professional specialization mask the importance of personalities in shaping and implementing policy. While it is too soon to evaluate the long-term effects of reorganization on the management of state lands in general, one can discern the direction in which the current administration has been taking the park system.

When Carrier joined the agency in 1985, he inherited a park system widely perceived as mediocre and stagnated by controversy. Moreover, the situation could not be remedied by resorting to an old pattern of increasing the number of parks. Carrier, in fact, felt that "there were too many areas classified as state parks and recreation areas, more than the state could manage."[1] Additionally, the division carried five unfinished projects as nearly dead weight. One of his goals, therefore, was to resolve the development issues that had brought these projects to a standstill. Although it took nearly ten years to accomplish, this goal has been realized. Pleasant Creek State Park has been developed to meet outdoor recreation demands in the urbanized Cedar Rapids area. Plans for developing Badger Creek Recreation Area (Madison County) have been dropped. Master development plans for the Mines of Spain and the Volga River State Recreation Area have been completed. Perhaps most important, the Brushy Creek impasse was resolved and development of this area began in 1993.

As a group, these five projects reflect a decided trend toward greater management efficiency, with the state developing and maintaining larger parks and county conservation boards or other local entities managing smaller areas that serve mostly local users. Currently, the number of parks, preserves, and recreation areas managed by Carrier's division is closer to seventy-five. More than thirty areas owned by the state are actually managed by nonstate entities (see appendix).

Management efficiency implies a detached, cost-benefit accounting approach. To be sure, budget constraints are as real now as they have ever been, and there is no doubt that funds often are channeled into projects that will affect the greatest number of users. However, it is also true that, for the first time in decades, a clear management philosophy has taken hold. Again, Brushy Creek is instructive. When the project was revived in 1986, the parks division set out to produce a new development concept that reflected a greater sensitivity to the inherent environmental features of the area. The new plan, approved by the Natural Resource Commission in 1988, called for a smaller, 690-acre lake sited in the northern area in order to avoid disturbing more highly valued natural and cultural resources located in the southern area, where a 260-acre tract was designated as a state preserve to protect a variety of plant communities and a burial mound group. At the same time, the legislature approved the acquisition of an additional 1,750 acres for equestrian trails, offsetting the loss of trails in the lake development area. At nearly 6,500 acres, Brushy Creek is now the largest park in the system.

The lessons of Brushy Creek are several. First, it taught Carrier not to "measure progress year to year," but "to stand away from the controversy . . . and look at it for the long term." Interest groups will continue to demand a voice in the policymaking process, yet it is difficult, and sometimes impossible, to build consensus in environmental matters. Brushy Creek became a testing ground to see if the agency could break political gridlock. "One of the things I felt strongly about," he continues, "was that the agency had lost a tremendous amount of respect and credibility because it had allowed itself to get mired in the Brushy Creek issue, mired in indecision."[2] The 1988 development plan recognized a wide range of user interests, but it did not try to accommodate every demand without due regard for inherent resource limitations. In other words, the 1988 plan was not another attempt to reach compromise simply for the sake of appeasing competing special interests. A new environmental impact statement, completed in 1992, roused opponents and supporters for one last skirmish, but the development plan emerged intact.[3] Construction of the lake began in 1993, seven years after the agency revived the project, five years after it had a plan to offer for review.

It is important that the 1988 plan reflected a heightened respect for environmental resources and, at the same time, responded to contemporary outdoor recreation interests. "Building a 1000-acre lake, as originally planned, would have been a travesty in view of some of the potential for land-based recreation," according to Carrier. At the same time, "to build the smaller lake [470 acres], which was the 1981 compromise, would have been to simply repeat the mistakes of the past and build one more low-quality artificial lake. It was gratifying," he continues, "to find a solution that hadn't been looked at before [but] that would accomplish both the goals of not building one more hard-to-manage low-quality lake and yet protect the area that needed protection."[4]

The 1988 solution began to reverse a long-standing tendency to acquire and manage state lands for maximum use or consumptive use. In this respect, Brushy Creek suggests a return to the multiple-use concept in much the way that W J McGee and Gifford Pinchot conceived it. For many people, "multiple use" is a term that has been freighted with excess use, both figuratively and literally. Yet Carrier is among those who hold that multiple use does not mean stretching resources to, or beyond, their limits in order to provide as many human benefits as possible. His park management approach is one that attempts to integrate the strategies of sustained yield, recreational use development, and preservation based on the inherent environmental attributes of each park.[5] Thus, as in the case of Brushy Creek, some areas would concentrate outdoor recreation while others might produce forest products or produce game. Still other areas might, through preserve status, provide maximum protection for rare plants, endangered animals, unusual geological formations, significant archaeological deposits, or important historical remains. This approach, of course, is much easier to implement in new parks and recreation areas; it is more difficult to restructure management practices in established parks.

Brushy Creek's development also recalls a road not taken in 1958, when the Wildlife Management Institute recommended that the State Conservation

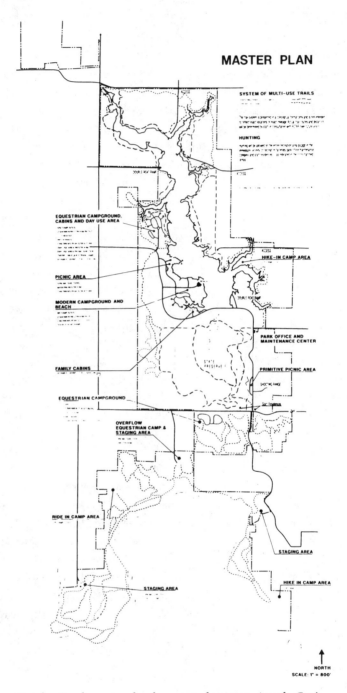

Brushy Creek master development plan, 1993, Angela Corio.
Courtesy Iowa Department of Natural Resources.

Commission concentrate on creating an "ideal" state park image by channeling expenditures into five geographically strategic areas. As a nonnative, Carrier's perception of Iowans is that "they are very utilitarian, somewhat Spartan, especially in their approach to public investment." In that respect, he notes, "Brushy Creek probably represents the spirit of Iowa because it is utilitarian. It provides development for what the recreation demand is today and expected to be in the future."[6] Still, the entire development project carries an estimated $15 million price tag, with lake construction representing half the cost. The other half covers associated amenities, including roads, more than forty miles of equestrian trails, three campgrounds, boat ramps, swimming beaches, and a new park headquarters complex.

Many people have been critical of Brushy Creek's preferential funding status, yet the legislature has been supportive. This, too, represents a shift of sorts because in the mid-1980s the legislature was using Brushy Creek as an example of why park spending should be curtailed. For so many years Brushy Creek expenditures had produced nothing but public controversy. A policy of maintaining "good economy" throughout the division has been critical to winning legislative support for capital improvements at Brushy Creek, and elsewhere for that matter. This policy has meant keeping operations "as lean as possible," including holding the line on personnel and administrative costs. Fiscal austerity of course has been a mandate from above. However, by convincing the legislature that his division can deliver professional park management without increasing staff and administrative costs, Carrier finds that it has been much easier to get new appropriations for capital improvements.[7]

Some of the money for capital improvements has come from the Resource Enhancement and Protection fund. The 1989 legislation creating REAP was greeted as one of the most progressive pieces of environmental legislation enacted by any state in recent years. REAP's preamble declares that it is state policy "to protect [Iowa's] natural resource heritage of air, soils, waters, and wildlife for the benefit of present and future citizens with the establishment of a resource enhancement program. The program shall be a long-term integrated effort to wisely use and protect Iowa's natural resources through the acquisition and management of public lands; the upgrading of public park and preserve facilities; environmental education, monitoring, and research; and other environmentally sound means. The resource enhancement program shall strongly encourage Iowans to develop a conservation ethic, and to make necessary changes in our activities to develop and preserve a rich and diverse natural environment."[8]

Passage of REAP committed the state legislature to a $300 million, ten-year program to implement the act's sweeping mandate. So far, annual allocations have fallen far short of the legislative commitment. Initially, appropriations from the general fund and from lottery proceeds financed REAP, although lottery proceeds have since dropped from the equation. During the first year of operation, 1990, the total appropriation was $15 million, which rose to $20 million in 1991. From that high, still only two-thirds of the anticipated $30 million per year, the annual appropriation has fallen to approximately $10 million. Of this amount, an established allocation formula directs nine percent, less than $1 million, to state land management. Because the legislature rescinded the park user fee in 1989, anticipating that REAP would provide a more stable source of funding for state parks, the department has placed the "state land management" share of REAP in the Parks, Recreation and Preserves Division budget. Even so, this amount is less than park user fees were generating before 1989. It remains to be seen whether the legislature will honor the $300 million commitment and extend REAP beyond its 2001 sunset provision.

Although the controversy over Brushy Creek has been laid to rest, seemingly irreconcilable points of view on public lands management assure that there will be future development battles and complex negotiated settlements. Recent comments reported in the press signal what is to come. As work progressed on Brushy Creek lake in summer 1995, Myron Groat of Fort Dodge, cochair of Brushy Creek for All Iowans, a prolake development group, began planning a grand party as part of the lake dedication. Attorney Wallace Taylor, chair of the Iowa Sierra Club, could see little cause for celebration. To the Sierra Club, Brushy Creek "define[s] what is wrong with the current DNR policy on public land management. They over-manage public lands and try to rearrange nature." Donna Schildberg of Menlo, a spokeswoman for equestrian groups and a member of the Brushy Creek Trails Advisory Board, stands somewhere in between. Although equestrian trails had to be sacrificed for lake development, Schildberg reluctantly came to accept the compromise because "there will be something for everyone."[9]

Opposing points of view and the embers of controversy aside, Brushy Creek may yet emerge as a place that embodies, in Iowa, an elusive "ideal." Without doubt, the development concept was a definite break with management practices that had prevailed for nearly forty years. Brushy Creek "isn't overdeveloped. It has an appropriate regard for protection of those parts that should be protected and not developed." What Carrier hopes Brushy Creek will come to symbolize is that "protection of unique natural areas and recreational development are not incompatible. . . . It is one of the few areas that really captures the dualism of park management."[10]

Fiscal austerity has meant that some areas and some programs have not received the attention and funding they deserve. Management of state preserves is among them. Carrier's greatest disappointment is that mandated budget cuts and reductions in staffing have hit the Preserves Advisory Board's programs the hardest, while, at the same time, the volume of work associated with preserves management has easily doubled. As a result, the staff struggles to keep up with the workload. It also means that the division is behind in placing all the state preserves under professional management plans, a commitment Carrier made when he joined the agency.[11]

These frustrations aside, it is also true that relations between the Preserves Advisory Board and the Department of Natural Resources have never been better. Carrier notes that the board "finally has a strong identity within the department whereas it felt like this unwanted foster child for a long time." Ten years ago, "there was real hostility in the department toward anything related to the preserve program. . . . It was derided as a bunch of preservationists. Now the board has developed much more credibility, even authority."[12] The change has been a two-way street. In part, the board's composition has changed, and in part, more staff members are involved in preserve activities. A spirit of cooperation is reflected in a new initiative to write ecosystem management plans for all park and recreation areas.[13] The working relationship between the department and the academic community is still fragmented and uneasy, but the Preserves Advisory Board once again provides a forum for rebuilding these bridges. Considering the origins of Iowa's park system and the long tradition of academic involvement in resource management issues, this change is promising.

Brushy Creek and the Preserves Advisory Board highlight the direction in which the Parks, Recreation and Preserves Division has moved during the past decade. In the process, Carrier has developed a corresponding vision that gives priority to resource management over expansion, balancing protection with use, and living within fiscal limits. In the foreseeable future, his vision includes four goals: first, to raise the level of support for the preserves program, including staffing and maintenance of the natural resources inventory; second, to maintain good staff morale with a high regard for public service; third, to write park resource management plans that include long-term acquisition strategies and that address all the resources, not just vegetation or wildlife; and four, to raise capital improvement budgets both for rehabilitating existing park facilities, including the 1930s CCC structures, and for developing "appropriate" new facilities.[14]

Carrier's vision and goals fit comfortably with the department's 1981 mission statement. They also complement director Larry Wilson's strong belief in multiple use within reasonable limits, by which he means that multiple use should not sacrifice resources, nor should multiple use damage the sometimes

delicate social relationships that exist among resource users and resource pro-
tectors. With respect to those who make up the latter group, Wilson was criticized in 1986 for recommending that the Preserves Advisory Board be eliminated as part of the reorganization. That recommendation was a mistake, he
now admits. Even though the statutory authority provides a clear mandate for
protecting resources with preserve status, something he felt his staff was professionally trained to carry out, Wilson did not understand, at the time, how
strongly environmentalists felt about the board's advisory role. In the past ten
years, he has accepted, even welcomed, greater public participation in the
decision-making process. For preserves, this means that citizens who serve on
the Preserves Advisory Board play an important role in managing the areas
that need the greatest protection.[15]

Wilson also has been criticized for not fighting to keep the State Conservation Commission from being merged into a new umbrella agency in 1986.
However, he maintains that, from the time he took the job as director of the
State Conservation in 1981, he understood the inevitability of a merger; the
idea had already gained considerable political currency. In fact, he would have
preferred a department served by one commission rather than two. His only
real choice was whether to entrench himself as a stand-patter, and then "go
packing," or to stay and help restructure the new Department of Natural Resources. He chose the latter. In so doing, he had to face those who looked at
the new arrangement and repeatedly asked, "How is that going to work, environmental over here and a bunch of hunters and fishermen over there want-
ing to exploit the trees and the forest and the prairies?"[16] There is, of course,
no answer that will satisfy those who ask this question. Time alone will tell.

It is fair to say that many, many people still question the wisdom of placing
an agency responsible for enforcing environmental regulations under the
same tent with an agency that had acquired a reputation for catering to resource users. If they do not repudiate the 1986 merger as a misguided attempt
to mix oil and water, they see the resulting superagency as a dangerous con-
centration of governmental power. Thus, Wilson has spent much of the past
decade trying to build cooperation and trust between the two sides of the
department. "There are times when we stumble," he admits, because the department's mandate is broad and the staff is large. Administratively, the department is a challenge; however, the merger works, in his opinion, precisely
because overlapping or conflicting policy matters do not have to be reconciled
through interagency negotiations. In that respect, "we are just an arm's length
away from an answer or a decision or professional advice."[17]

Of equal concern to those who watch the agency is the threat of political
interference from the governor's office. Wilson acknowledges that both com-

missions lost considerable authority with reorganization. They are still important, inasmuch as they set budgets, administrative rules, and general policy
guidelines, but, as he notes, "the two big winners in reorganization were the
governor and the department directors." Governor Terry Branstad has generally refrained from heavy-handed interference, but there is no assurance that
in the future the directors of any and all state departments will not become
strictly political appointments. For Wilson, the question becomes not which
party or personality is in power, but how to maintain continuity and profes-
sionalism in the department when political administrations and directors
change. The answer to this, in Wilson's estimation, is up to the agency's
constituencies.

"Let's face it, we always talk about environment, about parks and trees, but
what generally drives these agencies are the fishing and hunting activities. . . .
While everyone of the three million people in our state should be concerned
about the environment, and I hope that they are, as far as [sustained, organized political] activity is concerned, well, there isn't a whole lot of it." [18] A
political constituency is something that Iowa's state parks have not had since
the Iowa Park and Forestry Association dissipated in the 1920s. State parks
have not attracted a powerful group of defenders that speak with one voice,
perhaps because parks are many things to many people. Contrast this with
roughly thirteen million park visitations annually. Iowa's state parks are still
among the most heavily used in the nation. Yet even though Iowans enjoy state
parks and preserves, and use them in record numbers, they feel little vested
ownership in them.

What then, is the key to maximizing public support for and minimizing
partisan political interference in the state park system? The as-yet-unrealized
potential of REAP is one possibility. Although yearly appropriations lag far
behind the promise, the law also created an entity known as the REAP Congress, a biennial assembly of interested citizens whose sole purpose, once every
two years, is to caucus and make recommendations to the governor, the legislature, and the Natural Resource Commission regarding the resource concerns that the program addresses: open space acquisition, state land management, county conservation, city parks, historic resource protection, soil and
water enhancement, roadside vegetation, and conservation education. The
democratic forum of a REAP Congress is a bold concept, although in some
respects, it complements the essence of the "conservation park" idea: "conser-
vancies" without the elitism the latter term implies.

Other strategies to build public support for parks include more outreach
programs and greater collaboration with environmental organizations that
have a broad membership base, such as the Iowa Natural Heritage Founda-

tion. Some have looked to the success of the Missouri sales tax initiative as a possible means of cultivating greater public support, both monetary and political. In reality, there is no clear answer, and probably no single answer, to the question of how to build a solid constituency for parks. Still, it is an important question to ponder if our parks and preserves are to remain special public assets.

State Lands in Parks and Preserves, 1918–1996

Park, Preserve, Site, or Recreation Area	Map #	County	Earliest Land Acquisition	Year Ownership Trans-ferred	1931 Classification
Anderson Prairie	5	Emmet	1980		—
Backbone	93	Delaware	1918		State Park
Badger Creek	124	Madison	1968		—
Barkley Memorial	101	Boone	1929		Preserve
Beaver Meadow	79	Butler	1934	1961	—
Beeds Lake	77	Franklin	1934		—
Bellevue	110	Jackson	1925		State Park
Big Creek	114	Polk	1972		—
Bixby	30	Clayton	1926		State Park
Black Hawk Lake	85	Sac	1934		—
Bluffton Fir Stand	11	Winneshiek	1946		—
Bobwhite	141	Wayne	1947		—
Brown's Lake (Bigelow)	84	Woodbury	1946		—
Brush Creek Canyon	26	Fayette	1936		—
Brushy Creek	88	Webster	1968		—
Brushy Creek Preserve	21	Webster	1968		—
Ambrose A. Call	70	Kossuth	1925		State Park
Cameron Woods	50	Scott	1977		—
Casey's Paha	40	Tama	1974		—
Cayler Prairie	4	Dickinson	1958		—
Cedar Rock	25	Buchanan	1983		—
Cheever Lake	6	Emmet	1937		—
Theo F. Clark	105	Tama	1921	1960	State Park
Clear Lake	73	Cerro Gordo	1924		State Park
Cold Springs	122	Cass	1935		—
Coldwater Cave	12	Winneshiek	1943		—
Dolliver Memorial	87	Webster	1920		State Park
Doolittle Prairie	38	Story	1979		—
Eagle Lake	71	Hancock	1923		State Park
Echo Valley	81	Fayette	1934		—
Elk Rock (Red Rock)	126	Marion	1969[2]		—

☐ Under jurisdiction of Fish and Game (now Fish and Wildlife) Division.

— Not yet owned by state OR no longer owned by state.

* Under management agreement with county conservation board, municipality, or organization.

1941 Classification	1964 Classification	Year Preserve Dedicated Under 1965 SP Act	Current Classification(s) (owned or managed by, if not parks)
—	—	1980	Preserve (DNR F&W)
tate Park	State Park		State Park, State Forest within
—	—		Recreation Area
eo-Bio Area[1]	Preserve		State Forest (DNR Forestry)
Vayside	—		(o-City of Parkersburg)
ec. Reserve	State Park		State Park
ec. Reserve	State Park		State Park
—	—		State Park
eo-Bio Area	State Park	1979	Preserve
ec. Reserve	State Park		State Park
—	☐	1969	Preserve (DNR F&W)
—	State Park		State Park
—	State Park		State Park*
orest Res.	State Park	1971	Preserve
—	—		Recreation Area and
—	—	1988	Preserve within
ec. Reserve	State Park		State Park
—	—	1977	Preserve*
—	—	1985	Preserve*
—	Preserve	1971	Preserve (DNR F&W)
—	—		Historic Site
☐	☐	1978	Preserve (DNR F&W)
Vayside	—		(o-Tama Co.)
ake Reserve	State Park		State Park
ec. Reserve	State Park*		State Park*
—	?	1970	Preserve
tate Park	State Park		State Park
—	—	1980	Preserve*
ake Reserve	State Park		State Park*
ec. Reserve	State Park		State Park*
—	—		State Park

. Geologic-Biologic Area.
. Land leased from U.S. Army Corps of Engineers.

(*continued*)

Park, Preserve, Site, or Recreation Area	Map #	County	Earliest Land Acquisition	Year Ownership Transferred	1931 Classificatio
Fairport	129	Muscatine	1968[3]		—
Fallen Rock	24	Hardin	1974		—
Farmington	146	Van Buren	1919	1949	State Park
Fish Farm Mounds	14	Allamakee	1935		—
Flint Hills	137	Des Moines	1925	1940	State Park
Fort Atkinson	8	Winneshiek	1921		State Park
Fort Defiance	58	Emmet	1923		State Park
Margo Frankel Woods	115	Polk	1946		—
Galland School	56	Lee	1926		?
A. Gardner Sharp Cabin	3	Dickinson	1943		—
Geode	136	Henry–DM	1937		—
Gitchie Manitou	1	Lyon	1926		State Park
Frank A. Gotch	75	Humboldt	1942		—
Green Valley	132	Union	1950		—
Gull Point–Okoboji	57	Dickinson	1934[5]		—
Hardin City Woodland	23	Hardin	1959		—
Hayden Prairie	7	Howard	1945		—
Heery Woods	78	Butler	1935		—
Holst Forest	102	Boone	1939		—
Honey Creek (Rathbun)	142	Appanoose	1969[6]		—
Indian Fish Trap	48	Iowa	sovereign[7]		
Kalsow Prairie	20	Pocahontas	1948		—
Kearny (Medium Lake)	68	Palo Alto	1940		—
Kish-Ke-Kosh	46	Jasper	1980		—
Lacey-Keosauqua	145	Van Buren	1919		State Park
Lake Ahquabi	125	Warren	1934		—
Lake Anita	121	Cass	1962		—
Lake Darling	128	Washington	1947		—
Lake Keomah	127	Mahaska	1934		—
Lake Macbride	118	Johnson	1933		—

3. Land leased from U.S. Army Corps of Engineers.
4. Historic-Archaeologic Area.
5. Gull Point serves as park headquarters for several parks, recreation areas, and access points, including Arnold's Park Pier (1930), Emerson Bay-Lighthouse (1980), Lower Gar Access, Marble

1941 Classification	1964 Classification	Year Preserve Dedicated Under 1965 SP Act	Current Classification(s) (owned or managed by, if not parks)
–	—		Recreation Area
–	—	1978	Preserve*
ec. Reserve	—		(o-City of Farmington)
ist-Arch⁴	Preserve	1968	Preserve (DNR F&W)
–	—		(o-City of Burlington)
ist-Arch	Preserve	1968	Preserve
ec. Reserve	State Park		State Park
–	State Park		State Park*
other"	Preserve		Historic Site*
–	Preserve*		Historic Site (SHSI)
tate Park	State Park		State Park
ist-Arch	Preserve	1969	Preserve (DNR F&W)
–	State Park*		State Park*
–	State Park		State Park
ake Reserve	State Park		State Park
–	☐	1968	Preserve*
–	☐	1968	Preserve (DNR F&W)
ec. Reserve	State Park		State Park*
orest Res.	State Forest		State Forest (DNR Forestry)
–	—		State Park
–		1976	Preserve*
–	Preserve	1968	Preserve
ake Reserve	State Park*		"Park"*
–	—	1981	Preserve*
tate Park	State Park		State Park
ec. Reserve	State Park		State Park
–	State Park		State Park
–	State Park		Recreation Area
ec. Reserve	State Park		State Park
tate Park	State Park		State Park

Beach (1942), Mini-Wakan (1933), Orleans Beach (1956), Pike's Point (1932), Pillsbury Point (1928), Templar Park (1987), Trapper's Bay (1933), and Triboji Beach (1982).
 Land leased from U.S. Army Corps of Engineers.
 Meandered stream included in state sovereign lands.

(continued)

Park, Preserve, Site, or Recreation Area	Map #	County	Earliest Land Acquisition	Year Ownership Trans-ferred	1931 Classification
Lake Manawa	120	Pottawattamie	1932		—
Lake of Three Fires	139	Taylor	1935		—
Lake Wapello	144	Davis	1932		—
Ledges	103	Boone	1920		State Park
Lennon Mills	113	Guthrie	1943		—
Lepley	89	Hardin	1920		State Park
Lewis and Clark	96	Monona	1924		State Park
Little Maquoketa Mounds	34	Dubuque	1981		—
Lost Island	67	Palo Alto	1924	1964	State Park
Malchow Mounds	54	Des Moines	1975		—
Maquoketa Caves	109	Jackson	1921		State Park
McIntosh Woods	72	Cerro Gordo	1943		—
Melanaphy Springs	9	Winneshiek	1947		—
Mericle Woods	39	Tama	1985		—
Merritt Forest	31	Clayton	1967		—
Mill Creek	65	O'Brien	1935		—
Mines of Spain	95	Dubuque	1980		—
Catfish Creek	35	Dubuque	1980		—
Montauk	27	Fayette	1975		—
Mossy Glen	29	Clayton	1978		—
Nine Eagles	140	Decatur	1940		—
Oak Grove	64	Sioux	1924		State Park
Oakland Mills	135	Henry	1920		State Park
Okamanpedan	59	Emmet	1923		State Park
Palisades-Kepler	107	Linn	1922		State Park
Palisades-Dows	41	Linn	1922		State Park
Pammel	123	Madison	1923		State Park
Pecan Grove	51	Muscatine	1973		—
Pellet Memorial Woods	43	Cass	1984		—
Pikes Peak–Point Ann	82	Clayton	1936		—
Pilot Grove	47	Iowa	1980		—
Pilot Knob	61	Hancock-Winn	1921		State Park
Pilot Knob Preserve	16	Hancock-Winn	1921		State Park

8. Originally established as private preserve by Frank Chapman Pellet.

1941 Classification	1964 Classification	Year Preserve Dedicated Under 1965 SP Act	Current Classification(s) (owned or managed by, if not parks)
Rec. Reserve	State Park		State Park
Rec. Reserve	State Park		State Park
State Park	State Park		State Park
State Park	State Park		State Park
—	Preserve		State Park*
Wayside	?		State Park*
Rec. Reserve	State Park		State Park
—	—	1981	Preserve*
Lake Reserve	State Park*		(o-Palo Alto Co.)
—	—	1978	Preserve (DNR F&W)
Geo-Bio	State Park		State Park
—	State Park		State Park
—	☐	1994	Preserve
—	—	1986	Preserve (DNR F&W)
—	—	1969	Preserve (DNR Forestry)
Rec. Reserve	State Park		State Park*
—	—		Recreation Area and
—	—	1991	Preserve within
—	—	1984	Preserve
—	—	1979	Preserve (DNR F&W)
State Park	State Park		State Park
Rec. Reserve	State Park*		State Park*
Rec. Reserve	State Park		State Park*
Lake Reserve	State Park		State Park
State Park	State Park		State Park and
State Park	State Park	1980	Preserve adjacent
State Park	State Park		State Park
—	—	1978	Preserve (DNR F&W)
—	—	1978[8]	Preserve*
State Park	State Park		State Park
—	—	1980	Preserve*
State Park	State Park		State Park and
State Park	State Park	1968	Preserve within

(*continued*)

Park, Preserve, Site, or Recreation Area	Map #	County	Earliest Land Acquisition	Year Ownership Transferred	1931 Classification
Pilot Mound	100	Boone	1939		—
Pine Lake	90	Hardin	1920		State Park
Pioneer	62	Mitchell	1938		—
Pleasant Creek	106	Linn	1973		—
Plum Grove	49	Johnson	1941		—
Prairie Rose	111	Shelby	1953		—
Preparation Canyon	97	Monona	1934		—
Red Haw	133	Lucas	1936		—
Rice Lake	60	Winn-Worth	1924		State Park
Roberts Creek	28	Clayton	1988		—
Rock Creek	117	Jasper	1950		—
Sylvan Runkel	36	Monona	1974		—
Rush Lake	69	Palo Alto	1930		?
St. James Church	10	Winneshiek	1970		—
Searryl's Cave	42	Jones	1991		—
Sharon Bluffs	143	Appanoose	1931		—
Sheeder Prairie	44	Guthrie	1961		—
Shimek Forest	147	Lee-Van Buren	1936		—
Silver Lake	94	Delaware	1923	1964	State Park
Silver Lake Fen	2	Dickinson	1940		—
Slinde Mounds	13	Allamakee	1946		—
Springbrook	112	Guthrie	1926		State Park
Spring Lake	99	Greene	1949		—
Starr's Cave	55	Des Moines	1974		—
Steamboat Rock	91	Hardin	1936		—
Steele Prairie	18	Cherokee	1987		—
Stephens Forest	134	Lucas-Monroe	1936		—
Nestor Stiles	19	Cherokee	1983		—
Stone	83	Woodbury	1935		—
Mt. Talbot	17	Woodbury	1935		—
Storm Lake	74	Buena Vista	1926	1961	State Park
Strasser Woods	45	Polk	1983		—
Swan Lake	98	Carroll	1933		—
Toolesboro	52	Louisa	1963		—

1941 Classification	1964 Classification	Year Preserve Dedicated Under 1965 SP Act	Current Classification(s) (owned or managed by, if not parks)
orest Res.	State Forest		State Forest (DNR Forestry)
ec. Reserve	State Park		State Park
ec Reserve	State Park		State Park*
—	—		Recreation Area
other"	Preserve		Historic Site (SHSI)
—	State Park		State Park
orest Res.	State Park		State Park
ec. Reserve	State Park		State Park
ec. Reserve	State Park		State Park
—	—	1990	Preserve (DNR F&W)
—	State Park		State Park
—	—	1996	Preserve
ake Reserve	State Park		State Park (DNR F&W)
—	—	1970	Preserve
—	—	1992	Preserve (DNR F&W)
ec. Reserve	State Park		State Park*
—	Preserve	1968	Preserve
orest Res.	State Forest		State Forest (DNR Forestry)
Vayside	State Park*		(o-Delaware Co.)
☐	☐	1972	Preserve (DNR F&W)
—	☐	1979	Preserve (DNR F&W)
tate Park	State Park		State Park
—	State Park		State Park*
—	—	1978	Preserve*
Vayside	State Park		Pt. Iowa River Greenbelt*
—	—	1987	Preserve*
orest Res.	State Forest		State Forest (DNR Forestry)
—	—	1982	Preserve*
tate Park	State Park		State Park and
tate Park	State Park	1989	Preserve adjacent
ake Reserve	—		(o-City of Storm Lake)
—	—	1982	Preserve (DNR F&W)
ake Reserve	State Park		State Park*
—	—	1981	Preserve

(continued)

Park, Preserve, Site, or Recreation Area	Map #	County	Earliest Land Acquisition	Year Ownership Transferred	1931 Classification
Turin Loess Hills	37	Monona	1974		—
Turkey River Mounds	32	Clayton	1939		—
Twin Lakes	86	Calhoun	1923		State Park
Union Grove	104	Tama	1940		—
Viking Lake	131	Montgomery	1953		—
Volga River	80	Fayette	1968		—
Wall Lake	76	Wright	1922		State Park
Walnut Woods	116	Polk	1925		State Park
Wanata	66	Clay	1934		—
Wapsipinicon	108	Jones	1921		State Park
Waubonsie	138	Fremont	1926		State Park
White Pine Hollow	33	Dubuque	1934		—
Wildcat Den	130	Muscatine	1926		State Park
Wilson Island	119	Pottawattamie	1960		—
Wittrock Indian Village	15	O'Brien	1936		—
Woodman Hollow	22	Webster	1927		Preserve
Woodthrush	53	Jefferson	1928	1962	Preserve
George Wyth Memorial	92	Black Hawk	1938		—
Yellow River Forest	63	Allama-Clay	1936		—

1941 Classification	1964 Classification	Year Preserve Dedicated Under 1965 SP Act	Current Classification(s) (owned or managed by, if not parks)
—	—	1978	Preserve (DNR F&W)
ist-Arch	Preserve	1968	Preserve
ke Reserve	State Park		State Park
ec. Reserve	State Park		State Park
—	State Park		State Park
—	—		Recreation Area
ate Park	☐		jurisdiction trans. to F&G 1933
ec. Reserve	State Park		State Park
ec. Reserve	State Park		State Park
ec. Reserve	State Park		State Park
ate Park	State Park		State Park
orest Res.	State Forest	1968	Preserve (DNR Forestry)
ate Park	State Park		State Park
—	Multiuse Area		Recreation Area
ist-Arch	Preserve	1968	Preserve
eo-Bio	Preserve	1970	Preserve
ayside	Preserve*	1975	Preserve (o-City of Fairfield)
arkway	State Park		State Park
orest Res.	State Forest		State Forest (DNR Forestry)

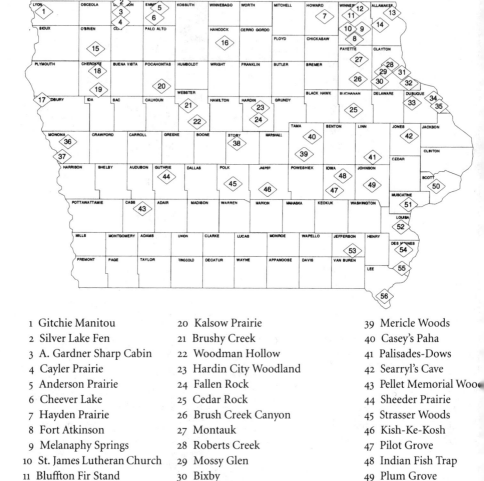

1 Gitchie Manitou
2 Silver Lake Fen
3 A. Gardner Sharp Cabin
4 Cayler Prairie
5 Anderson Prairie
6 Cheever Lake
7 Hayden Prairie
8 Fort Atkinson
9 Melanaphy Springs
10 St. James Lutheran Church
11 Bluffton Fir Stand
12 Coldwater Cave
13 Slinde Mounds
14 Fish Farm Mounds
15 Wittrock Indian Village
16 Pilot Knob
17 Mount Talbot [Stone SP]
18 Steele Prairie
19 Nestor Stiles Prairie

20 Kalsow Prairie
21 Brushy Creek
22 Woodman Hollow
23 Hardin City Woodland
24 Fallen Rock
25 Cedar Rock
26 Brush Creek Canyon
27 Montauk
28 Roberts Creek
29 Mossy Glen
30 Bixby
31 Merritt Forest
32 Turkey River Mounds
33 White Pine Hollow
34 Little Maquoketa River Mounds
35 Catfish Creek [Mines of Spain]
36 Sylvan Runkel
37 Turin Loess Hills
38 Doolittle Prairie

39 Mericle Woods
40 Casey's Paha
41 Palisades-Dows
42 Searryl's Cave
43 Pellet Memorial Wood
44 Sheeder Prairie
45 Strasser Woods
46 Kish-Ke-Kosh
47 Pilot Grove
48 Indian Fish Trap
49 Plum Grove
50 Cameron Woods
51 Pecan Grove
52 Toolesboro
53 Woodthrush
54 Malchow Mounds
55 Starr's Cave
56 Galland School

Iowa State Preserves and Historic Sites, 1918–1996. *Courtesy Iowa Department of Natural Resources.* Read map numbers *left to right*, top to bottom. Note: Only preserves located on state-owned land are listed. Consult the *Iowa State Preserves Guide* for a complete list of state preserves dedicated as of 1992.

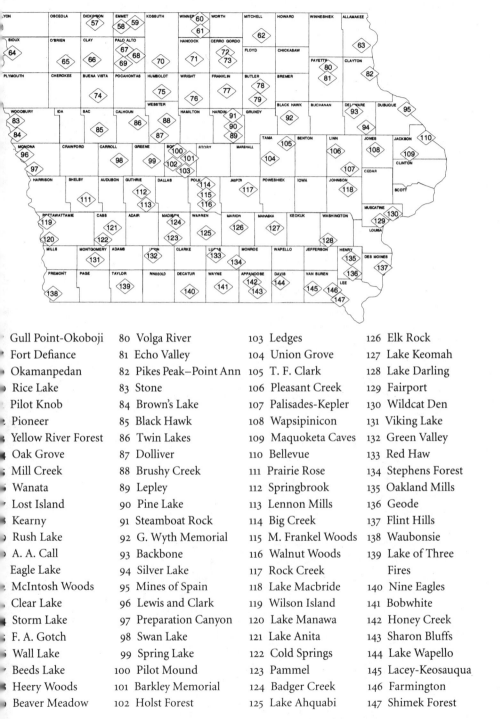

Gull Point-Okoboji	80 Volga River	103 Ledges	126 Elk Rock
Fort Defiance	81 Echo Valley	104 Union Grove	127 Lake Keomah
Okamanpedan	82 Pikes Peak–Point Ann	105 T. F. Clark	128 Lake Darling
Rice Lake	83 Stone	106 Pleasant Creek	129 Fairport
Pilot Knob	84 Brown's Lake	107 Palisades-Kepler	130 Wildcat Den
Pioneer	85 Black Hawk	108 Wapsipinicon	131 Viking Lake
Yellow River Forest	86 Twin Lakes	109 Maquoketa Caves	132 Green Valley
Oak Grove	87 Dolliver	110 Bellevue	133 Red Haw
Mill Creek	88 Brushy Creek	111 Prairie Rose	134 Stephens Forest
Wanata	89 Lepley	112 Springbrook	135 Oakland Mills
Lost Island	90 Pine Lake	113 Lennon Mills	136 Geode
Kearny	91 Steamboat Rock	114 Big Creek	137 Flint Hills
Rush Lake	92 G. Wyth Memorial	115 M. Frankel Woods	138 Waubonsie
A. A. Call	93 Backbone	116 Walnut Woods	139 Lake of Three
Eagle Lake	94 Silver Lake	117 Rock Creek	Fires
McIntosh Woods	95 Mines of Spain	118 Lake Macbride	140 Nine Eagles
Clear Lake	96 Lewis and Clark	119 Wilson Island	141 Bobwhite
Storm Lake	97 Preparation Canyon	120 Lake Manawa	142 Honey Creek
F. A. Gotch	98 Swan Lake	121 Lake Anita	143 Sharon Bluffs
Wall Lake	99 Spring Lake	122 Cold Springs	144 Lake Wapello
Beeds Lake	100 Pilot Mound	123 Pammel	145 Lacey-Keosauqua
Heery Woods	101 Barkley Memorial	124 Badger Creek	146 Farmington
Beaver Meadow	102 Holst Forest	125 Lake Ahquabi	147 Shimek Forest

Iowa State Parks, Recreation Areas, and Forests, 1918–1996. *Courtesy Iowa Department of Natural Resources.* Read map numbers *left to right*, top to bottom.

Notes

Prologue

1. Thomas H. Macbride, "The Present Status of Iowa Parks," 1922 address as reprinted in Macbride, *On the Campus* (1925), 177–178; the text of his speech also appears in *Iowa Conservation* 6 (1922), 41–42, 56–58.

2. This is the thesis that Thomas R. Cox advances in his book, *The Park Builders: A History of State Parks in the Pacific Northwest* (1988), one of the few scholarly treatments of state parks.

3. *A Study of the Park and Recreation Problem of the United States* (1941).

4. Two seminal articles reinforced this impression: Thomas P. Christensen, "The State Parks of Iowa," *Iowa Journal of History and Politics* [hereafter *IJHP*] 26 (1928), 331–414, and Louis H. Pammel, "The Arbor Day, Park and Conservation Movements in Iowa," *Annals of Iowa* 17 (1929–1930), 83–104, 189–232, 270–313.

5. My perspective on environmentalism has been shaped by works such as Samuel P. Hays, *Beauty, Health and Permanence: Environmental Politics in the United States, 1955–1985* (1987), and Robert Gottlieb, *Forcing the Spring: The Transformation of the American Environmental Movement* (1993); both analyze the strands of thought and action that anchor post–World War II environmental concerns to the earlier conservation movement.

CHAPTER 1   Conservation Parks

1. T[homas] H. Macbride, "County Parks," *Proceedings of the Iowa Academy of Science* [hereafter *Proceedings IAS*] 3 (1896), 95.

2. Thomas H. Macbride, "The Present Status of Iowa Parks," *On the Campus*, 166.

3. Ibid., 176.

4. [Edgar R. Harlan], "Iowa State Parks," *Annals of Iowa* 13 (1921), 144.

5. In 1905, California returned the land to the national government for inclusion in Yosemite National Park, but for thirty years the valley and nearby Mariposa Grove remained a state park. See Joseph H. Engbeck, Jr., and Philip Hyde, *State Parks of California from 1864 to the Present* (1980), 17-28; Natural Heritage Trust, *Fifty Years: New York State Parks, 1924-1974* (1975); Raymond H. Torrey, *State Parks and Recreational Uses of State Forests in the United States* (1926), 20-23; Beatrice Ward Nelson, *State Recreation: Parks, Forests and Game Reservations* (1928), 3-5; Roy W. Meyer, *Everyone's Country Estate: A History of Minnesota's State Parks* (1991), 1-22.

6. Cox, *The Park Builders*, especially chap. 3, "Ben Olcott's Crusade to Save Oregon's Scenery," 32-46, and chap. 5, "Asahel Curtis, Herbert Evison, and the Parks and Roadside Timber of Washington State," 57-78; Engbeck and Hyde, *State Parks of California*, 29-56; Eugene James O'Neill, "Parks and Forest Conservation in New York, 1850-1920" (1963), especially chap. 4, "The Rise of a Separate Park Movement," 96-113; Wisconsin State Planning Board and Wisconsin Conservation Commission, *A Park, Parkway and Recreational Area Plan* (1939), 33-36.

7. Glory-June Greiff, "People, Parks, and Perceptions: Eighty Years of Indiana State Parks" (work in progress, 1995), 19-20, cited with author's permission; see also Greiff, "New Deal Resources in Indiana State Parks," National Register of Historic Places Multiple Property Documentation, 1991.

8. Bohumil Shimek, "In Memoriam," *Proceedings IAS* 41 (1934), 34.

9. Macbride, "County Parks," 91; Thomas H. Macbride, "The President's Address," *Proceedings IAS* 5 (1898), 17-23.

10. A. Hunter DuPree and Samuel P. Hays were among the first to explore the role of scientists in fashioning resource conservation policies at the federal level; see DuPree, *Science in the Federal Government: A History of Policies and Activities to 1940* (1957), 232-255, and Hays, *Conservation and the Gospel of Efficiency: The Progressive Conservation Movement, 1890-1920* (1959). More recently, Michael L. Smith has detailed the role of scientists in California; see *Pacific Visions: California Scientists and the Environment, 1850-1915* (1987). Taking Hays to task, John F. Rieger argued that sportsmen "were the real spearhead of conservation," initiating the movement in the 1870s; see Rieger, *American Sportsmen and the Origins of Conservation* (1975), 21. It is true that in Iowa, as elsewhere, the rapid depletion of game birds and fish led to the adoption of restrictive game laws beginning in 1857; see Henry Arnold Bennett, "Fish and Game Legislation," *IJHP* 24 (July 1926), 335-444. It is also true that Iowa was in the forefront of wildlife conservation. However, wildlife conservationists tended to focus on bag limits and artificial propagation until well into the twentieth century. Serious concern for habitat protection came much later, after botanists, geologists, zoologists, foresters, and others began to advocate an integrated approach to resource conservation.

11. Biographical information has been compiled from the following published sources: Shimek, "In Memoriam," as noted above; Ruth Gallaher, "The Man," and Robert B. Wylie, "The Scholar," in a memorial issue of *Palimpsest* 15 (1934), 161-192; Mary Winifred Conklin Schertz and Walter L. Myers, *Thomas Huston Macbride* (1947); Stow Persons, *The University of Iowa in the Twentieth Century: An Institutional History*

(1990); and Debby J. Zieglowsky, "Thomas Macbride's Dream: Iowa Lakeside Laboratory," *Palimpsest* 66 (1985), 42–65. Additional information from primary sources is noted below.

12. During Macbride's academic career, the term "applied science" often referred to the emerging field of "engineering" as we understand it today. Thus, in 1903, when the State University concentrated various engineering courses into one area, it did so by establishing a new School of Applied Science; see Persons, *The University of Iowa in the Twentieth Century*, 48. The term as used here denotes the application of science to the understanding and solution of practical problems, a current definition that accurately characterizes Macbride's professional career.

13. In *The University of Iowa in the Twentieth Century*, Persons divides the university's institutional development into three distinct phases: a "provincial university" from its founding in 1847 to the early twentieth century, superseded by an "era of creative anarchy" that stretched from 1908 to midcentury, when the third, and present, "age of institutional inertia" began to emerge. During the first phase, Calvin and Macbride were among those who contributed to establishing the principal colleges, schools, and departments of the university as a whole, a process essentially complete by 1908. The University of Iowa's academic development reflected a widespread movement toward "multipurpose universities" with faculty divided into separate departments. During the 1870s the number of academic scientists increased dramatically in U.S. colleges and universities, although their relative influence in faculty composition declined. At the same time, schools of science emerged in many universities. See, for instance, Stanley M. Guralnick, "The American Scientist in Higher Education, 1820–1910," in *The Sciences in the American Context: New Perspectives*, ed. Nathan Reingold (1979), 99–141; Frederick Rudolph, *The American College and University: A History* (1962); Laurence R. Veysey, *The Emergence of the American University* (1965).

14. Macbride, "President's Address: The Present Status of Iowa Parks," *Proceedings of the Iowa Park and Forestry Association* [hereafter *Proceedings IPFA*] 1 (1902), 4–5.

15. Ibid., 5; Hays, in *Conservation and the Gospel of Efficiency*, discerned a "wide difference in attitude" separating those in the conservation movement who promoted "rational and comprehensive planning" (the Roosevelt-Pinchot-McGee element) and those who saw conservation as a moral crusade to "save resources from use rather than to use them wisely" (141–146). This seminal work in environmental history, which has influenced more than a generation of historians, fostered the notion that the "utilitarians" and the "moral conservationists" (or preservationists) were two separate camps motivated by distinct philosophies. In reality, the "differences of attitude"—and there were many—made the conservation movement "highly elastic," as Hays also notes (176). Moreover, Thomas Macbride provides a case in point demonstrating that goals and motives others would later see as divisive and competing were not, at the time, so easily distinguished. Macbride certainly was motivated by his moral convictions, and he often sprinkled his conservation writings with Biblical allusions, but he did not see resource conservation as strictly a moral problem.

16. Macbride, "County Parks," 91.

17. Macbride, "President's Address" (1898), 14–15.

18. "The Citizen vs. The City: Difficulties in Municipal Problems," delivered before the League of Iowa Municipalities in 1906; reprinted in *On the Campus*, 145–162.

19. Gov. Albert Cummins to Macbride, 7 September 1908, Thomas Huston Macbride Papers, University of Iowa [hereafter cited as Macbride Papers]; Macbride to Louis H. Pammel, 19 September 1908, Louis H. Pammel Papers, Iowa State University [hereafter cited as Pammel Papers]. Gifford Pinchot, head of the U.S. Forestry Department and the first chair of the National Conservation Commission, requested the governors of all states to appoint state commissions. Other members of the Iowa Forestry Commission included Louis Pammel of Iowa State College, Eugene Secor of Forest City, William Louden of Fairfield, I. M. Earle of Des Moines, and Wesley Green of Des Moines.

20. Thomas H. Macbride, "A Forestry Commission for Iowa," *Report of the Iowa State Horticultural Society* 43 (1909), 169–170.

21. A. C. Miller to Macbride, 15 September 1910, Macbride Papers. Miller, president of the Home Savings Bank in Des Moines, chaired the Iowa State Drainage, Waterways and Conservation Commission and attended the 1910 National Conservation Congress, held in St. Paul, where the events and issues of the Ballinger-Pinchot controversy dominated the convention.

22. The authors of individual sections of the commission's report are not identified. However, Miller to Macbride, 14 September 1910, Macbride Papers, specifically mentions that Macbride was assigned the first two topics. Louis Pammel, in addition, identifies Macbride as the author of the section on lakes and streams in "The Arbor Day, Park and Conservation Movements in Iowa," 282. Several letters in the 1910 exchange between Macbride and Miller reveal that Macbride was centrally involved in the investigation of lakes and streams, which tends to confirm Pammel's information.

23. *Proceedings of the Third National Conservation Congress, 1911* (1912), 29, 277. Perfunctory correspondence sent to Macbride from various officers of the National Rivers and Harbors Congress in 1910, Macbride Papers, list Macbride as the Iowa vice president at that time. Governor Carroll additionally named him an official delegate to the 1911 national convention, but without funds to cover expenses. There is no indication that Macbride attended. See Hays, *Conservation and the Gospel of Efficiency*, 219–240, on the competing proposals of U.S. senators Ransdall (Louisiana) and Newlands (Nevada) for rivers and harbors legislation during the 1910s.

24. Harriet Bell Carlander, "The American School of Wildlife, McGregor, Iowa, 1919–1941," *Proceedings IAS* 68 (1961), 294–300.

25. Macbride to Shimek, 7 October 1931, Macbride Papers.

26. Gallaher, 168.

27. Zieglowsky, "Thomas Macbride's Dream: Iowa Lakeside Laboratory," 62–64.

28. W J McGee to Macbride, 6 June 1898, Macbride Papers.

29. Smith, *Pacific Visions*, 159–163.

30. W J McGee to Samuel Calvin, 16 May 1898, Macbride Papers.

31. A series of exchanges in late 1898 and early 1899, for instance, shows that Macbride arranged for Bohumil Shimek to assist Pinchot as a "special agent" to supervise the government's forestry work; Pinchot to Macbride, 13 December 1898, 10 January 1899, 19 January 1899, and 27 January 1899. After leaving his position as chief forester

and becoming president of the National Conservation Association, Pinchot enlisted Macbride's help in order to strengthen membership; Pinchot to Macbride, 16 March 1910. Likewise, after Macbride had retired from academic life and Pinchot had moved on to become the Governor of Pennsylvania, Macbride sought his assistance to promote better forestry practices in Florida; Pinchot to Macbride, 8 February 1923, and Macbride to Pinchot, 5 May 1923, Macbride Papers.

32. Robert B. Wylie, "The Scholar," 174.

33. Thomas H. Macbride, "The Present Status of Iowa Parks," *On the Campus*, 176–177.

34. Macbride, "President's Address: The Present Status of Iowa Parks" (1902), 8.

35. John F. Sears, *Sacred Places: American Tourist Attractions in the Nineteenth Century* (1989), 100–102, 116–121.

36. Macbride, "A Forestry Commission for Iowa," 168–169.

37. [Thomas Macbride], "The Conservation of Our Lakes and Streams," in *Report of the Iowa State Drainage, Waterways and Conservation Commission* (1911), 188.

38. Pammel sent out notice of the meeting through newspapers; see *Proceedings IPFA* 1 (1902), 4. The meeting was held in historic Agricultural Hall, also known as Old Botany (and currently known as Carrie Chapman Catt Hall), on the Iowa State campus. To the list of states with active conservation organizations should be added California, where the Sierra Club was founded in 1892.

39. "Constitution and By-Laws," *Proceedings IPFA* 1 (1902), 4–5.

40. Smith, *Pacific Visions*, chap. 7, describes a similar phenomenon in California, where scientists affiliated with the University of California Berkeley, Stanford University, the U.S. Geological Survey, the U.S. Coast & Geodetic Survey, and the California Academy of Sciences dominated the charter membership of the Sierra Club, which, according to the author, "provided a forum for the social concerns they had derived from their work."

41. Macbride to Pammel, 16 October 1902, Pammel Papers.

42. Richard A. Overfield, *Science with Practice: Charles E. Bessey and the Maturing of American Botany* (1993), 17–20; 38–39.

43. G. B. MacDonald, "State Forestry: The Early Period," *Iowa Conservationist* 4 (1945), 123.

44. Overfield, *Science with Practice*, 21–22; Jacob A. Swisher, "The Iowa Academy of Science," *IJHP* 29 (1931), 315–337.

45. Titles are taken from the *Proceedings IAS*, 9 (1902), 10 (1903), 12 (1905), 13 (1906), 14 (1907), 17 (1910), and 20 (1913).

46. Swisher, "Iowa Academy of Science," 343.

47. Macbride to Pammel, 10 December 1902, Pammel Papers.

48. Macbride to Pammel, 4 April 1903, Pammel Papers.

49. See the tables of contents for the *Proceedings IPFA*, 1902–1907.

50. *Report of the Iowa State Horticultural Society* 35 (1901); *Proceedings IAS* 8 (1901).

51. Macbride to Pammel, 4 April 1903; Macbride to Pammel, 12 May 1903; Macbride to Pammel, 21 December 1903; Pammel Papers.

52. *Proceedings IPFA* 2 (1903), 55–56; 3 (1904), 19–21; 4 (1905), xiii; 5 (1906), 11–13; see also Christensen, "The State Parks of Iowa," 339–340.

53. B. Shimek, "Report of Committee on Legislation," *Proceedings* IPFA 5 (1906), 11–12.

54. "Report of Committee on Legislation," *Proceedings* IPFA 6 (1907), 11–15.

55. IPFA, "Report of the Committee on Resolutions," in *Report of the Iowa State Horticultural Society* 43 (1909), 134–135.

56. Charles Mulford Robinson, *City Planning Report for Des Moines, Ia.* (1909), 71–95.

57. Edgar R. Harlan, "Proposed Improvement of the Iowa State Capitol Grounds," *Annals of Iowa* 11 (1913), 96–114.

58. Louis H. Pammel, "The Arbor Day, Park and Conservation Movements in Iowa," 231–232.

59. At the 1909 meeting, for instance, Shimek spoke on forestry matters in the state and Pammel delivered some exhortatory remarks on "Our Neglected Opportunities," urging those assembled to write letters supporting federal legislation for national parks. With no discussion, at least none that was recorded, the program then proceeded to the "Municipal Control of Street Trees," "The Catalpa in Iowa as Post Timber," and "Grading and Packing of Fruit for Market." See *Report of the Iowa State Horticultural Society* 44 (1910), 65–79.

60. Pammel to Macbride, 10 November 1913; Macbride to Pammel, 14 November 1913; Pammel to Macbride, 19 November 1913, Pammel Papers.

61. Biographical information on Lacey comes from the John Fletcher Lacey Papers, State Historical Society of Iowa Library and Archives Division, Des Moines [hereafter Lacey Papers], and Annette Gallagher, C.H.M., "Citizen of the Nation: John Fletcher Lacey, Conservationist," *Annals of Iowa* 46 (1981), 9–22.

62. John F. Lacey, "Autobiography," bound typescript compiled by Berenice Lacey, 123, Lacey Papers.

63. For a history of U.S. national monuments, see Hal Rothman, *Preserving Different Pasts* (1989).

64. William T. Hornaday, "John F. Lacey," *Annals of Iowa* 11 (1915), 582.

65. *Major John F. Lacey Memorial Volume*, ed. Louis H. Pammel (1915).

66. Pammel to Harlan, 27 November 1914, Pammel Papers.

67. Evelyn Fox Keller, *Reflections on Gender and Science* (1985), 76–77, 162–176.

68. Carolyn Merchant, "Women of the Progressive Conservation Movement, 1900–1916," *Environmental Review* 8 (1984), 57–67; Mildred White Wells, *Unity in Diversity: The History of the General Federation of Women's Clubs* (1953), 192–197; Mary I. Wood, *The History of the General Federation of Women's Clubs for the First Twenty-Two Years of Its Organization* (1912), 118; Hays, *Conservation and the Gospel of Efficiency*, 142.

69. *Proceedings of the Third National Conservation Congress, 1911*, 130.

70. Mrs. Hugh Buffum and Louella Thurston, *History of the Iowa Federation of Women's Clubs, 1893–1968* [1968], 74–75.

71. "Civic Improvement and Forestry," *Year Book 1908–09* [1908]; Buffum and Thurston, *History of the Iowa Federation of Women's Clubs*, 74–75. The IPFA entered the campaign for an Appalachian National Park in 1908; see "Report of Committee on

Resolution," *Proceedings, IPFA* 7 (1908), 10; "Report of Committee on Resolutions" [IPFA], in *Report of the Iowa State Horticultural Society* 43 (1909), 134.

72. Alfred Runte, *National Parks: The American Experience* (1987), 112–118.

73. Philip V. Scarpino, *Great River: An Environmental History of the Upper Mississippi, 1890–1950* (1985), chap. 4; Jill York O'Bright, *The Perpetual March: An Administrative History of Effigy Mounds National Monument* (1989), 47–48; Rebecca Conard, "The Conservation Movement in Iowa, 1857–1942," National Register of Historic Places Documentation (1991), 34–35, 96–98; John O. Anfinson, "Commerce and Conservation on the Upper Mississippi River," *Annals of Iowa* 52 (1993), 395–399.

74. "Report of Conservation Committee," [Eleventh Biennial Period] *Year Book, 1915–1916* [Iowa Federation of Women's Clubs] (August 1915), 159–160.

75. Cora Call Whitley Papers, Iowa Women's Archives [hereafter Whitley Papers]; see also Mrs. Francis Edmund Whitley, "Chapter XII: Twelfth Biennial Period, 1915–1917," *History of the Iowa Federation of Women's Clubs*, 93–95.

76. Whitley, "Twelfth Biennial Period, 1915–1917," 93–95.

77. Whitley to Pammel, 21 November 1915; Pammel to Whitley, 23 November 1915; Whitley to Pammel, 2 December 1915; Pammel to Whitley, 7 December 1915, Pammel Papers.

78. "Minutes of the Business Meeting of the Iowa Forestry and Conservation Association, February 2, 1916," Pammel Papers.

79. Karen J. Blair, *The Clubwoman as Feminist: True Womanhood Redefined* (1980).

80. Whitley, "Twelfth Biennial Period, 1915–1917," 100.

81. Mrs. Francis E. Whitley, "Some Aspects of Conservation in Iowa," *Proceedings of the Iowa Forestry and Conservation Association, 1914–1915* [hereafter *Proceedings IFCA*] (1916), 186, 188.

82. Whitley, "Twelfth Biennial Period, 1915–1917," 94.

83. Hays, *Conservation and the Gospel of Efficiency*, 176; Charles R. Van Hise, *The Conservation of Natural Resources in the United States* (1910); Richard T. Ely, et al., *The Foundations of National Prosperity: Studies in the Conservation of Permanent Natural Resources* (1917). Van Hise, a geologist by training, was president of the University of Wisconsin, where Taylor and Pammel studied.

84. Mrs. H. J. [Rose Schuster] Taylor, "Conservation of Life Through City Parks," *Iowa Conservation* 1 (1917), 13.

85. Jane Parrott to G. B. MacDonald, 12 February 1917; MacDonald to Parrott, 16 February 1917; Pammel to MacDonald, 5 March 1917, Pammel Papers.

86. T. C. Stephens, "Needed Changes in Our Game Laws" and "Game Protection in Iowa," *Proceedings IFCA* (1916), 14–24, 61–69; O. A. Byington, "The Passage of the Quail Bill," *Iowa Conservation* 1 (1917), 1, 35–36; T. C. Stephens, "A Review of Wildlife Protection in Iowa," *Iowa Conservation* 1 (1917), 49–52, 60, 64–68; Bruce F. Stiles, "Last Installment of History of Iowa Wildlife Legislation," *Iowa Conservationist* 2 (1943), 86.

87. A[rthur] R. Carhart, "A System of Parks—National, State and County," *Report of the Iowa State Horticultural Society* 51 (1917), 79.

88. Ibid., 81.

89. "Memorial to the Twenty-sixth General Assembly from the Iowa Academy of Science" [drafted by Thomas Macbride, L. H. Pammel, and B. Find], *Proceedings IAS* 3 (1895), 15.

90. "Constitution and Bylaws," *Proceedings IPFA* 1 (1902), 4–5.

91. [Macbride], "The Conservation of Our Lakes and Streams," 197–198.

92. Frank Edward Horack, "Legislation of the Thirty-sixth General Assembly," *IJHP* 13 (1915), 501. In 1913 and 1915 the legislature also passed various laws enabling cities, towns, and rural townships to establish publicly supported parks and playgrounds; see Horack, 498–501; Horack, "The Work of the Thirty-fifth General Assembly," *IJHP* 11 (1913), 565.

93. H. C. Ford, "What the State is Doing Toward the Conservation of Iowa Lakes," *Proceedings IFCA* (1916), 176–179.

94. *Report of State Highway Commission, Part One* (1917); R. W. Clyde, "Lake Survey of Iowa," *Iowa Conservation* 3 (1919), 44–45.

95. G. B. MacDonald, "The Value of Iowa's Lakes for the People," *Proceedings IFCA* (1916), 180–184.

96. Louis H. Pammel, "The Preservation of Iowa's Lakes from the Standpoint of a Botanist," *Proceedings IFCA* (1916), 173–175.

97. MacDonald, "The Value of Iowa's Lakes for the People," 183–184.

98. C. F. Curtiss, "Forest and Game Reserves in Iowa," *Proceedings IFCA* (1916), 109–110.

99. Louis H. Pammel, "The Preservation of Natural History Spots in Iowa," *Proceedings IFCA* (1916), 108.

100. Louis H . Pammel, "Report of Executive Committee," "The Park Movement in Iowa," and "Dedication of Twin Lake[s] State Park," undated typescripts, Pammel Papers.

101. See, for instance, "The Preservation of Natural History Spots in Iowa," 106, where he states, "It is our duty as a state to preserve for the future these plants that tell us the story of the development of our land." Richard Lieber voiced a similar argument at about the same time, calling for state parks as historical monuments to Indiana's past; see Greiff, "People, Parks, and Perceptions," 19–20.

102. Pammel, "Report of Executive Committee," Pammel Papers.

103. Chap. 236, *Laws of the Thirty-seventh General Assembly*, 1917.

104. Louis H. Pammel, "What the Legislature Did with Reference to State Parks in Iowa," *Iowa Conservation* 3 (1919), 14–15; Christensen, "The State Parks of Iowa," 339–340.

105. Charles C. Deering, "The Telephone in Iowa," *Annals of Iowa* 23 (1942), 287–308; "Death Claims Holdoegel, 70: Former Teacher and State Senator," *Des Moines Register*, 1 August 1940; biography of Perry C. Holdoegel in *Past and Present of Calhoun County, Iowa*, vol. 2 (1915), 253–255.

106. Louis H. Pammel, "Dedication of Twin Lake[s] State Park," undated typescript, Pammel Papers.

107. Thomas H. Macbride, "Parks and Parks," *Bulletin: Iowa State Parks* [hereafter *Bulletin ISP*] 2 (1922), 13–14.

108. Macbride, "Parks and Parks"; see also "What's What in Parks," undated type-script [c. 1925], Macbride Papers .

CHAPTER 2   Pammel's Way

1. Louis H. Pammel, "Rural Parks," *Iowa State Park Bulletin* 2 (1924), 5.

2. The inspiration for this chapter's title comes from Marjorie Conley Pohl's brief biography of Pammel, "Louis H. Pammel, Pioneer Botanist: A Biography" published in *Proceedings IAS* 92 (1986), 1–50. In it she relates an exchange between geologist George McKay and Pammel regarding the latter's "feistiness" and his insistence "on the right attitude toward things." Pohl notes that "the right way was usually Pammel's way" (11). Much of the biographical information about Pammel in this chapter is based on Pohl's article, although some dates have been corrected.

3. Donald Worster provides a concise summary of Humboldt's place in the history of natural science and his influence on Charles Darwin in *Nature's Economy: A History of Ecological Ideas* (1977), 131–138.

4. Overfield, *Science with Practice*, see chapters 1–3 for the history of Bessey's contributions to botany education at Iowa State University; quotation, 9.

5. Pammel to Members of the State Board of Conservation, 29 April 1922, Pammel Papers.

6. Both controversies and the 1918 campaign are well documented in the William Lloyd Harding Papers, Sioux City Public Library; see also John E. Briggs and Cyril B. Upham, "The Legislation of the Thirty-eighth General Assembly," *IJHP* 17 (1919), 490; Leland L. Sage, *A History of Iowa* (1974), 251–252.

7. E. C. Hinshaw, *Report of the State Fish and Game Warden for the Biennial Period Ending June 30, 1918* (1918), 12–13. Missouri was the first state to succeed in piggybacking state park development onto an existing fish and game department. When Missouri's state park program was begun in 1917, the initial percentage set aside for land acquisition and park development was far too little, however, and Missouri did not acquire its first park until seven years later. Nonetheless, by the late 1920s, state park "experts" viewed fee-generated revenue as a much more secure means of park funding, since it was not tied to the politics of the legislative appropriation process or dependent upon the goodwill of donors. See Susan Flader, ed., *Exploring Missouri's Legacy: State Parks and Historic Sites* (1992), 4–5; Nelson, *State Recreation: Parks, Forests and Game Preserves*, 11–14, 142.

8. The Executive Council comprised the governor, secretary of state, state auditor, state treasurer, and secretary of agriculture.

9. Senator O. A. Byington, "The Passage of the Quail Bill," *Iowa Conservation* 1 (1917), 35–36; Stephens, "A Review of Wild Life Protection in Iowa," 50–52, 60–63; E. C. Hinshaw, "Quail in Iowa," *Twenty-first Report of the State Fish and Game Warden, 1913–1914* (1915), 14; Bennett, "Fish and Game Legislation in Iowa," 419, notes that although the Quail Bill was controversial, petitions supporting the bill outweighed those in opposition.

10. Frank Edward Horack, "The Legislation of the Thirty-seventh General As-

sembly," *IJHP* 15 (1917), 548; Chaps. 111, 202, 233, *Acts of the Thirty-seventh General Assembly*, 1917.

11. "Dr. T. C. Stephens Honored by the Directors of the Permanent Wild Life Protective Fund," *Iowa Conservation* 1 (1917), 1.

12. Stephens, "A Review of Wild Life Protection in Iowa," 49–50.

13. Hinshaw, *Twenty-first Report of the State Fish and Game Warden, 1913–1914*, 10–12; E. C. Hinshaw, *Report of the State Fish and Game Warden for the Biennial Period Ending June 30, 1916* (1916), 5–9, 13–14.

14. Pammel to Gov. N. E. Kendall, 20 June 1921, Pammel Papers.

15. Stephens to G. B. MacDonald, 9 May 1918, enclosing draft resolution by Bohumil Shimek; MacDonald to Pammel, 13 May 1918; Pammel to MacDonald, 16 May 1918, Pammel Papers.

16. Pammel suggested C. F. Curtiss, dean of agriculture at Iowa State College; George Kay, state geologist and professor of geology at the University of Iowa; and archaeologist Ellison Orr, then serving as president of the Iowa Conservation Association. See Pammel to Harlan, 22 September 1917; Pammel to Senator P. C. Holdoegel, 16 October 1917, Pammel Papers.

17. Harlan to Pammel, 31 January 1920, Pammel Papers.

18. Briggs and Upham, "The Thirty-eighth General Assembly," 541; Chap. 368, *Acts of the Thirty-eighth General Assembly*, 1919.

19. Minutes, Board of Conservation [hereafter BC Minutes], 5 September 1919, 14 May 1920, 28 May 1920, 18 June 1920, 28 July 1920, 24–25 September 1920, 19 November 1920, 3 December 1920.

20. John E. Briggs, "The Legislation of the Thirty-ninth General Assembly," *IJHP* 19 (1921), 570; Chap. 135, *Acts of the Thirty-ninth General Assembly*, 1921.

21. John E. Briggs and Jacob Van Ek, "The Legislation of the Fortieth General Assembly," *IJHP* 21 (1923), 615–616; Chap. 33, *Acts of the Fortieth General Assembly*, 1923.

22. BC Minutes, 14 May 1929.

23. See, for instance, BC Minutes, 23 April 1920, 14 September 1923, 4 October 1927, 4 August 1928, 5–6 February 1929, 5 March 1929, 6 August 1929, 7–8 January 1930, 4 March 1930.

24. Pammel to Kendall, 10 October 1922. See also "Meeting for Purpose of Presenting Petition, Etc., to Conservation Board, 3/8/1919," at which Pammel stated that "On this Board there is to be a representative of Iowa State College, State University of Iowa, the fish and game warden, the curator of the State Horticultural [Historical?] Society, a member of the state highway commission and two business men, with business ability." Pammel to R. S. Herrick, 5 November 1921, contains a lengthy explanation of Pammel's foiled attempt to establish a representative of the State Horticultural Society on the board. Without naming names, Pammel makes it clear that "an influential member of the legislature" blocked the attempt. Pammel Papers.

25. Pammel to S. A. Beach, 8 January 1921; Pammel to Albert, 17 January 1921, Pammel Papers.

26. Pammel, "The Arbor Day, Park and Conservation Movements in Iowa," 293.

27. Pammel to Kendall, 27 January and 1 February 1921, 2 October 1922, Pammel Papers.

28. Pammel, untitled and undated report to the members of the State Board of Conservation [contents reveal that it was written shortly after April 1921], Pammel Papers. Runte, in *American Parks: The National Experience*, argues that only areas seen as "wastelands" have been set aside as national parks in the United States. There is no evidence that the economic value (for agriculture, mining, etc.) of proposed state park lands in Iowa was a routine topic of public debates or private correspondence, undoubtedly because practically all arable land in Iowa was already under cultivation. Consequently, there was little or no public land to squabble over.

29. Pammel to Morda V. Coleman, 27 May 1919, Pammel Papers.

30. Pammel, typescript of remarks for introducing Governor Harding at Backbone State Park dedication, 28 May 1920, Pammel Papers.

31. "Meeting for Purpose of Presenting Petition, 3/8/1919," Pammel Papers.

32. The Iowa Federation of Women's Clubs established a committee on good roads in 1913 or 1914 as one of three subcommittees under the Committee on Conservation. Helen Taylor of Bloomfield served as chairman of this subcommittee, and also as vice president of the Iowa Good Roads Association. It is difficult to determine what this relationship initially meant in terms of shaping IFWC activities regarding parks and conservation. This is so because the IFWC's archival record is thin up to 1922, when the federation began publishing a quarterly newsletter. Activities reported in the newsletter show that during the 1920s various subcommittees would focus on highway beautification and promoting "outdoor manners" among the traveling public. See Buffum and Thurston, *History of the Iowa Federation of Women's Clubs*, 85–86, and *Iowa Federation News*, passim 1922–1927.

33. Cox, *The Park Builders*, 28–30.

34. Between 1913 and 1924, sixty-four named highways were officially registered with the Iowa Highway Commission, although many of these routes were nothing more than dirt or graveled roads. They included several "highways" that led to national park or tourist destination points: the National Parks Pike (present-day Highway 18), which ran through northern Iowa on its way from Madison, Wisconsin, to Yellowstone National Park; the Black Hills Highway (also Highway 18), which began in Chicago and ended in Denver; the Custer Battlefield Highway and the Glacier Trail, both of which would carry travelers from several Iowa locations west to Glacier National Park in Montana; and the Atlantic-Yellowstone-Pacific Highway (present-day Highway 9) through northern Iowa. See "Iowa Registered Highway Routes, 1914–1925," in William H. Thompson, *Transportation in Iowa: A Historical Summary* (1989).

35. "Meeting for Purpose of Presenting Petition, 3/8/1919."

36. Ibid.

37. Macbride to Crane, 7 February 1933, Macbride Papers.

38. Samuel Calvin, "The Devil's Backbone," *Midland Monthly* 6 (July 1896), 26.

39. BC Minutes, 27 December 1918, as reported in State Board of Conservation, *Iowa Parks: Conservation of Iowa Historic, Scenic, and Scientific Areas*, (1919), 11–14.

40. BC Minutes, 1 January 1919, as reported in *Iowa Parks*, 14–17.

41. "Meeting for Purpose of Presenting Petition, 3/8/1919." Bohumil Shimek's list of desirable park areas, published in *Iowa Conservation* 1 (1917) 16–17, seems to have been the basic working list.

42. *Iowa Parks*, xiii–xv and passim.

43. *Iowa Parks*, ix–xi.

44. In this regard, Pammel and others certainly were influenced by Frederic Clements' seminal work on "organic," or "dynamic" ecology; see *Plant Succession: An Analysis of the Development of Vegetation* (1916). In it Clements argued that plants grew and matured until a state of equilibrium (the climax stage) emerged. Importantly, Clements did not adequately address the complicating factor of human activity. Thus, then-current scientific thought gave some credence to the notion that parks could serve both conservation and recreation demands, although numerous statements by Pammel and other natural scientists indicate that they remained skeptical about the results and, hence, about the efficacy of Clements's theory of plant succession. Worster details Clements's role in the history of environmental thought in *Nature's Economy*, chap. 11, "Clements and the Climax Community."

45. *Iowa Parks*, 4–5.

46. Harlan to State Board of Conservation, 22 July 1919; "Board of Conservation resolution approving purchase of park at Keosauqua, Iowa," 17 October 1919; Harlan interview with J. Henry Strickling, transcript, 2 July 1921, Pammel Papers and Edgar R. Harlan Papers, State Historical Society of Iowa, Des Moines [hereafter Harlan Papers]; "The State Park to Embrace 1126 Acres," *State Line Democrat*, 24 October 1919.

47. "Use of Area for Recreation," undated typescript of a report on Lacey-Keosauqua State Park by L. H. Pammel, Pammel Papers; BC Minutes indicate that this report may have been presented to the board on 13 September 1922.

48. Lacey-Keosauqua dedication program, 26–27 October 1920; "Lacey-Keosauqua Park Dedicated," *Des Moines Register*, 31 October 1920; "Fine Tract at Keosauqua is Dedicated," *Van Buren Barometer*, 27 October 1920, Pammel Papers.

49. Harlan interview with S. W. Manning, transcript, 2 July 1921, 2–4; Harlan interview with H. E. Blackledge, transcript, 1 July 1921, 2–4, 6, Pammel Papers and Harlan Papers.

50. Blackledge interview, 3, 6–13; Strickling interview, 2–6.

51. BC Minutes, 28–29 January 1921, 22 April 1921, 17–18 June 1921, 19 July 1921, 15 October 1921.

52. Strickling interview, 12–15.

53. BC Minutes, 13 September 1922.

54. Pammel, "Use of Area for Recreation," n.p.; E. A. Piester, "Report on the Proposed Traffic System within Back Bone State Park," report to the Board of Conservation, 11 June 1923, Pammel Papers.

55. BC Minutes, 11 June 1926.

56. "Use of Area for Recreation," n.p.; Peter J. Schmitt in *Back to Nature: The Arcadian Myth in Urban America* (1990) explores the complex idealistic and xenophobic motives of those who promoted outdoor education and outdoor recreation for youth during the early twentieth century.

57. BC Minutes, 1–2 May 1928.

58. Pammel, "Use of Area for Recreation," n.p.

59. Blackledge to Pammel, 25 April 1924; Pammel to Blackledge, 28 April 1924, Pammel Papers.

60. Pammel to Members of the State Board of Conservation, 28 April 1924; Saunders to Pammel, 29 April 1924, Pammel Papers.

61. Harlan to Pammel, 1 May 1924, Pammel Papers.

62. Pammel to W. E. Albert, 18 June 1924; Pammel to Harold E. Pammel, 24 June 1924; Harlan to Pammel, 5 August 1924; Pammel to Harlan, 24 September 1924; Pammel to W. C. Merckens, 4 October 1924; Pammel to P. H. Elwood, 19 November 1924, Pammel Papers.

63. BC Minutes, 21 July 1924.

64. Pammel to Blackledge, 18 June 1924; Blackledge to Pammel, 21 June 1924; Harlan to Pammel, 5 August 1924, Pammel Papers. Letters indicate that as many as eight copies of the plan were produced, but none of them seems to have survived.

65. Pammel to Members of the Board, 14 November 1924; Pammel to Rees, 24 November 1924; Pammel to Harold Pammel, 6 January 1924; Pammel to Harold Pammel, 8 January 1925; "Petition to Members of the State Board of Conservation," 25 February 1925; Blackledge to Pammel, 2 March 1925; Pammel to Rees, 10 March 1925, Pammel Papers .

66. D. R. Therme, Lacey-Keosauqua State Park Golf Club, to Pammel, 14 May 1925; Blackledge to Pammel, 15 May 1925; Pammel to Blackledge, 18 May 1925; C. W. Clarke, Centerville Association of Commerce, to Pammel, 29 May 1925; Pammel to Clarke, 2 June 1925; Pammel to W. C. Merckens, 11 June 1925; Pammel to Rees, 17 July 1925, Pammel Papers.

67. Arthur J. Secor, County Agricultural Agent, to Pammel, 27 February 1925; Pammel to B. S. Pickett, ISC Horticulture Department, 10 March 1925; Pickett to Pammel, 12 March 1925; Pammel to Pickett, 14 March 1925, Pammel Papers.

68. Pammel to Rees, 24 March 1925; Pammel to W. M. Rockel, Possum Hollow Game Farm, 31 March 1925; Pammel to Rees, 26 October 1925; Pammel to Rees, 29 October 1929; Pammel to C. H. Diggs, ISC Landscape Architecture Department, 12 November 1925, Pammel Papers; BC Minutes, 4 October 1927.

69. Secor to Pammel, 29 August 1925; Pammel to Blackledge, Rees, and Secor, 31 August 1925; notice of "Big Outdoor Natural Life and Conservation Program, September 10th," Pammel Papers; "Lacey Conservation Day at Keosauqua," Bulletin ISP 3 (1925), 41–43.

70. Pammel to Blackledge, 30 June 1925; Pammel to Rees and Blackledge, 15 October 1925; Merckens to Pammel, 14 October 1925; Blackledge to Pammel, 28 October 1925; Pammel to Blackledge, 29 October 1925; Blackledge to Pammel, 3 December 1925; Pammel to Clifford L. Niles, 3 December 1925; Niles to Pammel, 4 December 1925; Niles to Blackledge, 4 December 1925; Blackledge to Merckens, 7 December 1925; Blackledge to Gov. John Hammill, 7 December 1925; Blackledge to Niles, 7 December 1925, Pammel Papers.

71. Pammel to Berenice Lacey Sawyer, 30 July 1926; Pammel to Hammill, 23 August 1926, Pammel Papers.

72. MacDonald to Pammel, 2 December 1926; Pammel to Merckens, 3 December 1926, Pammel Papers.

73. State Board of Conservation, *Administration of Iowa Parks, Lakes, and Streams* (1931), 19.

74. Deer nonetheless were reintroduced in Backbone and Ledges. In both locations, they proved to be so prolific that by the mid-1930s the Board of Conservation was forced to take steps to reduce their numbers; see BC Minutes, 8 February 1935.

75. Louis Pammel, "The Arbor Day, Park and Conservation Movements in Iowa," 293.

76. Frankel to Pammel, 16 June 1930; Pammel to Darling, 1 July 1930; Pammel to Trelease, 7 July 1930, Pammel Papers.

77. BC Minutes, 8 April 1931.

78. Torrey, *State Parks and Recreational Uses of State Forests in the United States*, passim; "Raymond Torrey Visits State Parks in Iowa," *Bulletin ISP* 3 (1925), 44–46.

79. Nelson, *State Recreation*, 5.

80. *Administration of Iowa Parks, Lakes, and Streams*, 4.

81. Jean C. Prior, *Landforms of Iowa* (1991), is a well-illustrated, accessible overview of Iowa's geology; see especially the map on page 110 of state parks, state preserves, state forests, national monuments, and national wildlife refuges where one may catch exemplary views of the state's varied landforms, followed by brief descriptions of what to look for.

82. Compiled from information contained in the 1919 and 1931 Board of Conservation reports and from current information available from the Iowa Department of Natural Resources, county conservation boards, and city park departments.

83. "Attendance at State Parks During the Season of 1923," *Bulletin ISP* 1 (1923), 11–12; *Administration of Iowa Parks, Lakes, and Streams*, 30–31.

CHAPTER 3   Reshaping Park and Conservation Goals in the 1920s

1. [Louis Pammel], "Conservation," *Bulletin ISP* 3 (November–December 1925), 66.

2. Donald C. Swain, *Federal Conservation Policy, 1921–1933* (1963).

3. Carol A. Buckmann, *The First 50: The Story of the Iowa Division, Izaak Walton League of America, 1923–1973* (1973), 10.

4. George Bennett, "The National Park Conference at Des Moines, Iowa, January 10–11–12, 1921," *Iowa Conservation* 5 (1921), 14–25.

5. "Minutes and Resolutions of the Iowa Conservation Convention at Ames, Jan 7 and 8, 1921," *Iowa Conservation* 5 (1921), 43.

6. Minutes of the Business Meeting Iowa Conservation Association, 7–8 January 1921, Pammel Papers.

7. MacDonald to Pammel, 22 May 1918; E. A. Bump to MacDonald, 1 April 1919; MacDonald to Bump, 10 April 1919; MacDonald to Pammel, 10 April 1919; Pammel to H. S. Conard, 28 April 1920; MacDonald to Conard, 27 May 1920; Pammel to Charles R. Keyes, 12 December 1921; Pammel to Rev. LeRoy Titus Weeks, 14 March

1922; Minutes, Meeting of Executive Board of the Iowa Conservation Association, 12 January 1924; MacDonald to Pammel, 16 January 1924, Pammel Papers.

8. Program, Iowa Conservation Association, 7–8 March 1924; Pammel to J. A. Rawlings, Associated Press, 2 February 1925, Pammel Papers.

9. The *Bulletin* did, however, set a precedent. In 1942, the State Conservation Commission, successor to the Board of Conservation, began publishing a similar newsletter, *Iowa Conservationist*, which continues to the present.

10. *Wildways* 8 (January–March 1928), 1.

11. Jacob A. Swisher, "The Iowa Academy of Science," *IJHP* 29 (1931), 358, 366.

12. Swisher, 368; Robert W. Hanson, "The Iowa Academy of Science: 1875–1975," *Proceedings IAS* 82 (1975), 5–6.

13. James H. Lees, "Some Geologic Aspects of Conservation," *Proceedings IAS* 24 (1917), 133–154.

14. Samuel W. Beyer, "Some Problems in Conservation," *Proceedings IAS* 26 (1919), 37–46.

15. Bohumil Shimek, "Drainage in Iowa," *Proceedings IAS* 31 (1924), 149–155.

16. George Bennett, "Keeping Iowa's Water Pure," *Proceedings IAS* 31 (1924), 431–435.

17. A. H. Wieters, "Status of Stream Pollution in Iowa," *Proceedings IAS* 35 (1928), 63–67.

18. *Proceedings IAS* 27 (1920), 20–21.

19. "Report of the Resolutions Committee, *Proceedings IAS* 31 (1924), 25–26.

20. "Report of the Committee on Biological Survey," *Proceedings IAS* 31 (1924), 21–22; O. H. Smith [IAS president] to Pammel, 31 January 1925; Pammel to Smith, 2 February 1925; undated draft of legislative bill for Natural History Survey, Pammel Papers.

21. "Report of the Committee on Biological Survey," *Proceedings IAS* 36 (1929), 27; Hanson, "The IAS, 1875–1975," 7.

22. "Report of the Committee on Conservation in Connection with Public Schools," *Proceedings IAS* 31 (1924), 22–23.

23. Bohumil Shimek, "Report of the Committee on Conservation," *Proceedings IAS* 34 (1927), 31–33.

24. BC Minutes, 5 March 1929.

25. BC Minutes, 1–2 May 1928, 11 December 1928, 8 January 1929, 5–6 February 1929, 5 March 1929, 6 April 1929, 6 August 1929, 3 October 1929, 5 November 1929, 7–8 January 1930, 4 February, 1930, 4 March 1930, 1 April 1930, 5 August 1930.

26. Pammel to McNider, 2 April 1923, Pammel Papers.

27. Armstrong to Pammel, 21 March 1927, Pammel Papers.

28. Whitley states in a 1923 article that "more than twice as many women are now members of state Forestry or Conservation Boards as there were two or three years ago." See "Conservation of Natural Resources," *Iowa Federation News* 4 (September– October 1923), 15. State studies, however, are lacking to confirm her observation and to determine what influence this gave women over policy formation. Sandra Jeanne Johnson, "Early Conservation by the Arizona Federation of Women's Clubs from 1900

to 1932" (M.S. thesis, 1993), is a pioneering study of clubwomen in conservation at the local and state levels. Her research indicates that, in Arizona, clubwomen took credit for influencing certain legislation, but they did not participate in policy formation. Additional state-level histories are needed before any generalizations can be made.

29. Carolyn Merchant, "Women of the Progressive Conservation Movement, 1900–1916," *Environmental Review* 8 (1984), 57–85, argues that as utilitarian conservationists gained a foothold in the federal bureaucracy and in professional organizations such as the American Forestry Association, they had less need for support from women's organizations. Women, for their part, began to see themselves as ideologically estranged from the utilitarian conservation approach, which increasingly accommodated commercial and material values. However, because the women of the conservation movement all enjoyed the benefits of comfortable middle-class lifestyles, there was no incentive to challenge the basic tenets of the economic system on which utilitarian conservation thought rested. Merchant's observations, in turn, are the basis of Sally Ranney's argument that after 1913 women had no real leadership roles in the conservation movement for about fifty years, until the 1960s, when they once again took a position in grassroots environmental leadership. See Sally Ann Gumaer Ranney, "Heroines and Hierarchy: Female Leadership in the Conservation Movement," in *Voices from the Environmental Movement*, ed. Donald Snow (1992), 41–42.

30. See, for instance, "Conservation of Natural Resources" (1920–1922), "A Plea for the Wild Flowers" [c.1922–1924], "Forestry and Natural Scenery" [c.1925–1926], and similar, though less substantive, printed reports and circulars for the period 1920–1930, General Federation of Women's Clubs Archive, Washington, D.C.; "The Second National Convention of the Izaak Walton League of America," *Outdoor America* 2 (May 1924), 54–55; Program for the Annual Meeting of the American Forestry Association and the Southern Forestry Congress, 6–7 January 1926, Whitley Papers.

31. "The Great Mississippi Valley Enterprise," *Iowa Conservation* 7 (1923), 63; Whitley, "Department of Applied Education, Conservation of Natural Resources," *IFWC Year Book 1925–26*, 251.

32. ICA Annual Meeting programs for 7–8 March 1919, 2–4 March 1922, 7 March 1924.

33. Mrs. Francis E. Whitley, "Forestry in State and Nation," WHO Radio Talk, 29 April 1927, printed text distributed by the General Federation of Women's Clubs, Whitley Papers.

34. Mrs. Francis Edmund Whitley, "Women's Clubs and Forestry: A Record of Active and Consistent Support of Forest Conservation by the Organized Women of America," *American Forests and Forest Life* 32 (1926), 221.

35. Cora Whitley, "Outdoor Good Manners: A Campaign of Education," circular for GFWC clubwomen, Whitley Papers; see also "Committee on Conservation of Forestry and Wild Life Refuges," *Iowa Federation News* 5 (January–February 1925), 7.

36. Mrs. Francis E. Whitley, "Outdoor Good Manners," Bulletin No. 20, American Nature Association, Whitley Papers.

37. "Mrs. Whitley Offers Prize to School Children," *The Iowa Clubwoman* 7

(March–April 1927), 11. Unfortunately, the IFWC archives contain no further information or materials concerning this contest.

38. Outdoor Good Manners promotional materials, Whitley Papers.

39. Ibid.; also "The Nature Lovers," *Nature Magazine*, July 1925, 32–33.

40. Jardine to Whitley, 3 January 1927, published in *The Iowa Clubwoman* 7 (March–April 1927), 11.

41. "Etiquette of the Outdoors Invoked Throughout Nation," *Christian Science Monitor*, 4 May 1925; "Good Manners Save Forests," *Des Moines Register*, 5 May 1925.

42. Swain, *Federal Conservation Policy, 1921–1933*, 12–17.

43. Schmitt, *Back to Nature*, especially chap. 15, "The Search for Scenery."

44. Mrs. Francis Edmund Whitley, "Our State Parks," *Bulletin ISP* 1 (April 1924), 15–16.

45. Pammel to Harlan, 18 December 1920, Pammel Papers.

46. Ibid.

47. G. B. MacDonald to Pammel, 7 May 1921, Pammel Papers.

48. Erwin to Pammel, 5 August 1919; R. A. Pearson to Pammel, 6 August 1919; Pammel to Pearson, 25 August 1919, Pammel Papers.

49. Pammel to Erwin, 25 August 1919, Pammel Papers.

50. BC Minutes, 9 March 1923. Pammel appears to have polled park custodians earlier than that, sometime in 1921, to find out what improvements were desired; undated typescript, "Improvements for State Parks (Suggested by Park Custodians)," Pammel Papers.

51. Robinson to McNider 16 May 1923; Robinson to Pammel, 5 June 1923; McNider to Pammel, 6 June 1923; "Agreement for the Services of [Pearse-Robinson] Landscape Architects on the Property of Board of Conservation," Pammel Papers.

52. Pammel to P. H. Elwood, Jr., Landscape Architecture Department, 4 October 1923; Elwood to Pammel, 5 October 1924; Pammel to C. L. Niles, 1 December 1923, Pammel Papers.

53. BC Minutes, 9 March 1923, 11 May 1923, 8 June 1923.

54. BC Minutes, 13 April 1923.

55. BC Minutes, 23 April 1920 and 14 September 1923; "Report of H. E. Pammel, Landscape Engineer, on Location of Permanent Features in Backbone State Park," 21 September 1923, Pammel Papers.

56. Pammel to H. E. Blackledge, 18 June 1924; Pammel to P. H. Elwood, 19 November 1924, Pammel Papers.

57. BC Minutes, 14 March 1924; W. E. G. Saunders to Pammel, 13 June 1924; R. E. Johnson, secretary of the Executive Council, to members of the Board of Conservation, 24 June 1924; Saunders to Pammel, 28 June 1924; Pammel to Saunders, 2 July 1924; R. K. Bliss, director, State Extension Service, to Pammel, 13 October 1924; Pammel to Bliss, 7 October 1924, Pammel Papers.

58. BC Minutes, 14 March 1924.

59. BC Minutes, 21 July 1924.

60. BC Minutes, 13 February, 12 June, and 14 August 1925.

61. Fitzsimmons's designs, for instance, include the converted custodian's resi-

dence (1925), the barn (1925), and a 1931 stone auditorium at Backbone State Park; the remodeled lodge (c. 1926) adjacent to the golf course in Lacey-Keosauqua; custodians' houses for Maquoketa Caves (1931, not built) and Waubonsie (1931); and shelters for Ft. Defiance and Twin Lakes (1931), Iowa Department of Natural Resources and Fitzsimmons Collection, Iowa State University.

62. Fitzsimmons Collection.

63. Linda Flint McClelland, *Presenting Nature: The Historic Landscape Design of the National Park Service, 1916–1942* (1993), 7–9, 80.

64. *Administration of Iowa Parks, Lakes, and Streams*, 15–25, 29; BC Minutes, 6 August 1929. Lodges were located at Ambrose A. Call, Pammel, Dolliver (2), Ledges, Bellevue, and Lacey-Keosauqua.

65. Pammel to Members of the State Board of Conservation, 2 February 1927; Pammel to G. B. MacDonald, 10 February 1927, Pammel Papers.

66. *Administration of Iowa Parks, Lakes, and Streams*, 31.

67. G. Perle Wilson Schmidt, "Preservation of Places of Historic Interest in Iowa," in *Iowa Parks*, 211.

68. Ibid.

69. "Our Lakes and State Parks," *Bulletin ISP* 3 (January–February 1926), 120.

70. J. A. Swisher, "Historical and Memorial Parks," *Palimpsest* 12 (1931), 202–204.

71. BC Minutes, 28 May 1920.

72. BC Minutes, 10–12 January 1921, 14 July 1922 and 10 November 1922; "Iowa State Parks," *Annals of Iowa* 13 (1921), 142.

73. BC Minutes, 4 September 1928, 5–6 February 1929, 5 May 1931.

74. Swisher, "Historical and Memorial Parks," 207–218.

75. Torrey, *State Parks and Forests*, 21–24, 94–96, 235–236; Meyer, *Everyone's Country Estate*, 1–22; Nelson, *State Recreation*, 205–208.

76. Swain, *Federal Conservation Policy, 1921–1933*, 142–143, argues that the rise of nativism and nationalism in the 1920s contributed to the success of the National Park Service. He cites, in particular, the activities of such groups as the Daughters of the American Revolution.

77. *Administration of Iowa Parks, Lakes, and Streams*, 15–24.

78. Harlan interview with H. E. Blackledge, 1 July 1921.

79. Chap. 236, *Acts of the Thirty-seventh General Assembly*, 1917.

80. Chap. 52, *Laws of Iowa*, 1906.

81. MacDonald to Pammel, 1 December 1920; MacDonald to Pammel, 19 January 1921, Pammel Papers.

82. Pammel to S. A. Beach, 5 January 1921; Beach to Pammel, 13 January 1921 [Beach was chief of Horticulture and Forestry at Iowa State College and, at the time, also president of the SHS]; Pammel to E. R. Harlan, 8 April 1921, Pammel Papers; see also BC Minutes, 3 December 1920.

83. Pammel to Foskett, 17 January 1921, Pammel Papers.

84. Pammel to Horchem, 31 January 1921, Pammel Papers.

85. Sherman to Pammel, 1 February 1921; Sherman to W. P. Dawson, 1 February 1921, Pammel Papers.

86. MacDonald to Pammel, 14 February 1921, Pammel Papers.

87. Pammel to Sherman, 3 February 1921; Beach to Pammel, 3 February 1921; Sherman to Pammel, 9 February 1921; Pammel to Sherman, 11 February 1921; S. A. Beach to Francis Sestier, 18 January 1921, Pammel Papers.

88. Pammel to W. P. Dawson, 3 February 1921, Pammel Papers.

89. Senate file 505, "A Bill for an act to encourage the planting and conservation of trees," 14 February 1921, Pammel Papers.

90. Horchem to Pammel, 12 April 1921; Pammel to Newberry, 13 April 1921; Pammel to Newberry, 16 April 1921, Pammel Papers.

91. Pammel to Herrick, 25 April 1921, Pammel Papers.

92. That rift continues to the present; the board's successor agencies, the Iowa Conservation Commission and the Iowa Department of Natural Resources, have been at odds with the Iowa Association of Nurserymen ever since.

93. G. B. MacDonald, "Reforestation in Iowa," text of address delivered at the Iowa Conservation Association meeting on July 15, 1921, as published in *Iowa Conservation* 5 (1921), 101–102.

94. L. H. Pammel, "The State Park Movement and Reforestation," undated typescript of an address delivered before the Iowa Conservation Association, probably the March 2–4, 1922, meeting, Pammel Papers.

95. MacDonald to Pammel, 19 January 1923; Newberry to Pammel, 9 February 1923; MacDonald to members of the Iowa Conservation Association [general letter], 21 February 1923, Pammel Papers.

96. Herrick to MacDonald, 28 February 1923, Pammel Papers.

97. Minutes of the Business Meeting of the Iowa Conservation Association, February 27, 28, 1923, typescript, Pammel Papers.

98. Briggs and Van Ek, "The Legislation of the Fortieth General Assembly," 528–529, 552, 615.

99. Pammel to Stoddard, 3 April 1923; Stoddard to Pammel, 5 April 1923; Berenice L[acey] Sawyer to Pammel, 9 April 1923; Pammel to Sawyer, 11 April 1923, Pammel Papers.

100. BC Minutes, 3 April 1920, 26 April 1922, 9 May 1922.

101. G. B. MacDonald, "Forestry and the Iowa Farmer," *Ames Forester* 10 (1922), 59–62.

102. Pammel to C. L. Niles, 1 December 1923, Pammel Papers.

103. General letter from MacDonald to members of the Iowa Conservation Association, 11 February 1924, Pammel Papers.

104. MacDonald to U.S. Forester [William Greeley], 13 December 1924, Pammel Papers.

105. MacDonald to E. A. Sherman [USDA], 5 August 1924, Pammel Papers; G. B. MacDonald, "Reforestation Work Under Clarke-McNary Act in Iowa," *Ames Forester* 14 (1926), 21; G. B. MacDonald, "Forestry Progress in Iowa," *Ames Forester* 29 (1941), 10–12.

106. Michael Williams, *Americans and Their Forests: A Historical Geography* (1989), 440–454; Swain, *Federal Conservation Policy, 1921–1933*, 9–29.

107. Runte, *National Parks*, 82–105, quotation 91.

108. Pammel to Stoddard, 3 April 1923, Pammel Papers; "Attendance at State Parks During the Season of 1923," *Bulletin ISP* 1 (July 1923), 11–12.

109. *Administration of Iowa Parks, Lakes, and Streams*, 17–18, 20–22.

110. BC Minutes, 14 May 1929, 3 October 1929, 7–8 January 1930, 4 March 1930, 1 April 1930, 5 August 1930, 2 June 1931.

111. [Louis Pammel], "Report of State Board of Conservation," in *Journal of the House*, Thursday, 17 February 1921, 554.

112. Harlan to Pammel, 31 January 1920; Pammel to Harlan, 2 February 1920; Pammel to Sterling, 18 February 1921, Pammel Papers.

113. Pammel to Sterling, 18 February 1921; see also HF 842 by House Committee on Conservation; SF 783 [substitution bill] by Senate Committee on Conservation; Pammel to unidentified recipient, 13 May 1921, Pammel Papers.

114. Briggs, "The Legislation of the Thirty-ninth General Assembly of Iowa," 570; Chap. 135, *Acts of the Thirty-ninth General Assembly*.

115. Pammel to Charles Rhinehart, Chair, House Committee on Conservation of Natural Resources, 13 March 1926, Pammel Papers. A "meandered" lake or stream refers to those bodies of water that at the time of the original Government Land Office surveys were so surveyed as to plat and compute the acreage of adjacent fractional sections of land. The line of meander typically was, and is, the mean high water mark.

116. Hays, *Conservation and the Gospel of Efficiency*, 114–121.

117. The name changed to Iowa Light and Power in 1923.

118. "Before the State Board of Conservation: In the Matter of the Overflowing of Certain Lands in the State Park Near the City of Boone, Iowa, by Reason of the Proposed Hydro-Electric Project of the Iowa Traction Company," filed with the Board of Conservation on 17 May 1922, Pammel Papers.

119. BC Minutes of meeting held at Boone, 14 July 1922; Harlan to Pammel, 3 October 1922; Pammel to Harlan, 5 October 1922; Harlan to Pammel, 20 December 1922; Pammel to Harlan, 23 December 1922; Pammel to Harlan, 1 February 1923; Pammel to Harlan, 14 February 1923, Pammel Papers.

120. W. G. Haskell to Attorney General Ben J. Gibson, 2 September 1922, Pammel Papers.

121. "A Park and a Dam," *Boone News-Republican*, 13 February 1923.

122. "A Report on the Effect of the Proposed Location of a Power Dam in the Des Moines River on the Park Values of the Ledges," presented to Board of Conservation, 21 February 1923, Pammel Papers.

123. "Abstract of Conference between Board of Conservation & Parties Interested in the erection of a dam by the Iowa Light & Power Company to affect The Ledges State Park" [9 February 1923], 7–8, Pammel Papers.

124. Ibid., 19–20.

125. BC Minutes, 9 February 1923.

126. Pammel to McNider, 2 April 1923; Pammel to Berenice L. Sawyer, 11 April 1923, Pammel Papers.

127. Pammel to McNider, 10 April 1923; F. C. Sampson [Rep. Audubon County]

to Pammel, 14 April 1923; Pammel to McNider, 14 April 1923, Pammel Papers; *Journal of the House*, 13 April 1923, 1718–1719.

128. McNider to Pammel, 3 April 1923; Pammel to McNider, 4 April 1923, Pammel Papers.

129. Briggs and Van Ek, "Legislation of the Fortieth General Assembly of Iowa," 615–616; Pammel to McNider, 2 April 1923, Pammel Papers.

130. "Synopsis of Engineer's Report: Muscatine Hydro-Electric, Muscatine, Iowa" [c. December 1924]; Louis Pammel, "Report on the Dam Project at Muscatine—Directing Water from the Cedar River," 23 December 1924; Pammel to W. C. Merckens, 24 December 1924; M. L. Hutton, "Report on Water Diversion and Water Power Project—Cedar River, Muscatine County, Iowa [January 1925], Pammel Papers. In an unrelated matter, Pammel expressed the opinion that the Board of Conservation should "provide ample reservoirs to hold the water back in natural and artificial lakes." See Pammel to E. W. Nelson, chief, Bureau of Biological Survey, Washington, D.C., 6 October 1921, Pammel Papers.

131. C. V. Findlay to Mrs. E. F. Armstrong, 30 September 1927; Armstrong to Pammel [c. 20 October 1927], Pammel Papers.

132. BC Minutes, 4 October 1927; open letter from Mary C. Armstrong to the Citizens of Webster County, 18 October 1927, Pammel Papers.

133. The Pammel Papers contain numerous letters and documents pertaining to the proposed dam. Cited here are Pammel to Hammill, 24 October 1927; Pammel to the secretary of state, the state auditor, the state treasurer, and the secretary of agriculture, 25 October 1927; "A Proposed Dam in the Dolliver Memorial Park, Fort Dodge," interview statement written by Pammel, 25 October 1927; Pammel to members of the State Board of Conservation, 25 October 1927; Pammel to L. D. Weld [Coe College, Iowa Academy of Science], 9 November and 15 November, 1927; Pammel to Mrs. F. E. Whitley, 28 November 1927; Pammel to A. L. Bakke, I. E. Melhus, J. E. Guthrie, C. J. Drake, E. F. Smith [Iowa State College], 28 November 1927; Pammel to James H. Lees, 2 December 1927; Pammel to H. S. Conard, 3 December 1927; Pammel to Bohumil Shimek, 3 December 1927.

134. BC Minutes, 6 December 1927.

135. BC Minutes, 6 and 8 December 1927; Minutes, Executive Council, 12 December 1927; George Bennett, "Dam at Dolliver Park?" *Des Moines Register*, 16 December 1927.

136. BC Minutes, 11 December 1928.

137. See BC Minutes 15 July 1929, 6 May 1930, 3 February 1931, and 9 February 1932 for board action on subsequent proposals.

138. This matter never got beyond discussion, however. The idea was abandoned after consulting landscape architect R. J. Pearse pointed out that such a structure would require expensive floodgates and, even with such protection, would expose large areas of the park to flooding. Pammel to McNider, 24 November 1922; E. I. Leighton, president, Fort Dodge Chamber of Commerce, to Pammel, 13 February 1923; Pammel to Leighton, 14 February 1923; R. J. Pearse, ASLA, "Report on the Existing Conditions and Proposed Landscape Development on the Tract of Land Near LeHigh, Iowa, Known as Dolliver Park," 19 June 1923, Pammel Papers.

139. BC Minutes, 6 October 1931.

140. [Macbride], "The Conservation of Our Lakes and Streams," 197.

141. Frank Edward Horack, "The Work of the Thirty-fifth General Assembly of Iowa," *IJHP* 11 (1913), 572–575.

142. Minutes of the Business Meeting, Iowa Conservation Association, 7–8 March 1919, Pammel Papers.

143. See Martin V. Melosi, *Garbage in the Cities: Refuse, Reform, and the Environment, 1880–1980* (1981); Gottlieb, *Forcing the Spring*, 51–59.

144. Joel A. Tarr, James McCurley, and Terry F. Yosie, "The Development and Impact of Urban Wastewater Technology: Changing Concepts of Water Quality Control, 1850–1930," in *Pollution and Reform in American Cities, 1870–1930*, ed. Martin V. Melosi (1980), 73–77.

145. Harris F. Seidel, ed., *Iowa's Heritage in Water Pollution Control* (1974), 4–12, 479.

146. Briggs and Van Ek, "The Legislation of the Fortieth General Assembly of Iowa," 613; Chap. 37, *Acts of the Fortieth General Assembly*.

147. Buckmann, *The First 50*, 19–20, 37–38.

148. Seidel, *Iowa's Heritage in Water Pollution Control*, 331–332, 457–458; Minutes, Business Meeting of the Iowa Conservation Association, 7–8 March 1924, Pammel Papers; Jacob A. Swisher, "The Legislation of the Forty-first General Assembly of Iowa," *IJHP* 23 (1925), 581–582.

149. BC Minutes, 1 November 1927.

150. Seidel, *Iowa's Heritage in Water Pollution Control*, 335–338, State Department of Health, *Report on the Investigation of Pollution of the Des Moines River From Estherville to Farmington, 1928–1934* (Des Moines, 1934).

151. BC Minutes, 2 September 1930, 9 December 1930.

152. *Administration of Iowa Parks, Lakes, and Streams*, 5.

CHAPTER 4   Toward a Resource Agency

1. From J. N. Darling's untitled speech before the 1930 Annual Convention of the Izaak Walton League of America, Iowa Division, August 15, 1930, n.p., IWLA–Iowa Division Records, Iowa State University [hereafter IWLA–Iowa].

2. These New Deal acronyms stand for the Civilian Conservation Corps (CCC), which operated from 1933 to 1942; the Civil Works Administration (CWA), a short-term program to relieve distress during the winter of 1933–34; the Public Works Administration (PWA), funded under the 1933 National Industrial Recovery Act and operated from then until 1935; the Works Progress Administration (WPA), created in 1935 and terminated in 1943; and the National Youth Administration (NYA), a division of the WPA for jobless youth and college students.

3. David L. Lendt's biography, *Ding: The Life of Jay Norwood Darling* (1979), covers the entire sweep of Darling's life, placing his involvement in conservation issues within the perspective of his career in journalism.

4. Scarpino, *Great River*, 114–129; Buckmann, *The First 50*, 3–9.

5. Lendt, 37, cites evidence that Darling was among the founders of the Iowa

division. Carol Buckmann, however, does not mention Darling in her account of the Iowa division's founding and membership growth (10, 13–15, 100), although elsewhere she acknowledges Darling's many contributions to conservation and the IWLA.

6. Scarpino, *Great River*, 129–139; George Bennett, "The National Park of the Middle West," *Iowa Conservation* 2 (1918), 43–46; Anfinson, "Commerce and Conservation on the Upper Mississippi River," 295–299; Harry C. Oberholser, "The Winneshiek Bottoms Drainage Project," *Iowa Conservation* 7 (1923), 9–10; [George Bennett], "The American School of Wild Life Protection, McGregor Heights—1923," *Iowa Conservation* 7 (1923), 40–44; House Committee on Agriculture, *Hearings on H.R. 4088, A Bill to Establish the Upper Mississippi River Wild Life and Fish Refuge*, 68th Cong., 1st Sess., 1924; Peter T. Harstad and Bonnie Lindemann, *Gilbert N. Haugen: Norwegian-American Farm Politician* (1992), 179–181; Haugen chaired the House Committee on Agriculture from 1919 until his defeat in the 1932 election; he supported the wildlife refuge bill in deference to his constituents.

7. Anfinson, "Commerce and Conservation," 399–410; Raymond H. Merritt, *The Corps, the Environment, and the Upper Mississippi River Basin* (1984), 45, 54–59; Swisher, "The Legislation of the Forty-first General Assembly of Iowa," 582–584; "Large Wild-Life Refuge Being Established on Upper Mississippi," *Bulletin ISP* 3 (1925), 91–93.

8. Ira N. Gabrielson, *Wildlife Refuges* (1943), 8, 14–15, 23.

9. *Hearings on H.R. 4088*, 3.

10. Donald Swain, *Federal Conservation Policy in the 1920s*, 31–48.

11. Scarpino, *Great River*, 119.

12. Scarpino, *Great River*, 141–150, views the fight for the Upper Mississippi River Wildlife and Fish Refuge Act as the "first modern environmental campaign" because it was the first time that organized grassroots pressure was successfully combined with expert testimony and direct lobbying of key government officials.

13. Buckmann, *The First 50*, 19–23, 28–35, 100; Henry S. Conard, "Notes for a Biography of Bohumil Shimek" (1946), 32–34, Bohumil Shimek Papers, University of Iowa.

14. "Izaak Waltons Rap Iowa Fish and Game Men," *Des Moines Register*, 5 February 1927; "Waltons Urge Change in Law," *Des Moines Register*, 12 February 1929; L. B. Sample, Sioux City Rod and Reel Club, to C. F. Weaver, Mason City, 7 June 1930; Ed H. Prior, Mason City, to E. S. Gage, Mason City, 29 September 1930, IWLA-Iowa.

15. J. N. Darling to R. G. Townsend, 7 March 1959, IWLA-Iowa. This letter is one of several collected in the late 1950s and early 1960s when the Iowa Division began to document its own history. Republican Dan Turner, a former state senator and a reform-minded politician, served only one term as governor, 1931–1933. Although he made government reorganization and tax reduction top priority issues, deteriorating economic conditions and farm distress created enormous public sentiment for political change. Turner lost his bid for reelection (though it was a narrow defeat) in the great Democratic landslide of 1932; Sage, *A History of Iowa*, 274–286.

16. Chap. 26, *Acts of the Forty-fourth General Assembly*; Jacob A. Swisher, *The Legislation of the Forty-fourth General Assembly of Iowa*, Iowa Monograph Series, no. 3, ed. Benjamin F. Shambaugh (1932), 68–69.

17. Swisher, *Legislation of the Forty-fourth General Assembly*, 73; Chap. 337, *Acts of the Forty-fourth General Assembly*.

18. J. N. Darling untitled speech, August 15, 1930.

19. Buckmann, *The First 50*, 37.

20. The other commissioners were Dr. W. C. Boone of Ottumwa, Dennis Goeders of Algona, Dr. J. F. Walter of McGregor, and A. E. Rapp of Council Bluffs.

21. Minutes, Fish and Game Commission [hereafter FG Minutes], 7 May 1931.

22. BC Minutes, 2 June 1931.

23. BC Minutes, 4 August 1931, 22 September 1931.

24. Darling to Crane, 20 November 1941, Jay N. Darling Papers, University of Iowa [hereafter Darling Papers]. So thorough was Darling's recasting of the origins of the state conservation plan and the State Conservation Commission that the key players are also misreported in his biography; see Lendt, 60.

25. Herbert Evison, "Plans for State Park Systems," in *American Civic Annual*, ed. Harlean James (1932), 135–139; Stiles, "History of Wildlife Legislation, 86; see also Lendt, 60, 158.

26. Jacob L. Crane, Jr., and George Wheeler Olcott, *Report on the Iowa Twenty-five Year Conservation Plan* (1933), 15–17. Leopold's "Report of the Iowa Game Survey" was published in four installments in *Outdoor America* 11 (1932–1933). Technically, Leopold's survey was principally funded by the Sporting Arms and Ammunition Manufacturers' Institute, which had engaged Leopold to conduct a more extensive game survey of North America; Iowa, however, paid Leopold a small consulting fee to finish the report in 1932 when SAAMI funding dried up. See Curt Meine, *Aldo Leopold: His Life and Work* (1988), 278–290.

27. Jacob L. Crane, Jr., "Preparation of the Iowa Conservation Plan," in *American Civic Annual* (1932), 140–141.

28. Crane and Olcott, *Twenty-five Year Conservation Plan*, 1–2.

29. Ibid., 8–9, 119–143.

30. Ibid., 3–11, 163–167.

31. BC Minutes, 2 June 1931, 15 June 1931.

32. Darling to Crane, 10 August 1941, Darling Papers.

33. James B. Trefethen, *Crusade for Wildlife* (1961), 264–272; Lendt, 63–77; Susan L. Flader, *Thinking Like a Mountain* (1974), 27, 132.

34. FG Minutes, 17 August 1931, 6 October 1931, 22 January 1932. The commission and the board met in joint session for the first time on 3 November 1931. The minutes of both agencies record that between then and December 1934 the two agencies met jointly twelve times. On a more regular basis, the superintendent of parks, M. L. Hutton, attended Fish and Game Commission meetings; likewise, the executive secretary of the Fish and Game Commission, I. T. Bode, often attended Board of Conservation meetings.

35. FG Minutes, 8 February 1932, 3 March 1932, 5 April 1932, and 8 December 1932; "Iowa Conservation Plan Refuge Policy," approved by both agencies December 1932 and filed with Fish and Game Commission minutes; BC Minutes, 9 February 1932, 30 November 1932.

36. FG Minutes, 10 December 1931 (joint meeting), 2 January 1933, 1 June 1933,

17 June 1933, 3 October 1933; BC Minutes, 13–14 October 1933, 28 November 1933, 29 January 1934, 5 March 1934, 29 March 1934, 12 April 1934, 8–9 August 1934.

37. *Report of the State Fish and Game Commission* (1934), 14–15.

38. FG Minutes, 23 May 1931, 27 June 1931, 10 December 1931 (joint meeting), 4 August 1932 (joint meeting), 10 October 1932, 7 February 1933, 19 July 1934, 26 August 1934; BC Minutes, 13 May 1932, 5 October 1932, 3–4 July 1934, 8 September 1933, 30 April–1 May 1934, 2–3 July 1934, 8–9 August 1934, 5 October 1934, 8 February 1935, 29 March 1935.

39. FG Minutes, 19 July 1934.

40. *Report of the Fish and Game Commission* (1934), 43.

41. BC Minutes, 3 November 1931, 5 January 1932, 6 April 1932, 7 September 1932, 5 October 1932, 7 March 1933, 4 April 1933.

42. BC Minutes, 19 May 1933; FG Minutes, 1 June 1933, 13 February 1934 (joint meeting), 18 February 1934. Several members of both agencies, however, also were members of the State Planning Board, including Margo Frankel, Grace Gilbert King, William P. Woodcock, Dr. W. C. Boone, and A. E. Rapp. This ensured a high degree of coordination for park planning and development at both state and municipal levels.

43. Iowa State Planning Board, *The Second Report* (1935),125–138.

44. *The Second Report*, 148–152; quotation 150. The State Historical Society was established as a private entity; the State Historical Department was a state agency; in 1986, the two merged under the Department of Cultural Affairs.

45. Chap. 13, *Laws of the Forty-sixth General Assembly*. To isolate the State Conservation Commission from partisan politics as much as possible, the law required that no more than four members could be of the same political party and that all appointments must be approved by the State Senate.

46. Minutes, Iowa State Conservation Commission [hereafter ICC Minutes], 17 May 1935, 22 May 1935, 6 June 1935, and 13 June 1935. Shortly afterward, Bode resigned to take as position working under Jay Darling in the U.S. Bureau of Biological Survey. This move led to his selection, in 1937, as director of the new Missouri Conservation Commission; see Flader, *Exploring Missouri's Legacy*, 10–11.

47. Whatever its initial merits, some commissioners felt that the district system would give way to turf wars, a reasonable concern; commissioner districts were abolished by majority vote on 28 November 1938. The system as a whole, however, was reorganized by districts during World War II because personnel shortages forced economies of travel and labor for maintenance and administration.

48. Fitzsimmons to Iowa Conservation Commission, transcript of remarks at 3–5 November 1941 meeting, IDNR files.

49. Iowa State Planning Board, *A Preliminary Report of Progress* (1934), 4–5, 72–85; *The Second Report*, 18–21.

50. G. B. MacDonald, "Report on the Status of Emergency Conservation Work in Iowa," *Thirty-sixth Annual Year Book* (1935), 182–183; Joyce McKay, "Civilian Conservation Corps Properties in Iowa State Parks, 1933–1942" (1989), Section E, 51–52 (Table 4), 71–72.

51. John Salmond, *The Civilian Conservation Corps, 1933–1942: A New Deal Case Study* (1967), remains the only scholarly study of the CCC.

52. *Biennial Report of the Iowa State Conservation Commission* [hereafter *Report ICC*] (1940), 19.

53. MacDonald, "Report on the Status of Emergency Conservation Work in Iowa," 183.

54. *Report ICC* (1936), 123−124; *Report ICC* (1938), 16, 116.

55. Summary information extrapolated from McKay, Table 4, 51. McKay identified a total of 713 extant CCC structures in 1990. While a few of them have since been destroyed or removed, the Iowa Department of Natural Resources now maintains and restores CCC structures as historic properties.

56. George D. Butler, *Municipal and County Parks in the United States, 1935* (1937), 4.

57. BC Minutes, 3−4 July 1933.

58. "Mrs. Frankel Given Medal for Park System Service," *Des Moines Register* 30 May 1933.

59. McKay, "Civilian Conservation Corps Properties in Iowa State Parks," 69−70.

60. McKay, 79, 82; BC Minutes, 2−3 July 1934.

61. ICC Minutes, 18 July 1935.

62. McClelland, *Presenting Nature*, 252−254.

63. McClelland, *Presenting Nature*, 253−262; Albert Good, *Park Structures and Facilities* (1935); Albert H. Good, ed., *Park and Recreation Structures*, 3 vols. (1938). Good credited Iowa with originating the hearth ring, and he called attention to the partially enclosed picnic shelter as an "agreeable and vigorous" regional design; see Good, *Park and Recreation Structures*, vol. 2, 40, 58−59.

64. BC Minutes, 26 April 1922, 14 August 1923 through 10 October 1924 passim, 11 June 1926, 2 May 1928, 5 June 1928; E. A. Piester, "Report on Proposed Traffic System within Back Bone State Park," 11 June 1923; Pammel's park dedication speech, undated typescript; Louis Pammel, "Report on the Backbone State Park" [c. 14 August 1923], Pammel Papers.

65. BC Minutes, 23 April 1920, 14 September 1923, 13 February, 12 June 1925, 14 August 1925; map of the existing fish hatchery site plan prepared for Job 713, "Obliteration," by CCC Camp SP-17, c. 1933; IDNR; "Suggested Sketch for the Proposed Reconstruction of the Lodge, Backbone State Park" and "Proposed Barn Plan, Backbone State Park," prepared by the Iowa Extension Service at Ames, John R. Fitzsimmons, Landscape Architect, IDNR; "Report of H. E. Pammel, Landscape Engineer, on Location of Permanent Features in Backbone State Park," 21 September 1923; Louis Pammel, "Report on the Backbone State Park [14 August 1923]; Byron W. Newberry to Pammel, 17 December 1924; Clifford L. Niles to Pammel, 10 December 1924; Pammel to Gov. John Hammill, 15 October 1926; Hammill to Pammel, 11 October 1926, Pammel Papers. The converted barn/custodian's residence now houses a museum that interprets the work of the Civilian Conservation Corps in the state parks.

66. "Backbone Sate Park Auditorium" and "Plan for Nature Study Headquarters, Backbone State Park," both by John R. Fitzsimmons, 1931, IDNR.

67. Macbride to Fitzsimmons, undated letter c. 1932, Macbride Papers; emphasis in the original.

68. Macbride to Fitzsimmons, 7 February 1933, Macbride Papers. During the late 1920s, there was considerable debate in Washington state over how much development should be allowed in Mt. Rainier National Park. The Macbrides, retired in Seattle, obviously followed the debate.

69. See, for example, Laurie D. Cox, "Use Areas in State Parks," in *American Planning and Civic Annual*, ed. Harlean James (1940), 163–169.

70. Backbone State Park is listed on the National Register of Historic Places.

71. McKay, "CCC Properties in Iowa State Parks, 1933–1942." Minutes of 29 October 1937 show that the Conservation Commission turned down a request from the Strawberry Point Lions Club to have an enclosed shelter house, with seating capacity for 250, erected near the auditorium. In rejecting the petition, the commission noted that Backbone State Park had been "treated very generously in the way of improvements." The artificial lake would later be considered a mistake because no thought had been given to land-use practices in the surrounding watershed and soil erosion caused the lake to start silting in rapidly; see F. T. Schwob, "Parks and Recreation from the State's Point of View," in *American Planning and Civic Annual*, ed. Harlean James (1943), 136.

72. *Report of the State Fish and Game Commission* (1934), 43; A. A. Welt of Iowa City handled fundraising and negotiated most of the land options. He raised money to purchase approximately 800 acres by selling more than 100 cottage sites located on a 40-acre tract of what would become a lake peninsula.

73. Joyce McKay, "Survey Report: CCC Properties in Iowa Sate Parks," (1990), 59.

74. *Report ICC* (1938), 103, 105, 128; visitors were reported at 92,672 in 1940 and 145,289 in 1942; see *Report ICC* (1940), 173; *Report ICC* (1942), 127.

75. *Report ICC* (1940), 154–191 passim; *Report ICC* (1942), 107–141 passim.

76. V. W. Flickinger, "Use Areas in Iowa State Parks," in *American Planning and Civic Annual*, ed. Harlean James (1940), 174–175.

77. F. T. Schwob, " Parks and Recreation from the State's Point of View," 135–139.

78. *Report ICC* (1938), 110; *Report ICC* (1940), 159; *Report ICC* (1942), 114–115.

79. ICC Minutes, 7 January 1936.

80. *Report ICC* (1940), 18; *Report ICC* (1942), 12, 35–37, 127–129, 138–139.

81. Iowa State Planning Board, *The Second Report*, 18–21; G. B. MacDonald, "Multiple Use of State Forests," *Iowa Conservationist* 5 (1946), 9, 11–12.

82. *Twenty-five Year Conservation Plan*, 76–79.

83. *Report ICC* (1938), 118.

84. *Report ICC* (1936), 114, 133–136; Malcolm K. Johnson, "White Pine Hollow," *Iowa Conservationist* 19 (1960), 93–94. The size was eventually increased to 712 acres, and it is now a designated National Natural Landmark; see John Fleckenstein, comp., *Iowa State Preserves Guide* (1992), 188–189.

85. *Report ICC* (1938), 117–118; MacDonald, "Forestry Progress in Iowa," 10–13. Commercial nurserymen succeeded in attaching a rider to the ICC's appropriation that forced the commission to limit reforestation efforts; see Mrs. Addison [Louise] Parker, "Iowa State Parks," in *American Planning and Civic Annual*, ed. Harlean James (1941), 186.

86. *Report ICC* (1940), 160–167; *Report ICC* (1942), 117.

87. *Report ICC* (1940), 166; *Report ICC* (1942), 121; MacDonald, "Forestry Progress in Iowa," 9–10.

88. ICC Minutes, 7 January 1936, contain a lengthy resolution supporting land acquisition for national forests, presented by MacDonald and passed by the commission.

89. G. B. MacDonald, "The Beginning of a National and State Forestry Program in Iowa," *Ames Forester* 23 (1935), 17–19.

90. ICC Minutes, 12 March and 6 October 1936, 12–13 March 1940; G. B. MacDonald, "Progress of the Forest Land Acquisition Program in Iowa," *Ames Forester* 25 (1937), 51–53.

91. MacDonald, "Forestry Progress in Iowa," 15–16; G. B. MacDonald, "Present State Forests," *Iowa Conservationist* 4 (1945), 167. In 1938, the federal government revised its purchase plans down to 210,000 acres; see ICC Minutes, 14 June 1938.

92. *Report ICC* (1942), 129, 137; MacDonald, "The Beginning of a National and State Forestry Program in Iowa," 20.

93. ICC Minutes, 6–7 May 1941, 18–19 August 1941, 3 December 1941, 6–7 January 1942.

94. MacDonald, "Forestry Progress in Iowa," 12–13.

95. BC Minutes, 2 September 1930, 9 December 1930, 9–10 July 1931.

96. *Twenty-five Year Conservation Plan*, 47–68.

97. *The Second Report*, 45–67.

98. BC Minutes, 5 October 1932, 5 October 1934; FG Minutes, 4 August 1932, 10 October 1932.

99. BC Minutes, 13 May 1932, 30 April–1 May 1934; FG Minutes, 19 July 1934, 26 August 1934.

100. "Spirit Lake Council Gives Answer to Criticism," *Spencer Daily Reporter*, 3 September 1937.

101. ICC Minutes, 18 June 1936.

102. ICC Minutes, 7 May 1937, 26 August 1937; "Surveys Are Being Made for Lakes Sewer Project," *Spencer Daily Reporter*, 15 June 1937; "State Conservation Officials Discuss Lakes Sewer," *Spencer Daily Reporter*, 2 July 1937; "Foresee Lakes Towns Agreement on Sewer Soon," *Spencer Daily Reporter*, 21 July 1937; "Failure of Spirit Lake City Council to Agree With Other Towns Holds up Sewer Plans," *Spencer Daily Reporter*, 19 August 1937.

103. "Governor Urges Sewage Action," *Spencer Daily Reporter*, 27 August 1937; "An Editorial to Spirit Lake," *Spencer Daily Reporter*, 28 August 1937; "Spirit Lake Council Gives Answer to Criticism," *Spencer Daily Reporter*, 3 September 1937; "Kraschel Warns Spirit Lake on Sewer," *Spencer Daily Reporter*, 4 September 1937; "Gentlemen of Spirit Lake, It Is Your Next Move in the Sewer Matter!" *Spencer Daily Reporter*, 4 September 1937; "Action Comes Following Conference With High State Officials of Iowa," *Spencer Daily Reporter*, 9 September 1937; ICC Minutes, 15 September 1937.

104. "Begin Work on Lakes Sewer Job," *Spencer Daily Reporter*, 10 November 1937; "Complete First Mile of Digging for Lakes Sewer," *Spencer Daily Reporter*, 8 December 1937; "Lack of Funds Stops Work on Sewer For Time," *Spencer Daily Reporter*, 6 June

1938; "Senator Herring and Keller Inspect Lakes Sewer for Completion," *Spencer Daily Reporter*, 12 July 1938.

105. "Action Taken to Get WPA Cash for Lakes Sewer Work," *Spencer Daily Reporter*, 11 March 1938; "Move to Get WPA Cash for Lakes Sewer," *Spencer Daily Reporter*, 12 March 1938; "Sanitary Sewer Board Organizes in Lakes Region," *Spencer Daily Reporter*, 21 May 1938; "Will Require $2,500 to Get $350,000 More," *Spencer Daily Reporter*, 13 July 1938; "Solicit Funds to Complete Sewer System," *Spencer Daily Reporter*, 16 July 1938; "Fate of Finish of Lakes Sewer Depends Upon More Money for Survey," *Spencer Daily Reporter*, 23 July 1938; "Another $400,000 for Lakes Sewer Awarded by WPA," *Spencer Daily Reporter*, 17 October 1938; "Lakes Sewage System Gets $7,800 from IERA," *Spencer Daily Reporter*, 5 November 1938; "Lake Sewer Project to Extend Next to the Methodist Camp Grounds on Okoboji," *Spencer Daily Reporter*, 27 November 1939; "Sewage and Water Lines Are Finished," *Spencer Daily Reporter*, 22 July 1940.

106. ICC Minutes, 15–18 July 1940, 20–21 August 1940.

107. ICC Minutes, 7–8 January 1941, 11–12 March 1941, 10–11 September 1941, 3–4 March 1942, 17 June 1942.

108. ICC Minutes, 24–25 August 1942, 16–17 November 1942, 7–8 December 1942, 22 February 1943, 7–8 June 1943, 12 July 1943, 21 August 1944, 13 November 1944, 11 December 1944, 11 June 1945.

109. ICC Minutes, 23–25 July 1945, 27–28 August 1945, 8 October 1945, 20–21 November 1947.

110. ICC Minutes, 11–13 July 1948 (quotation), 10–11 January 1949, 19 July 1951.

111. For instance, Commissioner Fred Mattes of Odebolt expressed his view "that the Great Lakes Sewer project is entirely a lake pollution problem" during a 1945 debate over how the commission should respond to the legislative act that transferred authority over the system to the State Conservation Commission; see Minutes 23–25 July 1945; see also ICC Minutes, 15 January 1950, 22–23 May 1950, 4–5 August 1950, 28–29 August 1950, 15 January 1951.

112. Chap. 13, *Acts of the Fifty-fourth General Assembly* (1951).

113. ICC Minutes, 27 September 1951, 10 April 1952, 22–23 May 1952.

114. ICC Minutes, 25 May 1953, 22–23 July 1953, 3–4 February 1959, 1 November 1961.

115. Seidel, *Iowa's Heritage in Water Pollution Control*, 335–337, 479.

116. Darling to Crane, 10 August 1941, Darling Papers.

117. Fitzsimmons to State Conservation Commission, transcript of remarks made at 3–5 November 1941 meeting, 11, IDNR files.

118. Ibid., 4–13.

119. Schwob memorandum to employees, 15 May 1942; E. B. Speaker, "Reappraisal of Iowa 25 Year Plan 1942," undated typescript; and James R. Harlan memorandum to Schwob, 27 May 1942, IDNR files.

120. ICC Minutes, 25 February 1938, 29 April 1938, 5 July 1938.

121. National Park Service, *Study of the Park and Recreation Problem of the United States*, 12–13.

122. *Study of the Park and Recreation Problem of the United States*, 132–271 (state-

by-state statistics). These figures must be used with caution, since states may not have reported statistics uniformly or completely. Iowa, for instance, only reported 24,828 acres in parks, preserves, wildlife refuges, and forests.

123. *The Second Report*, 152.

124. *Report ICC* (1942), 12.

CHAPTER 5   Seeking Balance

1. *Report ICC* (1944), 7, 127–135 passim; ICC Minutes, 6–7 January 1942, 21–22 April 1942, 27–28 May 1942, 6 October 1942, 7–8 December 1942, 16–17 August 1943.

2. *Report ICC* (1944), 134; *Report ICC* (1946), 149; ICC Minutes, 16–17 November 1942, 22 February 1943, 12 April 1943, 12 July 1943, 16–17 August 1943.

3. *Report ICC* (1944), 137–140; *Report ICC* (1946), 157; ICC Minutes, 6 October 1942, 15 January 1945.

4. ICC Minutes, 21 August 1944; F. T. Schwob, "Iowa Plans Comprehensive Postwar Conservation Program," *Iowa Conservationist* 3 (1944), 73, 75; State Conservation Commission to Postwar Rehabilitation Committee, "Recommendation to the Postwar Rehabilitation Committee" [c. October 1944], Papers of Governor Robert Blue, State Historical Society of Iowa [published as "Schwob Makes Recommendations to Postwar Committee," *Iowa Conservationist* 3 (1944), 81–82].

5. *Report ICC* (1944), 11, 126–127; ICC Minutes, 10–11 September 1941, 27–28 May 1942, 17 June 1942, 6 October 1942, 12 July 1943, 6 December 1943.

6. *Report ICC* (1946), 143–144, 151.

7. ICC Minutes, 19 November 1945.

8. This is reflected in the 1942 biennial report, which includes, on page 111, a brief report on the commission's "historical program," the first time such a report was included.

9. She was always publicly known as Mrs. Addison Parker after her marriage.

10. Author interview with Addison M. Parker, Jr., 22 March 1995; biographical notes with Louise L. Parker Papers, Iowa State University.

11. Kautsky to Parker, 12 November 1947; Parker to Kautsky, 18 November 1947, Parker Papers.

12. ICC Minutes, 5 July 1938.

13. These impressions of her personality are gleaned from commission minutes and scattered correspondence as well as conversations with her son, Addison M. Parker, Jr.

14. ICC Minutes, 7 December 1948, 14–15 March 1949, 19–20 April 1949, 23 May 1949.

15. BC Minutes, 9 August 1924, 13 August 1926, 10 December 1926, 4 September 1928, 8 January 1929, 5–6 February 1929, 1 April 1930, 5 May 1931, 3 December 1934; Haugen to W. C. Merckens, Board of Conservation Secretary, 17 September 1928, Ft. Atkinson Research Files, SHSI.

16. Good, *Park and Recreation Structures*, vol. 2, 185–196; Greiff, "People, Parks, and Perceptions," 120–127.

17. Paul G. Hirschler, assistant architect, BBS, to W. C. Price, National Park Ser-

vice, 2 April 1936; MacDonald to Donald B. Alexander, National Park Service, 17 April 1936; Harry E. Curtis, National Park Service, to MacDonald, 23 April 1936; MacDonald to M. L. Hutton, 24 April 1936; Logan Blizzard to Hutton, 12 September 1936; MacDonald to Hutton, 29 September 1936, Ft. Atkinson Research Files.

18. ICC Minutes, 15 March 1938, 2 May 1939, 5 July 1939, 14 August 1939, 9 February 1940, 7 May 1940, 4 June 1940.

19. ICC Minutes, 18 September 1940, 7 January 1941, 17 June 1941, 6 January 1942, 21 April 1942, 20 July 1942; F. L. Carnes, architect, to Reque, 26 October 1942, Fort Atkinson Research Files.

20. ICC Minutes, 6 December 1943, 7 February 1944, 25 September 1944, 12 March 1945, 23 July 1945, 28 August 1945, 4 September 1946, 18 November 1946, 20 January 1947, 15 July 1947; "Fort Atkinson—Professor Reque," report to ICC by Mrs. Addison Parker, 29 June 1945, and S. S. Reque, "Ft. Atkinson Report" [1947], Fort Atkinson Research Files.

21. ICC Minutes, 2 June 1955, 27 March 1956, 10 April 1956, 5 July 1961, 7 March 1962; Merrill J. Mattes, "Historian's Report on Preservation of Fort Atkinson, Iowa," 10 December 1956, Fort Atkinson Research Files.

22. ICC Minutes, 15–18 July 1940; *Report ICC* (1942), 111; Jacob Swisher, "Plum Grove," *Palimpsest* 29 (1948), 26.

23. ICC Minutes, 6–7 January 1942, 3–4 February 1942, 3–4 March 1942, 21–22 April 1942, 27–28 May 1942, 17 June 1942.

24. ICC Minutes, 7–8 December 1942, 7–8 June 1943, 12 July 1943, 16–17 August 1943.

25. The commission's intent was to restore the house as nearly as possible to its original floor plan and appearance, although, for reasons unknown, this did not happen; instead, it was "completely rejuvenated and reconditioned," according to Swisher, "Plum Grove," 28. See also ICC Minutes, 7–8 February 1944; Jan R. Nash, "Plum Grove Historic Site Project: Evaluation of the Interpretation of Plum Grove and Recommendations"(1993), 3, 25.

26. ICC Minutes, 26 February 1946, 24–25 February 1947; Nash, 3–4. In 1994, the Department of Natural Resources transferred administrative control of the site to the State Historical Society of Iowa.

27. ICC Minutes, 3–5 November 1941, 3 December 1941, 6–7 January 1942, 21–22 April 1942, 12 April 1943, 16–17 August 1943, 17 June 1946, 17–18 July 1952; *Report ICC* (1944), 127; *Report ICC* (1946), 145; *Report ICC* (1948), 76.

28. O'Bright, *The Perpetual March*, 47–63.

29. ICC Minutes, 14–16 August 1939, 12–13 March 1940; *Report ICC* (1942), 130.

30. ICC Minutes, 15 May 1945, 11 June 1945, 23–25 July 1945, 27–28 August 1945, 15–16 April 1946, 22–23 September 1947, 5–6 January 1948.

31. O'Bright, *The Perpetual March*, 65–107, contains detailed discussions of land acquisitions and related issues that were negotiated from 1946 to 1984.

32. Florence Cowles Kruidenier was the daughter of Gardner Cowles, owner and publisher of the *Des Moines Register and Tribune*; her husband, David Kruidenier, also entered the family business and became the newspaper's publisher.

33. Aldo Leopold, "The Land Ethic," in *A Sand County Almanac* (1949), 243.

34. Daryl D. Smith, "Iowa Prairie—An Endangered Ecosystem," *Proceedings IAS* 88 (1981), 7–8.

35. John Madson, *Where the Sky Began: Land of the Tallgrass Prairie* (1995), 203.

36. Daryl Smith, "'Mystique' of the Prairie," in *Fifth Midwest Prairie Conference Proceedings*, ed. David C. Glenn-Lewin and Roger Q. Landers, Jr. (1978), 194–197.

37. Biographical information throughout this section is drawn from three articles: Lois Hattery Tiffany, "Reflections on Women Scientists and the Iowa Academy of Science," *Proceedings IAS* 82 (1975), 94–95; Jan Lovell, "She Fought to Save Iowa's Prairies," *Iowan* 36 (1987), 22–27, 56–57; Duane Isely, "Ada Hayden: A Tribute," *Journal IAS* [formerly *Proceedings IAS*] 96 (1989), 1–5.

38. Isely, 2; Lovell, 26, 56.

39. Ada Hayden, "The Ecologic Foliar Anatomy of Some Plants of a Prairie Province in Central Iowa," *American Journal of Botany* 6 (1919), 69–85; Hayden, "The Ecologic Subterranean Anatomy of Some Plants in a Prairie Province in Central Iowa," *American Journal of Botany* 6 (1919), 87–104; "Notes on the Floristic Features of a Prairie Province in Central Iowa," *Proceedings IAS* 25 (1918), 368–389.

40. Bohumil Shimek, "Prairie Openings in the Forest," *Proceedings IAS* 17 (1911), 16–19; "The Prairies," *Laboratories of Natural History*, Bulletin 6 (1911), 169–240; "The Plant Geography of the Lake Okoboji Region," *Laboratories of Natural History*, Bulletin 7 (1915), 1–90; "The Prairies of the Mississippi River Bluffs," *Proceedings IAS* 31 (1926), 205–212; "Papers on the Prairie," *Laboratories of Natural History*, Bulletin 11 (1925), 1–36.

41. Ada Hayden, "Conservation of Prairie," *Iowa Parks* (1920), 283–284.

42. Crane and Olcott, *Twenty-five Year Conservation Plan* (1933), 139–140.

43. As referenced in Ada Hayden, "The Selection of Prairie Areas in Iowa Which Should be Preserved," *Proceedings IAS* 52 (1945), 125–148; Smith, "Iowa Prairie—An Endangered Ecosystem," 8.

44. BC Minutes, 2–3 July 1934, 4 September 1934; ICC Minutes, 2–3 July 1935, 9 March 1937, 23 March 1937.

45. ICC Minutes, 24 May 1938.

46. ICC Minutes, 7–8 October 1941.

47. Isely, "Ada Hayden: A Tribute," 4.

48. ICC Minutes, 6 March 1944; Hanson, "The Iowa Academy of Science: 1875–1975," 14, 16.

49. ICC Minutes, 13 November 1944; Parker also issued a call for public assistance in identifying prairie remnants in 1944; see "Our Vanishing Prairie," *Iowa Conservationist* 3 (1944), 63.

50. Parker to Ewald Trost, 22 June 1963, Parker Papers; ICC Minutes, 15 January 1945, 12 March 1945, 15 May 1945; see also Ada Hayden, "Iowa Had a Coat of Many Colors," *Iowa Conservationist* 5 (1946), 25, 31; Robert M. Bliss, "A Baltimore Piano Maker's Investment Saved Some of Iowa's Native Prairie," *Iowa Conservationist* 8 (1949), 155.

51. Hayden, "The Selection of Prairie Areas in Iowa Which Should be Preserved," 127–148 [identified parcels were located in O'Brien, Emmet, Howard, Cherokee,

Dickinson, Ida, Pocahontas, Guthrie, Crawford, and Pottawattamie counties]; Ada Hayden, "A Progress Report on the Preservation of Prairie," *Proceedings IAS* 53 (1947), 45–82.

52. ICC Minutes, 27–28 August 1945, 8 October 1945, 14–15 January 1946, 26 February 1946, 18–19 August 1947, 5–6 April 1948, 15–16 November 1948; Hayden "A Progress Report on the Preservation of Prairie," 57. The minutes indicate that the commission first considered a 60-acre tract in Pocahontas County referred to as the Gunderson Tract. It is not clear whether this was the same as or part of the 160-acre Kalsow Tract. Hayden's report describes only the Kalsow Prairie.

53. Ada Hayden and J. M. Aikman, "Considerations Involved in the Management of Prairie Preserves," *Proceedings IAS* 56 (1949), 133–142; Edward T. Cawley, "The History of Prairie Preservation in Iowa," *Proceedings of the Second Midwest Prairie Conference*, 1970, ed. James H. Zimmerman (1972), 23.

54. ICC Minutes, 9 April 1945, 14 May 1945, 10–11 January 1949; Parker to Hayden, 17 January 1949, Parker Papers.

55. ICC Minutes, 25–26 July 1955, 1 September 1955, 18 February 1957, 9–10 April 1957, 6 May 1958. An odd footnote to this acquisition demonstrates how Cold War ideology permeated even the most unlikely corners of American life. The minutes of the 9–10 April 1957 meeting record Bruce Stiles as stating that "on his recent trip to Washington, D.C. he had a conference with Dr. John Wolf of the Atomic Energy Commission [who] explained to him how important the preservation of this type of land [native prairie] is . . . in order to provide information on original soil composition in case it is ever necessary to rebuild an area following an atomic explosion."

56. ICC Minutes, 12 February 1945, 15 May 1945, 25 July 1949; George L. Collins, National Park Service to Parker, 4 May 1945, Parker Papers.

57. Parker to Ewald Trost, 22 June 1963, Parker Papers.

58. Hanson, "The Iowa Academy of Science, 1875–1975," 16; ICC Minutes, 28–29 August 1950; Hayden Prairie was designated a National Natural Landmark in 1966.

59. ICC Minutes, 1 September 1955.

60. ICC Minutes, 13–14 October 1954.

61. Madson, *Where the Sky Began*, 287.

62. Paul Matthiae, chief of Natural Areas Section, Wisconsin Department of Natural Resources, telephone interview, 29 September 1995; Forest Stearns and Cliff Germain, "Natural Areas Preservation Council, 1951–1991," (1991); Thomas E. Toney, *Public Prairies of Missouri* [c.1994]; Donald M. Christisen, "A Vignette of Missouri's Native Prairie," *Missouri Historical Review* 61 (January 1967), 166–186; Ralph E. Ramey and E. Roger Troutman, "Ohio Prairies and the Ohio Prairie Survey Project," *Fifth Midwest Prairie Conference Proceedings*, 207–208; *Directory of Indiana's Dedicated Nature Preserves* (1991); Hank Hoffman, Indiana Department of Natural Resources, telephone interview, 9 May 1995; Don McFall and Jean Karnes, eds., *A Directory of Illinois Nature Preserves*, 2 vols. (1995); Jerry Steinauer, Nebraska Game and Parks Commission, telephone interview, 5 May 1995.

63. Worster, *Nature's Economy*, 205–220, 232–237.

64. Worster, *Nature's Economy*, 253.

65. Schwob, "Iowa Plans Comprehensive Postwar Conservation Program," 73; see

also V. W. Flickinger, "Twenty-Five Years of Iowa Parks," *Iowa Conservationist* 6 (1947), 118.

66. *Report of the State Fish and Game Commission* (1934), 42–45.

67. *Report ICC* (1936), 119; *Report ICC* (1938), 105–107.

68. ICC Minutes, 24–25 July 1944; Wilbur A. Rush, "Iowa State Parks in 1947," *Iowa Conservationist* 6 (1947), 188.

69. ICC Minutes, 21 August 1944, 26 February 1946, 24–25 February 1947, 5–6 August 1948, 7–8 September 1948. The papers of Ewald Trost (University of Iowa), a member of the ICC 1945–1957, contain preliminary reports prepared in 1945 and 1946 on proposed artificial lake sites in (1) Geode State Park, (2) Jefferson and Washington counties, (3) Poweshiek and Jasper counties, (4) Shelby County, and (5) Cass, Montgomery, and Page counties.

70. ICC Minutes, 13 November 1944.

71. *Report ICC* (1948), 108–115; ICC Minutes, 22–23 September 1947; Bruce Stiles, "New Laws," *Iowa Conservationist* 6 (1947), 145, 151; V. W. Flickinger, "Current Land and Waters Development," *Iowa Conservationist* 7 (1948), 17, 22.

72. *Report ICC* (1948), 115–117, 126; *Report ICC* (1950), 29–37, 98, 143.

73. John W. Henderson, "Legislature Votes Six New Artificial Lakes," *Iowa Conservationist* 6 (1947), 129, 133; H. W. Freed, "Iowa's Six New Lakes," *Iowa Conservationist* 6 (December 1947), 185, 191; "Allocate Funds for Iowa's Largest Man-Made Lake," *Des Moines Register*, 26 July 1949.

74. ICC Minutes, 17 October 1949.

75. ICC Minutes, 28–29 August 1950, 19 July 1951, 23–24 August 1951; Walter Shotwell, "Close Valve for New Lake," *Des Moines Register*, 18 September 1950; *Report ICC* (1954), 28; *Report ICC* (1956), 28.

76. The other seven states were Ohio, Indiana, Michigan, Illinois, Wisconsin, Minnesota, and Missouri.

77. Meine, *Aldo Leopold*, 259–268, 272–278.

78. Flader, *Thinking Like a Mountain*, 21–26.

79. Aldo Leopold, "Game and Wildlife Conservation," *Condor* 34 (1932), 103–104.

80. FG Minutes, 28 January 1932, 8 February 1932, 17 March 1932.

81. *Twenty-five Year Conservation Plan*, 91.

82. Before the merger, the Fish and Game Commission purchased the initial tracts of two wildlife refuges: Dewey's Pasture (Clay Co.) and Allen Green Slough (Des Moines Co.).

83. *Report ICC* (1942), 12, 20–23, 33–35 (quotation, 21).

84. Lonnie L. Williamson, "Evolution of a Landmark Law," in *Restoring America's Wildlife, 1937–1987* (1987), 1–12; the funding formula stipulates 75 percent federal money, 25 percent state money. The Iowa Cooperative Wildlife Research Unit, established at Iowa State University in 1932, was the prototype for the federal program. Darling had so much faith in this concept that he funded one-third of the cost personally; Iowa State and the Fish and Game Commission funded the other two-thirds; see Arnold O. Haugen, "History of the Iowa Cooperative Wildlife Research Unit," *Proceedings IAS* 73 (1966), 136–145.

85. John R. Langenbach, "Restoring a Land Base," in *Restoring America's Wildlife*, 69; see also "Saving the Wetlands," *Iowa Conservationist* 14 (1955), 116.

86. Stiles, "Brief Analysis of Iowa Fish and Game Policy," 178.

87. Acreage figures derived from ICC biennial reports 1940 through 1950; see also Lester F. Faber, "Iowa's 14 Years in the Pittman-Robertson Program," *Iowa Conservationist* 10 (1951), 180.

88. *Report ICC* (1950), 274.

89. *Report ICC* (1950), 274–276; *Report ICC* (1952), 72–77; *Report ICC* (1954), 64–68; *Report ICC* (1956), 67–70; *Report ICC* (1958), 66–70; *Report ICC* (1960), 80. Of the total 29,700 acres, approximately 2,400 acres were purchased under a second federally financed program set up by the Federal Aid in Sport Fisheries Restoration Act of 1950, commonly known as the Dingell-Johnson Act.

90. *Administration of Iowa Parks, Lakes and Streams*, 23; *Iowa Parks*, 32; *Report ICC* (1936), 122, 137; *Report ICC* (1938), 35; "Rice Lake Visioned as Iowa's Fourth Largest," *Des Moines Register*, 8 September 1935.

91. *Report ICC* (1942), 37; *Report ICC* (1950), 34; *Report ICC* (1956), 70–71, 205; ICC Minutes, 14–16 July 1947.

CHAPTER 6   The Center Does Not Hold

1. Darling, "Poverty of Conservation Your National Problem," *Iowa Conservationist* 4 (1945), 115, 118.

2. "Riaski Speaks for State Parks," *Iowa Waltonian* 10 (June 1955), 1.

3. "Resolutions of the Iowa Division, 33rd Annual Convention," *Iowa Waltonian* 11 (November 1956), 6; William A. Riaski, "Iowa Parks Are Being Neglected," *Iowa Waltonian* 11 (March 1957), 1, 9.

4. *Report ICC* (1954), 32, 35–36.

5. Ira N. Gabrielson, *Wildlife Conservation* (1941) and *Wildlife Refuges* (1943).

6. ICC Minutes, 22–23 September 1947. Gabrielson's report states that the investigation was jointly requested by the commission and by Gov. Robert Blue; however, Jay Darling explained the situation a bit differently. According to Darling, "the survey was not instigated by the Governor but was arranged by the members of the Conservation Commission for their own purposes and the good of the cause. . . . Quite characteristic of the Governor's present attitude he, having learned of the proposed survey, called in a few members of the Commission and said it was a fine idea, he would cooperate to the fullest extent and appoint a committee to work with Gabrielson, and then he stole the ball and announced it as his idea and his program." Darling to Richard C. Leggett, Fairfield, Iowa, 10 October 1947, with copies to newspaper columnist Ries Tuttle and select members of the Izaak Walton League, Darling Papers.

7. ICC Minutes, 12 February 1945.

8. J. N. "Ding" Darling, "Add Crime Wave" [cartoon title], *Des Moines Register* 14 August 1946; C. C. Clifton, "Ding Explains Cartoon Basis," *The Decorah Journal*, 21 August 1946; see also Lendt, *Ding*, 116–117.

9. Darling to Poyneer, 5 December 1947; Poyneer to Darling, 8 December 1947; Poyneer to Darling, 30 December 1947, Fred J. Poyneer Papers [ICC commissioner

1939–1951], University of Iowa. Gabrielson, incidentally, donated his services so as to avoid any obligation to the commission or to the governor.

10. Conservation Study Committee and Ira N. Gabrielson, "A Joint Report to Governor Robert D. Blue and the Iowa Conservation Commission" [December 1947], Trost Papers (also Poyneer Papers); also published as "Gabrielson, Committee Report," *Iowa Conservationist* 7 (1948), 1, 5–8. By law, the salaries of commission employees were set by the legislature and not by the commission, a situation that Darling repeatedly criticized; see, for instance, Darling to Poyneer, 21 September 1943; Darling to Don L. Berry, publisher of the *Indianola Record and Tribune*, 11 September 1946, Darling Papers.

11. Ira N. Gabrielson, "Report to the Iowa Conservation Commission by the Wildlife Management Institute," *Iowa Conservationist* 14 (March 1955 supplement), 1.

12. Gabrielson, "Report to the Iowa Conservation Commission," 2–3, 7. A 1953 investigation of ICC accounting practices and expenditures came to the same conclusion: the Lands and Waters Divisions "as a general condition has not had enough money to operate efficiently." See M. B. Bolsem, superintendent of State Audits, "Special Report, The Iowa Conservation Commission," 25 February 1953, Trost Papers.

13. ICC Minutes, 2 June 1955, 28 June 1955, 21–22 August 1956; *Report ICC* (1958), 40–41; *Report ICC* (1960), 47–49; *Report ICC* (1962), 43–44.

14. "The Status of Salaries of Employees of the State Conservation Commission," 9 October 1956, IWLA-Iowa. The league's study, as well as a related questionnaire sent to all legislators, was publicized in the *Des Moines Register*, 8 September 1956, 4. What neither the IWLA nor Gabrielson reported is that conservation officers were encouraged to log excessive mileage, for which they were reimbursed, as a means of subsidizing their salaries; see 1953 state audit report.

15. "New Conservation Laws Enacted," *Iowa Conservationist* 16 (1957), 137, 144; Bill Riaski, "Iowa Conservation Legislation in 1957," *Iowa Waltonian* 11 (July 1957), 2; ICC Minutes, 13 August 1957; ICC Minutes, 7 July 1959.

16. ICC Minutes, 28–29 August 1950, 7–8 May 1951, 29 October 1952, 27–28 August 1954, 14 February 1955, 6 January 1959.

17. "State Parks Most Popular With Residents of Iowa Towns," *Iowa Conservationist* 8 (1949), 140; Bruce F. Stiles, "Some Conservation Needs in Iowa as Expressed by a National and State-wide Survey," *Proceedings IAS* 64 (1957), 83.

18. ICC Minutes, 9–10 April 1957, 18 June 1957, 13 August 1957, 5 November 1957; Wildlife Management Institute, "A Ten Year Program for the Iowa State Conservation Commission" (1958).

19. ICC Minutes, 3–4 February 1959. In 1952, the Little Hoover Commission recommended that the ICC, Natural Resources Council (discussed later in this chapter), Soil Conservation Committee, and Geological Survey be consolidated into a proposed Department of Conservation and Natural Resources. All agencies involved opposed the move.

20. WMI, "A Ten Year Program for the Iowa State Conservation Commission," 83–137, 149–157.

21. *Report ICC* (1962), 13–18. The ICC minutes of 7 February 1962 contain a lengthy entry concerning the need to fill the position of Lands and Waters chief and the need

for two new staff positions, an administrative assistant to the director and a CPA for the administration division. The discussion suggests that few of the commissioners had any clear sense of how the agency functioned on a day-to-day basis. *Report ICC* (1964) contains the first mention of a planning section.

22. *Report ICC* (1960), 15–25.

23. Williamson, "Evolution of a Landmark Law," 13–14; ICC Minutes, 7–8 May 1951.

24. Figures compiled from biennial reports 1954 through 1960.

25. Stiles, "Some Conservation Needs in Iowa," 82–84.

26. Attendance statistics must be accepted with caution. The vast majority of states did not collect gate fees or have any other mechanism for tabulating day users accurately, so reported figures are usually estimates. Iowa, for instance, took actual counts on three different days of the summer season and extrapolated figures to arrive at its estimate. Presumably, other states employed a sampling technique of some sort to arrive at reasonable estimates.

27. National Park Service, *State Park Statistics—1952*, passim.

28. National Park Service, *State Park Statistics—1960*, passim.

29. National Park Service, *State Park Statistics—1957*, 2–5.

30. One would hardly know it, though, from reading Freeman Tilden's breezy 1962 book, *The State Parks: Their Meaning in American Life*, which is considered a standard in the meager historical literature on state parks. Tilden devotes a scant twelve pages to "Principles, Policies, and Problems," in which he manages to avoid discussing all three in any meaningful way. To be fair, Tilden no doubt understood just how difficult an assignment this would have been, given the fact that states administer their park systems through a variety of agencies, and, likewise, have devised a variety of funding sources. This may be the reason he chose to focus on the federal government's role in encouraging standardized administration of state park systems.

31. Gabrielson, "Report to the Iowa Conservation Commission," 7.

32. Women who served as commissioners between 1950 and 1965 were Florence Kruidenier, Des Moines (1949–1951); Mrs. Emmet Ryan, Underwood (1951–1952); Mrs. Emmet Hannan, Council Bluffs (1952–1955); and Mrs. John W. Crabb, Jamaica (1955–1961). Men who served were E. B. Gaunitz, Lansing (1939–1951); Fred J. Poyneer, Cedar Rapids (1939–1951); Ewald G. Trost, Fort Dodge (1945–1957); Arthur C. Gingerich, Wellman (1947–1951); James D. Reynolds, Creston (1948–1959); C. A. Dinges, Emmetsburg (1949–1955), Joe Stanton, Des Moines (1951–1957); Floyd S. Pearson, Decorah (1951–1957); William F. Frudeger, Burlington (1951–1953); George M. Foster, Ottumwa (1953–1959); George V. Jeck, Spirit Lake (1955–1961); Dr. Albert N. Humiston, Cedar Rapids (1957–1963); Clyde M. Frudden, Green (1957–1963); George H. Meyer, Elkader (1957–1963); Sherry R. Fisher, Des Moines (1959–1965); Earl E. Jarvis, Wilton Junction (1959–1965); Robert Beebe, Sioux City (1961–1967); Edward Weinheimer, Greenfield (1961–1967); Dr. N. K. Kinney, Ida Grove (1963–1969); Rev. Laurence H. Nelson, Bellevue (1963–1969); and Mike F. Zack, Mason City (1963–1969).

33. ICC Minutes, 3 March 1964.

34. As quoted in Lendt, *Ding*, 149.

35. Stiles actually was asked to resign as director a few days before he died from

complications associated with surgery to remove a malignant tumor. David Lendt has pieced together the story in *Ding*, 147–148. Stiles had been in ill-health for some time, ostensibly because of alcohol abuse, and he was no longer handling his job responsibly. When new commissioners came on board in July 1959, they succeeded in removing him as director. Stiles, however, was respected for his earlier contributions and well liked by his colleagues. His death effectively removed the cloud that attended his last years with the commission.

36. "Conservation Director Resigns; Bruce F. Stiles Takes Over Director Post," *Iowa Conservationist* 7 (1948), 50; Stan Widney, "Our Silver Anniversary," *Iowa Conservationist* 19 (1960), 38.

37. ICC Minutes, 1 May 1963.

38. Speaker Park, adjacent to Black Hawk Lake State Park, is named in his honor.

39. Biographical profile with Everett B. Speaker Papers, Iowa State University.

40. E. B. Speaker, "Notes for Interview with Mr. Dale Poel, July 11, 1968," Speaker Papers.

41. Gabrielson, "Report to the Iowa Conservation Commission," 7.

42. Chap. 203, *Laws of the Fifty-third General Assembly* (1949).

43. P.L. 534, 78th Congress, 1944.

44. ICC Minutes, 28 July 1939, 15 September 1939. This was the beginning of controversy over the Saylorville Dam; see chap. 7.

45. ICC Minutes, 6–7 November 1939, 5–6 December 1939, 16–18 January 1940.

46. Darling to Poyneer, 21 September 1943, Darling Papers.

47. Darling to C. R. Hallowell (president Iowa Division, IWLA), 7 October 1943, Darling Papers.

48. "Tentative Report of the Conservation of Resources Committee" [1944], Gov. Blue Papers.

49. H. Garland Hershey, "Formation of a Natural Resources Council: An Historical Outline" [c. 1950], personal files of Michael H. Smith, assistant attorney general, Iowa Department of Justice; H. Garland Hershey, "Water Supply," *Proceedings IAS* 52 (1945), 44–47.

50. Darling to Gov. Blue, 10 July 1946, Darling Papers.

51. Cover letter from Darling and list of 26 recipients, 17 September 1946; cover letter from Darling to Gov. Blue, 17 September 1946; "Iowa Natural Resources Council," 27 September 1946, Darling Papers.

52. *Report of the Interim Flood Control Committee* (1948), 10. Hershey was an appointed member of the committee; Darling served as an advisor, as did Bruce Stiles, Everett B. Speaker, and G. L. Ziemer.

53. Darling to Gov. Blue, 17 September 1946, Darling Papers.

54. "Text of Governor Blue's Speech" [probably 16 August 1948 meeting of Interim Flood Control Committee], Gov. Blue Papers.

55. Chap. 203, *Laws of the Fifty-third General Assembly* (1949).

56. In a verbal report to the State Conservation Commission, then-director G. L. Ziemer noted that the composition of the council had been a "controversial question" with members of the Interim Flood Control Committee. Some members favored representation by various state departments, as advocated in the Darling-IWLA proposal,

but this plan lost in the final vote; see ICC Minutes, 4−5 October 1948, 10−11 January 1949, 14−15 February 1949.

57. Darling to Henry L. Adams, 12 February 1947, copies to Ries Tuttle, G. L. Ziemer, and members of the State Conservation Commission; Darling to Governor-elect William S. Beardsley, 13 December 1948, Darling Papers.

58. ICC Minutes, 28−29 November 1949.

59. ICC Minutes, 28−29 August 1950, 9 October 1950 (quotation).

60. Chap. 229, *Laws of the Fifty-seventh General Assembly* (1957). These amendments followed recommendations contained in the 1956 *Report of the Iowa Study Committee of Water Rights and Drainage Laws*, and the changes were closely followed by the legal community; see *Iowa Law Review* 41 (Winter 1956), the entire issue of which is devoted to articles on water rights issues. Other midwestern states, notably Michigan, Wisconsin, and Minnesota, also were revising their water laws during this period to gain control over irrigation. Water quality, rather than water supply, has been the more critical resource issue for Iowa and other Upper Midwestern States. Supply, in fact, was never perceived as a problem in Iowa until there were temporary shortages between the flood years of 1947 and 1954. More important, though, studies conducted by the NRC showed that 85 percent of Iowa's water supply came from precipitation and runoff, with only 15 percent from aquifers. This information heavily influenced the legislative recommendations contained in the 1956 report. The rationale behind the 1957 law, as Judge John Tobin pointed out in "Irrigators' Challenge is Answered by Waltonian," *Iowa Waltonian*, 11 (August-September 1958), 1, 3, is that water "once in a stream [or natural lake], belongs to the public and is no longer an accessory to the land."

61. Some farmers began organizing for water rights immediately after the law was passed. The strategy of a group known as the Iowa Irrigators was to maximize water rights for agricultural use by securing as many irrigation permits as possible.

62. ICC Minutes, 13 August 1957; E. W. Rogers, LeMars Chamber of Commerce, to Justin J. Rogers, Izaak Walton League, 4 February 1958, IWLA-Iowa Division; Tobin, "Irrigators' Challenge is Answered by Waltonian," 1. Considering that George M. Foster, chair of the ICC 1955−1956, was also a member of the study committee that drafted the 1957 Water Resources Conservation Act, the breadth of its language is puzzling. Richard G. Bullard, the first state water commissioner, has stated that "the idea of regulating the state water resources in the public interest so appealed to the Iowa legislature that the bill swept through both houses of the 57th General Assembly without a dissenting vote"; see Bullard, "Iowa Water Laws," in *Water Resources of Iowa*, ed. Paul J. Horick (1970), 114. Given broad support for a law that empowered the state to appropriate water rights on the principle of public benefit, it is quite possible that the ramifications of the sweeping policy language were not thought through completely, or that the language was deliberately left vague in order to ensure quick passage.

63. ICC Minutes, 2−3 December 1957.

64. ICC Minutes, 7 January 1958; "Why Irrigate in Iowa," statement issued by Iowa Natural Resources Council, 28 February 1959, IWLA-Iowa Division.

65. Tobin, "Irrigators' Challenge is Answered by Waltonian," 3; see also Eugene Davis, "Water Rights in Iowa," *Iowa Law Review* 41 (1956), 216−236.

66. ICC Minutes, 8 April 1958, 6 May 1958, 5–6 August 1958, 4 September 1958, 6 January 1959, 10 March 1959, 2 September 1959, 7 October 1959. The Wildlife Management Institute, the National Wildlife Federation, and the Sport Fishing Institute all joined the ICC in its lawsuit against the NRC, which gave the commission some added leverage vis-à-vis a pending bill in the state legislature prohibiting one state agency from suing another.

67. ICC Minutes, 8–9 October 1957, 2–3 December 1957.

68. Walter Shotwell, "Close Valve for New Lake," *Des Moines Register*, 18 September 1950.

69. Ewald Trost, "The Birth of Lake Darling," remarks delivered at the Lake Darling gate-setting ceremony, 17 September 1950, Trost Papers; Shotwell, "Close Valve for New Lake."

70. Henderson, "Legislature Votes Six New Artificial Lakes," 129, 133.

71. ICC Minutes, 27 October 1950, 12–13 November 1950, 7–8 May 1951.

72. Dedication pamphlet for Rock Creek Lake and Park, 24 August 1952, Trost Papers.

73. H. W. Freed, "Preliminary Report in re: Proposed artificial lake sites in Shelby County," March 21, 1946; Freed, "Artificial Lake—Cass, Montgomery & Page Counties," April 2, 1946, Trost Papers; ICC Minutes, 26 February 1946, 24–25 February 1947, 22 August 1949, 27–28 February 1950, 23–24 August 1951, 10 April 1952.

74. "Finds Flaws in 4 State Dams," *Des Moines Register*, 24 August 1952; ICC Minutes, 12 August 1953, 21 September 1953, 30 November 1953, 4–5 December 1952.

75. ICC Minutes, 25–26 July 1956, 13 August 1957; *Report ICC* (1956), 29; *Report ICC* (1958), 34; "Forming Iowa's Newest Lake," *Iowa Conservationist* 15 (1956), 52; "Set Viking Lake Dedication," *Iowa Conservationist* 16 (1957), 161.

76. ICC Minutes, 17–18 July 1952, 23 July 1952, 14 August 1952; Don Allen, "Fight Loss of Fertile Farms to New Lake," *Des Moines Register*, 28 September 1952.

77. ICC Minutes, 30 November 1953, 23 May 1956, 18 June 1957, 7 October 1959, 1 November 1961; *Report ICC* (1954), 20, 144; *Report ICC* (1958), 142; *Report ICC* (1960), 43; "Prairie Rose Lake Dedication," *Iowa Conservationist* 21 (1962), 47.

78. *Report ICC* (1936), 116; *Report ICC* (1938), 121; *Report ICC* (1940), 186; *Report ICC* (1942), 139.

79. John Madson, "Iowa's State-Owned Playgrounds," *Iowa Conservationist* 11 (1952), 62; "An All-Time High in State Park Use," *Iowa Conservationist* 13 (1954), 12; "Look What Happened in the Parks!" *Iowa Conservationist* 14 (1955), 105; "Another State Park Attendance Record Set," *Iowa Conservationist* 15 (1956), 2; "Iowa Parks High in Visit Totals," *Iowa Conservationist* 16 (1957), 154.

80. "State Parks Most Popular with Residents of Iowa Towns," *Iowa Conservationist* 8 (1949), 140.

81. *Report ICC* (1938), 17–18.

82. ICC Minutes, 14 December 1938, 12–13 March 1940.

83. Iowa State Planning Board, *The Second Report* (1935), 136.

84. *The Second Report*, 133.

85. ICC Minutes, 2–3 July 1935, 6 October 1936.

86. A more complete account of Squirrel Hollow's development history appears in Rebecca Conard, "National Register of Historic Places Registration: Squirrel Hollow County Park Historic District" (1991).

87. *Report ICC* (1940), 155, 173–175; *Report ICC* (1950), 143.

88. ICC Minutes, 7–8 January 1941; J. Harold Ennis, "The County Conservation Program in Iowa," *Proceedings IAS* 69 (1962), 220.

89. ICC Minutes, 11 December 1944, 12 February 1945, 9 April 1945, 17 June 1946, 2 December 1946, 11–13 July 1948, 28–29 August 1950.

90. "State Parks in Need of Funds," *Iowa Waltonian* 9 (August 1954), 10; William A. Riaski, "Concerning State Parks," *Iowa Waltonian* 9 (December 1954), 4–5; "Riaski Speaks for State Parks," *Iowa Waltonian* 10 (June 1955), 1; Chap. 12, *Laws of the Fifty-sixth General Assembly* (1955); "The County Conservation Law," *Iowa Conservationist* 15 (1956), 65–66; "Fifteen Counties to Set Up Recreation Areas," *Iowa Conservationist* 16 (1957), 97, 104; "County Conservation Boards," *Iowa Conservationist* 17 (1958), 73, 77; ICC Minutes, 11–12 October 1956, 13–14 December 1956.

91. *Report ICC* (1960), 61–62.

92. ICC Minutes, 2 December 1958, 2 September 1959. The Cass County agreement, though, was not finalized until 1961 because of further negotiations to develop an artificial lake in Cass County. The commission agreed to collaborate with Cass County to develop Lake Anita on the condition that the county conservation board also take over its management and operation; see ICC Minutes, 7 December 1960, 6 December 1961.

93. ICC Minutes, 2 August 1962; "State Parks in Need of Funds," *Iowa Waltonian* 9 (August 1954), 10.

94. Ennis, "The County Conservation Program," 220.

95. Macbride, "County Parks," 91.

96. *Report ICC*, 1962, 61–62; Ennis, "The County Conservation Program," 222.

97. *Report ICC*, (1986), 61; telephone conversation with Don Brazelton, executive secretary, Iowa Association of County Conservation Boards, 17 June 1995.

98. Iowa State Conservation Commission, *Outdoor Recreation in Iowa* (1966), 60.

99. ICC Minutes, 3 April 1963.

100. *Report ICC* (1968), 9; *Report ICC* (1970), 20–21.

CHAPTER 7   The Lightning Rod Decades

1. ICC Minutes, 11–12 March 1980.

2. Author interviews with Marion Pike, 31 July 1995, and John Field, 10 August 1995; see ICC Minutes, 4–5 September 1979, 6 November 1979, and 7 December 1979, for representative accounts of Leonard's disruptive tactics.

3. Keith McNurlen, "Action for Outdoor Recreation—Iowa Division," remarks prepared for 42nd IWLA, Iowa Division, Annual Convention, 24–27, 1964, Keith Mc-Nurlen Papers, Iowa State University.

4. "$50 Million Recreation Plan," *Des Moines Sunday Register*, 7 June 1964.

5. Arnold O. Haugen, ed., *Report on Iowa's Outdoor Resources: Their Conservation and Use in Outdoor Recreation* (1964).

6. Chap. 135, *Laws of the Sixty-first General Assembly* (1965).

7. Hays, *Beauty, Health, and Permanence*, 52–57; Gottlieb, *Forcing the Spring*, 81–114.

8. Telephone conversation with S. Galen Smith, 25 September 1995; Myrle M. Burk, "Natural Areas Owned by the Iowa Chapter of the Nature Conservancy," *Proceedings IAS* 80 (1973), 175–177. The Nature Conservancy, a descendant of the Ecological Society of America, began acquiring lands and waters for preservation in 1954; the Iowa Chapter purchased its first site in 1965. According to Smith, two events sparked his interest in organizing the Iowa Chapter. Shortly after he joined the ISU faculty in 1960, Smith mounted a campaign to stop construction of a four-lane road and bridge through Pammel Woods. Then he was instrumental in organizing support to stop the destruction of a prairie patch located on the grounds of Ames High School, which ISU professors had long used as a field site for teaching purposes. Roger Landers also played an important role in establishing the Iowa Chapter.

9. One can trace the activities of the IAS Conservation Committee through its reports, included in the *Proceedings* from 1944 through 1960. In 1961 the committee organized as a "section" of the IAS, and for a few years after that published a brief summary of conservation activity statewide in the *Proceedings*. See especially "Report of the Committee on Conservation," *Proceedings IAS* 52 (1945), 30–47; J. M. Aikman, "What an Academy Can Do to Promote the Conservation of Natural Resources," *Proceedings IAS* 56 (1949), 29–37; "Conservation in Iowa, 1963, *Proceedings IAS* 71 (1964), 38.

10. At that time, the "geologic-biologic" areas were Bixby State Park, Barkley Memorial Preserve, Maquoketa Caves State Park, and Woodman Hollow. "Historic-archaeologic" areas were Fort Atkinson, Pillsbury Point State Park, Gitchie-Manitou, and the Wittrock Indian Village Tract; see ICC Minutes, 13–14 February 1941.

11. The fifteen areas, as listed in the ICC Minutes, 1 September 1955, were Barkley Memorial Preserve, Fish and Game [Fish Farm] Mounds, Fort Atkinson, Galland School, Abbie Gardner Sharp Cabin, Arnolds Park Pier, Pillsbury Point, Gitchie Manitou, [Wittrock] Indian Village, Kalsow Prairie, Lennon Mill [Panora], Plum Grove, Turkey River Mounds, Woodman Hollow, and Woodthrush Preserve.

12. ICC Minutes, 3 October 1962. The three state preserves not designated as wildlife refuges were Fort Atkinson, Kalsow Prairie, and Woodthrush Preserve.

13. Smith conversation, 25 September 1995; see also S. Galen Smith, "Natural Area Conservation in Iowa," *Proceedings IAS* 70 (1963), 191–196.

14. Haugen, *Report on Iowa's Outdoor Resources*, 36.

15. Smith to author, 30 October 1995; Edward T. Cawley telephone conversations, 30 June 1995, 23 September 1995; J[ohn]. T. Curtis, "Scientific Areas in Wisconsin," in Haugen, *Iowa's Outdoor Resources*, 118–119; Stearns and Germain, "Natural Areas Preservation Council: A Brief History and Record of Activity," 1. During his last four years of graduate school, Cawley was the arboretum botanist at the University of Wisconsin Arboretum, and his responsibilities included monitoring a restored prairie area.

16. Cawley telephone conversation, 30 June 1995; Smith conversation, 25 September 1995.

17. Cawley conversation, 30 June 1995; Edward T. Cawley, "The Iowa State Preserves System: A Progress Report," *Proceedings IAS* 76 (1969), 137; Edward T. Cawley, "The History of Prairie Preservation in Iowa," 22–24.

18. Cawley, "The Iowa State Preserves System," 138; "Iowa's New State Advisory Board for Preserves," *Iowa Waltonian* 7 (October 1965), 4; Dean Roosa, "Iowa's Preserves System," *Iowa Conservationist* 35 (1976), 13. Cawley was among those appointed to serve on the first Preserves Advisory Board. Others included ecologist Margaret Black of Drake University; George Knudson, Luther College Chemistry Department; state archaeologist Marshall McKusick of the University of Iowa; William Petersen, director of the State Historical Society; Robert Russell, representing the Izaak Walton League Iowa Division (which had supported the bill); and Everett B. Speaker of the State Conservation Commission.

19. ICC Minutes, 5–6 September 1967, 7–8 May 1968, 2–3 August 1968. Preserve status for Plum Grove was debated for several years, without resolution, because the board could not work out a management agreement with the Colonial Dames. As of this writing, Plum Grove still has not been dedicated as a state preserve. Ironically, this property was a major reason for writing the law to include cultural properties.

20. Roosa, "Iowa's Preserves System," 13.

21. Fleckenstein, *Iowa State Preserves Guide*, 3.

22. ICC Minutes, 13–14 December 1956; telephone conversation with Duane Anderson, state archaeologist 1975–1986, 28 August 1995. The Office of State Archaeologist was established in 1959 (Chap. 201, *Acts of the Fifty-eighth General Assembly*) as an unfunded position attached to the department of anthropology at the University of Iowa. Archaeologists Reynold J. Ruppé and David Stout of the University of Iowa lobbied for the position as a mechanism for involving academic archaeologists in the emerging enterprise of contract archaeology, which was initially associated with federal highway projects. Thus, the motivation for establishing the Office of State Archaeologist was distinct from the responsibilities and concerns facing the State Conservation Commission. Marshall McKusick replaced Ruppé as state archaeologist when the latter left Iowa in 1959.

23. ICC Minutes, 7 October 1959, 5–6 July 1960.

24. ICC Minutes, 2–3 August 1960.

25. Author interview with Adrian Anderson, former state historic preservation officer, 4 August 1995; conversation with Duane Anderson. Adrian Anderson's superior, state archaeologist Marshall McKusick, had not sought the historic preservation program, and he was no more interested in it than William Petersen had been. During the next three years, Anderson made the State Office of Historic Preservation operational, but administratively it became Iowa's orphan agency. Its second home proved also to be temporary. In 1971, Marshall McKusick asked the governor to remove it from the Office of State Archaeologist. Thus, Anderson's position was transferred, administratively, to the State Conservation Commission, although actual program operations remained in Iowa City. Moreover, the Office of Historic Preservation operated with its own state appropriation as well as with federal grants. As a result of this shuffling, Iowa was nearly the last state to have its historic preservation program certified by the fed-

eral government. The history of this confusing period and its implications for historic preservation in Iowa remain to be told, but I am grateful to Adrian Anderson and Duane Anderson (no relation) for unraveling the main threads.

26. Adrian Anderson interview.

27. Marion Clawson and Jack L. Knetsch, *Economics of Outdoor Recreation* (1966), see especially 304–317.

28. *Outdoor Recreation for America: The Report of the Outdoor Recreation Resources Review Commission* (1962). The Heritage Conservation and Recreation Service superseded the Bureau of Recreation during the Carter administration. Then, in 1981, the Reagan administration abolished HCRS, and the LWCF program was transferred to the National Park Service.

29. Phyllis Myers and Sharon Green, *State Parks in a New Era*, vol. 2 (1989), 1, 14–16.

30. Bill Brabham, "What the Land and Water Conservation Fund Means to Iowa," *Iowa Conservationist* 24 (1965), 67; Bill Brabham, "Land and Water Conservation Funds," *Iowa Conservationist* 25 (1966), 31; ICC Minutes, 2 May 1972.

31. *Outdoor Recreation in Iowa* (1966), 2–3.

32. ICC Minutes, 5–6 April 1966, 7 June 1966, 6–7 September 1966, 4–5 October 1966, 6–7 December 1966.

33. IDNR Budget and Grants Bureau.

34. At this writing, the LWCF still exists, but it is a pale shadow of the program that flourished in the 1970s, and funds for state projects have all but dried up. In recent years not more than $25 million annually has been allocated for all states. There have been repeated calls to reinvigorate the LWCF, but this seems unlikely given mounting fiscal problems at the federal level and current public attitudes regarding government spending; see, for instance, "New LWCF Plan Would Spread Funds Widely," *Common Ground* 5 (September–October 1994), 1, 7.

35. IDNR Budget and Grants Bureau.

36. *Report ICC* (1966), 41–44.

37. Acreage figure as reported in Myers and Green, *State Parks in a New Era*, vol. 1, 40; *Report ICC* (1984), 40–41, 68–74. Acreage figures for 1966 and 1985 include land only.

38. ICC Minutes, 27 August 1975. The state legislature no longer funds the Green Thumb program, although the Division of Parks, Recreation, and Preserves is still required to give priority to hiring seasonal workers who would have otherwise met the qualifications to work under the Green Thumb program.

39. ICC Minutes, 2 May 1978, 7 July 1981, 6–7 October 1981; *Report ICC* (1978), 8.

40. ICC Minutes, 3 December 1981, and passim until 1 May 1986, when the State Conservation Commission Minutes close.

41. ICC Minutes, 2–3 July 1984 to 1 May 1986, passim.

42. ICC Minutes, 6–7 October 1981, 5–6 January 1982, 6 August 1982.

43. Flader, *Exploring Missouri's Legacy*, 18–19; Myers and Green, *State Parks in a New Era*, vol. 2, 4–9.

44. ICC Minutes, 1–2 August 1985, 7–8 January 1986. The park user fee was re-

scinded in 1989, when the Resource Enhancement and Protection Act was passed; Chap. 236, *Laws of the Seventy-third General Assembly*.

45. Trost to Speaker, 7 June 1964 (draft copy), Trost Papers.

46. "Statement by Ewald Trost for the Fort Dodge Group meeting with the State Conservation Commission in Des Moines on July 7th, 1964"; G. W. Graalmann, Fort Dodge Chamber of Commerce to Trost, 23 June 1964; "Group Will Meet with Commission," *Fort Dodge Messenger* 6 July 1964, Trost Papers.

47. Don L. Russell, Fort Dodge Chamber of Commerce to E. B. Speaker, 14 October 1964; "Conservation Members Check Area Near City," *Fort Dodge Messenger*, 24 October 1964; "Start Surveys of Lake Sites," *Des Moines Register*, 9 November 1964; "Lizard Creek Area Passes First Test," *Fort Dodge Messenger*, 9 November 1964; Trost to State Conservation Commission, 10 November 1964, Trost Papers; ICC Minutes, 5 November 1964.

48. ICC Minutes, 2–3 February 1965, 4 May 1965, 2 November 1965, 7 June 1966.

49. "Senate Votes Outdoor Bill," *Des Moines Register*, 13 June 1967; "Conservation Bill Approved," *Des Moines Register*, 30 June 1967; ICC Minutes, 16–17 November 1967, 7–8 May 1968.

50. Pike interview; Field interview; Gerald Schnepf interview with author, 1 August 1995.

51. ICC Minutes, 6–7 December 1966, 5–6 December 1967, 5 September 1972, 5 February 1974, 1 April 1975, 2 December 1975.

52. ICC Minutes, 3 August 1976, 4 January 1977, 1 February 1977, 2–3 July 1979; Pike interview; Schnepf interview. At the 3–4 January 1968 meeting, the commission adopted a policy that it would not expend funds to impound waters for artificial lakes until the watershed had been "sufficiently protected against erosion to prevent a loss of over 25 percent of its recreational value during the first 100 years of operation."

53. ICC Minutes, 5–6 December 1967; Otto Knauth, "State Vows That 2 Lakes Won't Be 'Sewage Lagoons,' " *Des Moines Register*, 6 December 1967. A 1965 state water pollution law finally took pollution control out of the hands of the Department of Health. It authorized yet another agency, the Iowa Water Pollution Control Commission, to provide for prevention, abatement, and control of water pollution.

54. "Plans for Brushy Lake," *Fort Dodge Messenger*, 30 August 1967; "New Lake Makes Zoning Necessary," *Fort Dodge Messenger*, 31 August 1967; Ewald Trost's handwritten notes of 19 January 1968 meeting of Webster County Board of Supervisors; D. M. Hill (Superintendent of Construction-Maintenance, ICC) to Everett B. Speaker, 22 January 1968, summarizing the 19 January Board of Supervisors meeting, Trost Papers.

55. Trost to Hill, 11 April 1968, 23 April 1968; "Waltonians Plan Study of Zoning," *Fort Dodge Messenger*, 29 May 1968, Trost Papers. The IWLA-Iowa Division and several local chapters were very involved in water pollution control matters beginning in the late 1950s; see Buckmann, *The First 50*, 83–88.

56. ICC Minutes, 2–3 August 1968; Dick Doak, "Landowners Attack New State Lake," *Des Moines Register*, 3 August 1968; "Ask Condemnation Right for Land at Brushy Creek," *Fort Dodge Messenger*, 12 December 1969. Later commissioners would

recall that "horse-trading" was more-or-less standard operating procedure among commission members, and every commissioner had a local constituency anticipating most-favored-trading status. Commissioners typically benefited from the system, in terms of goodwill and community status, so it was highly unusual for sitting commissioners to air such personal opinions in public.

57. ICC Minutes, 4 November 1975; Brice, Petrides & Associates, Inc., *Brushy Creek State Recreation Area Environmental Impact Study* (1982), 2.1.

58. Quinn Johnson, Fort Dodge Chamber of Commerce, to Rep. Vince Mayberry, 24 January 1969; Johnson to Sen. Joe Coleman, 24 January 1969; Trost to D. M. Hill, 12 May 1969; Trost to Lloyd Bailey, ICC, 2 October 1969; Trost to William Boswell, assistant director, ICC, 12 February 1970; Johnson to Fred Priewert, 6 August 1970; Priewert to Johnson, 11 August 1970; Rep. Dale M. Cochran to Trost, 11 August 1970, Trost Papers.

59. ICC Minutes, 2 December 1975, 6 April 1976, 31 August 1976.

60. ICC Minutes, 4 October 1977.

61. ICC Minutes, 1 November 1977.

62. Marian Pike, in particular, "was not strong for a lake, *anywhere*" during her early years on the commission; Pike interview.

63. ICC Minutes, 6 February 1979; Brice, Petrides & Associates, *Brushy Creek EIS*, 2.1.

64. Brice, Petrides & Associates, *Brushy Creek EIS*, 2.1–2.2, 3.2–3.3.

65. ICC Minutes, 7–8 April 1981; Brice, Petrides & Associates, *Brushy Creek EIS*, 2.3, sections 5, 6.

66. Author interview with Michael Carrier, 2 August 1995.

67. See, for instance, Arthur Maass, *Muddy Waters: The Army Engineers and the Nation's Rivers* (1951).

68. ICC Minutes, 27–28 August 1945, 26 February 1946, 3–4 September 1946, 26–27 September 1949, 27–28 February 1950, 18 December 1950, 23–24 August 1951, 27 March 1956, 6 January 1960, 3 February 1960; C. W. Daly, "Will Expand Lake Macbride," *Iowa Conservationist* 9 (1950), 20; "A New Lake Macbride," *Iowa Conservationist* 16 (April 1957), 125; *Report ICC* (1958), 144.

69. ICC Minutes, 28 July 1939, 15 September 1939.

70. ICC Minutes, 15–16 November 1948.

71. Natural Resources Council, *Supplemental Report on Water Resources and Water Problems, Des Moines River, Iowa* (1955), 6.

72. ICC Minutes, 27–28 May 1954; Stiles to Natural Resources Council, 8 September 1955, Save the Ledges Coalition Papers, Iowa State University.

73. ICC Minutes, 19–20 July 1954; "Saylorville Dam Still Under Hot Debate," *Iowa Waltonian* 9 (October 1954), 5; "General notes on the Corps meeting September 13, 1954" and "General notes on the Natural Resources Council meeting September 27, 1955," Save the Ledges Coalition Papers. Also see Elmer Peterson, *Big Dam Foolishness* (1954).

74. "Conservation Bill Approved," *Des Moines Register*, 30 June 1967.

75. Iowa State Conservation Commission, *Outdoor Recreation in Iowa* (1968), 179; ICC Minutes, 2 December 1969.

76. ICC Minutes, 4 January 1972, 12 November 1974.

77. Gerry Schnepf, who was in charge of the ICC's planning section in the 1970s, now views the Hawkeye Naturama concept in a different light. "In retrospect," he says, "the man [Arnold Haugen] was on target. I think at the time, we, including myself, kind of pooh-poohed Naturama and held it down, didn't let that whole concept come to the forefront. But what it proposed was lifestyle training for youth and family"; Schnepf interview.

78. Buckmann, *The First 50*, 74–76.

79. ICC Minutes, 5–6 September 1967, 6 August 1971, 3 October 1972.

80. Otto Knauth, "Consider Flood Plans for Ledges," *Des Moines Sunday Register*, 14 September 1971.

81. Bob Davis, "Saylorville Dam Foes Form Alliance," *Boone News-Republican*, 30 November 1972.

82. ISU members included Clark Bowen, assistant dean of sciences and humanities and chairman of the Iowa Chapter of the Sierra Club; Robert Moorman, a wildlife conservationist with the ISU Extension Service; botanist Roger Landers; and James O'Toole of the ISU College of Veterinary Medicine, who also represented the Iowa Confederation of Environmental Organizations.

83. Bob Davis, "Court Issues Ledges Restraining Order to Corps," *Boone News-Republican*, 4 December 1972; *Iowa Citizens to Save Ledges State Park, Inc., et al. v. Robert Froehlke, et al.* [Secretary of the Army], U.S. District Court, 72-285-2, 1972. The lawsuit came on the heels of one of the more important environmental decisions of the twentieth century, *Sierra Club v. Morton*, 405 U.S. 727 (1972), in which the U.S. Supreme Court ruled that the Sierra Club, by virtue of its stated mission to protect the natural environment of the Sierra Nevada, had legal standing in court to defend the natural environment of Mineral King Valley against economic development interests, in this case, Walt Disney Enterprises. The decision prompted Justice William O. Douglas to take the legal analysis one step further and write a pathbreaking minority opinion in which he argued that Mineral King Valley *itself* should have standing in the case. Douglas's argument was influenced, in part, by Christopher Stone's 1972 essay, "Should Trees Have Standing?" For more on Douglas and the evolution of his legal thinking, see Roderick Nash, *The Rights of Nature: A History of Environmental Ethics* (1988), 130–131; James C. Duram, "Justice William O. Douglas and the Wilderness Mind," in *Great Justices of the U.S. Supreme Court*, ed. Pederson and Proviser (1994), 233–248.

84. Bonnie Wittenburg, "Saylorville's Impact on Local Lives," *Des Moines Sunday Register*, 13 May 1973.

85. Bob Davis, "Corps Offers Cash Settlement for Ledges Damage," *Boone News-Republican*, 25 July 1973.

86. ICC Minutes, 7 April 1970.

87. ICC Minutes, 4 September 1973. According to Gerry Schnepf, he and Dick Fleishman, representing the Corps, worked out a new settlement, which was eventually adopted. It carried a price tag of about $7 million to pay for additional land acquisition at Ledges and a river corridor trail from Des Moines to the reservoir. In addition, the agreement also included a stipulation to monitor the flood control system for ten

years, after which time the State Conservation Commission could request an additional sum to cover losses unanticipated when the project was complete.

88. "Saving the Ledges," *The Peoples Press* [ISPIRG], 6 September 1973; Otto Knauth, "Pack Room for Hearing on Saylorville Dam Plan," *Des Moines Register*, 27 September 1973.

89. U.S. Army Engineer District, Rock Island, *Final Environmental Impact Statement, Saylorville Lake Flood Control Project, Des Moines River, Iowa* (1974).

90. Bonnie Wittenburg, "Corps of Engineers Rejects Barrier Dam at Ledges Park," *Des Moines Register*, 30 May 1974.

91. See the lengthy comments and responses contained in section 8 of the final EIS. Environmental groups had reason to be skeptical of the Corps' conclusions. This section exemplifies the non sequitur responses that the COE and other federal agencies routinely served up to environmentalists in an attempt to minimize the effects of the National Environmental Policy Act on their entrenched philosophies and procedures.

92. Ibid., 71 – 72.

93. "Protecting the Ledges," *Des Moines Register*, editorial, 4 June 1974.

94. *Iowa Citizens to Save Ledges State Park, Inc., et al. v. Robert F. Froehlke, et al.* Motion to Dismiss and in the Alternative Motion for Summary Judgment (6 June 1974).

95. "Seek Help of Public Officials for Ledges," *Boone News-Republican*, 17 June 1974; Bonnie Wittenburg, "Environmentalists Ask Ray for Ledges Park Support," *Des Moines Register* 3 July 1974; "Groups Continue Efforts to Prevent Flooding of Ledges," *Ames Daily Tribune*, 9 July 1974; "Document Evaluates Ways to Save the Ledges from Flooding Waters," *Boone News-Republican*, 13 July 1974.

96. "Ray Asks Special Task Force to Study 9-Point Ledges Proposal," *Boone News-Republican*, 20 August, 1974; "Press release, 20 August 1974, Office of the Governor," Records, Iowa Citizens to Save Ledges State Park [hereafter Records CSL].

97. ICC Minutes, 10 September 1974.

98. Robert B. Smythe, COE, to Goeppinger, 25 November 1974 [form letter]; Goeppinger to "Chairman" Peterson, COE, 20 December 1974, Records CSL.

99. Mary Burke, "Nystrom Proposes Saving Ledges," *Boone News-Republican*, 24 March 1975; "Congressman Tom Harkin Reports from Washington," newsletter, 4 June 1975, Records CSL.

100. "See Ray Approval on Saylorville Dam Plan," *Des Moines Register*, 18 June 1975.

101. Otto Knauth, "Drop Fight for Ledges Barrier Dam," *Des Moines Register*, 22 July 1975 (quotation); Mary Burke, "Time a Factor in Ledges Policy Change," *Boone News-Republican*, 22 July 1975; "Iowa Citizens to Save the Ledges State Park Statement, Press Conference on the Ledges, 21 July 1975, presented by Hans Goeppinger," Records CSL.

102. See Hays, *Beauty, Health, and Permanence*, and "From Conservation to Environment: Environmental Politics in the United States Since World War II," *Environmental Review* 6 (1982), 14 – 29.

103. "Contract for Relocation and Modification of Ledges State Park Facilities,"

between Iowa State Conservation Commission and U.S. Army Corps of Engineers, 27 December 1976, IDNR.

104. And the flood of all floods came just sixteen years later, in 1993, making a mockery of Corps predictions by outstripping the capacity of nearly every flood control structure in the Upper Mississippi River Valley. Behind Saylorville Dam, the flood-pool rose to 892 feet and devastated Ledges State Park, depositing as much as ten feet of silt on the canyon floor. Based on recent reports emanating from the Corps of Engineers, the floods of '93 *appear* to have convinced Army engineers that it is not feasible to continue building ever larger structural works to hold and/or channel floodwaters and that taxpayer money would be better spent acquiring farms and homes located in floodplains; see "Bleed Rivers, Not Taxpayers," *Des Moines Register*, 9 July 1995. It is too early to tell whether this conclusion is based simply on the fiscal restraints that will shape federal spending into the foreseeable future, on a fear that Congress will vest more authority over floodplain management with the Department of Interior (as recommended by the Clinton administration's Interagency Floodplain Management Review Committee), or whether the Corps of Engineers has truly adopted a new environmental ethic. One positive outcome of the floods, however, was acquisition of the 2,600-acre Louisa No. 8 Levee District near the confluence of the Iowa and Mississippi rivers, which has been added to the Mark Twain National Wildlife Refuge. The 1994 Galloway Report, *Sharing the Challenge: Floodplain Management Into the 21st Century*, cited the Louisa Levee District project as a "national model for alternative floodplain management." Two private, nonprofit environmental organizations, the Iowa Natural Heritage Foundation and The Conservation Fund, played key roles in bringing the project to fruition, in cooperation with the U.S. Fish and Wildlife Service. And the lessons of Ledges continue. In 1995, the Department of Natural Resources spent just $128,000 to clear out the several feet of accumulated silt and to restore the canyon, a feat nearly everyone thought impossible in 1993, and surely one that could not be done so inexpensively.

105. *"Rediscovering" Ledges State Park: A Master Plan for Redevelopment* (1981), 1–3, 19–20.

106. Ibid., 4; original quotation, Rene Dubos, *The Wooing of Earth* (1980), 7.

107. Schnepf interview. Commissioners appointed by Ray were Jim D. Bixler, Council Bluffs (1969); Joan Geisler, Dubuque (1969); Les L. Licklider, Cherokee (1969); John G. Link, Burlington (1971); Thomas A. Bates, Bellevue (1972); Herbert T. Reed, Winterset (1973); John C. Thompson, Forest City (1973); Carolyn Lumbard Wolter, Des Moines (1973); Marian Pike, Whiting (1975); John C. Brophy, Lansing (1975); Richard T. Kemler, Marshalltown (1977); Donald E. Knudson, Eagle Grove (1979); John Field, Hamburg (1979); Baxter Freese, Wellman (1981); and Richard Thornton, Des Moines (1981).

108. Pike interview. Blue Lake is part of Lewis and Clark State Park.

109. Carolyn Lumbard, a foreign language instructor at Drake, was active in the Polk County Soil Conservation District and the Environmental Coordinating Council, one of the many environmental groups active during the 1970s.

110. Pike interview. Her perspective on the role she and Carolyn Wolter played in

commission deliberations, while it is all too briefly characterized, raises an interesting question about the role of women, generally, in American environmentalism. From the Progressive Era to the present, women's contributions to environmentalism largely have been analyzed through the activities of women's organizations, women-led mass protests, or the voices of prominent feminists. This has yielded a complex picture of women and environmentalism—see, for instance, Gottlieb, *Forcing the Spring*, 207–234—but it still does not adequately address the ways in which women have influenced environmental policy from within bureaucratic systems.

111. Pike interview.

112. Pike interview; Schnepf interview; Field interview.

113. Field interview.

114. Author interview with Larry Wilson, 2 August 1995.

115. ICC Minutes, 7 April 1970, 7 July 1981.

116. [Larry Wilson], "A Message from the Director of the Iowa Conservation Commission," *Iowa Conservationist* 40 (1981), 6–7; ICC Minutes, 7 July 1981. The Nature Conservancy still provides program direction for the natural areas inventory, which, among other things, enables the state to keep track of threatened or endangered species.

117. ICC Minutes, 2–3 July 1984, 2–3 October 1984, 7 February 1985, 1–2 October 1985.

118. Chap. 1245, sections 1801–1899, *Laws of the Seventy-first General Assembly* (1986). Reorganization of environmental agencies was part of a massive reorganization of state government in general.

119. ICC Minutes, 15 January 1951, 6 October 1955.

120. ICC Minutes, 3–4 January 1967, 6–7 February 1968, 1 October 1969.

121. ICC Minutes, 4 April 1972, 2 May 1972.

122. Chap. 1119, *Laws of the Sixty-fourth General Assembly, Second Session* (1972).

123. Chap. 1199, *Laws of the Sixty-ninth General Assembly* (1982).

124. Field interview. According to him, the commissioners met informally in the Wallace Office Building to debate which provisions of the draft reorganization bill they could accept and those they could not. Commissioner Richard Thornton, a seasoned lobbyist, served as liaison to key legislators, and in this fashion the reorganization bill was redrafted. For obvious reasons, there are no official records of these meetings, but Marian Pike confirms that such meetings took place, and documents in her personal records attest to a high level of communication among commissioners during early 1986.

125. See, for instance, Marian Pike's editorial, "Change Erodes Conservation Board," *Des Moines Register*, 14 September 1987, which ran shortly after she completed her second six-year term.

126. Wilson interview.

127. Macbride, "The Citizen vs. The City," 1906; Macbride, "A Forestry Commission for Iowa," 1909.

128. Macbride, "County Parks," 91.

129. Garrett Hardin, "The Tragedy of the Commons," *Science* 162 (1968), 1243–1248.

130. Kenneth Boulding, "Commons and Community: The Idea of a Public," in *Managing the Commons*, ed. Hardin and Baden (1977), 286–293.

Epilogue

1. Carrier interview.

2. Carrier interview.

3. ICC Minutes, 6 March 1986; Brice, Petrides-Donahue, *Proposed Dam and Lake, Brushy Creek State Recreation Area, Webster County, Iowa, Final Environmental Impact Statement* (1992) contains a summary of public participation from 1986 to 1990.

4. Ibid.

5. Carrier to author, 9 December 1995.

6. Carrier interview.

7. Ibid.

8. Chap. 236, *Acts of the Seventy-third General Assembly* (1989).

9. Larry Stone, "27 Years in Making, Lake is Taking Shape," *Des Moines Register*, 6 September 1995.

10. Carrier interview.

11. Ibid.

12. Ibid.

13. Minutes, State Preserves Advisory Board, 20 December 1994.

14. Carrier interview.

15. Wilson interview.

16. Ibid.

17. Ibid.

18. Ibid.

Bibliography

Manuscript and Special Collections

Blue, Robert D. Papers as Governor of Iowa, 1944–1948. State Historical Society of Iowa, Library and Archives Division, Des Moines.

Darling, Jay Norwood "Ding." Papers, 1909–1979. Special Collections, University of Iowa.

Fitzsimmons, John. Drawing Collection. Special Collections, Iowa State University.

Fort Atkinson Research Files. State Historical Society of Iowa, Manuscripts and Archives Division, Iowa City.

General Federation of Women's Clubs. Program Records 1920–1940. GFWC Archives, Washington, D.C.

Harding, William Lloyd (1877–1934) and the Harding/Lamoreux Families. Papers, 1830–1964. Sioux City Public Library.

Harlan, Edgar R. Papers, 1887–1937. State Historical Society of Iowa, Library and Archives Division, Des Moines.

Iowa Citizens to Save Ledges State Park. Records, 1972–1978. Special Collections, Iowa State University.

Iowa Department of Natural Resources, Des Moines. Engineering Division [architectural drawings and development plans]; Parks, Recreation, and Preserves Division [technical reports, master plans, administrative records]; Central Records [minutes, biennial reports].

Iowa Federation of Women's Clubs. Yearbooks and miscellaneous records. Iowa Federation of Women's Clubs, Des Moines, Iowa.

Izaak Walton League of America, Iowa Division, Records. Special Collections, Iowa State University.

Lacey, John Fletcher. Papers, 1847–1913. State Historical Society of Iowa, Library and Archives Division, Des Moines.

Macbride, Thomas Huston. Papers, 1848–1934 and 1845–1954. Special Collections, University of Iowa.

McNurlen, Keith A. Papers, 1958–1972. Special Collections, Iowa State University.

Pammel, Louis Hermann. Papers, 1882–1931. Special Collections, Iowa State University.

Parker, Louise L. Papers, 1944–1972. Special Collections, Iowa State University.

Poyneer, Fred J. Papers, 1939–1972. Special Collections, University of Iowa.

Save the Ledges Coalition. Records, 1948–1977. Special Collections, Iowa State University.

Shimek, Bohumil. Papers, 1890–1946 [includes finding aid to Bohumil Shimek Papers, 1878–1936, Smithsonian Institution]. Special Collections, University of Iowa.

Speaker, Everett B. Papers, 1930–1970. Special Collections, Iowa State University.

Trost, Ewald George. Papers, 1925–1952. Special Collections, University of Iowa.

Whitley, Cora Call [Mrs. Francis Edmund]. Papers. Iowa Women's Archives, University of Iowa.

Interviews and Personal Communications

Anderson, Adrian. Interview with author. Jewell, Iowa, 4 August 1995.

Anderson, Duane. School of American Research, University of New Mexico. Telephone conversation with author, 28 August 1995.

Brazelton, Don. Executive secretary, Iowa Association of County Conservation Boards. Telephone conversation with author, 17 June 1995.

Carrier, Michael. Interview with author. Des Moines, 2 August 1995. Letter to author, 9 December 1995.

Cawley, Edward T. Biology department, Loras College. Telephone conversations with author, 30 June 1995 and 23 September 1995.

Field, John D. Interview with author. Hamburg, Iowa, 10 August 1995.

Hoffman, Hank. Indiana Department of Natural Resources. Telephone conversation with author, 9 May 1995.

Matthiae, Paul. Chief, Natural Areas Section, Wisconsin Department of Natural Resources. Telephone conversation with author, 29 September 1995.

Parker, Addison M., Jr. Unrecorded conversation with author. Des Moines, Iowa, 22 March 1995.

Pike, Marian S. Interviews with author. Waverly, Iowa, 20 April 1995; Whiting, Iowa, 31 July 1995.

Schnepf, Gerald F. Interview with author. Des Moines, 1 August 1995.

Smith, Michael H. Assistant attorney general, Iowa Department of Justice. Unrecorded conversation with author, 13 January 1995.

Smith, S. Galen. Professor emeritus, University of Wisconsin-Whitewater. Telephone conversation with author, 25 September 1995. Letter to author, 30 October 1995.

Steinauer, Jerry. Nebraska Game and Parks Commission. Telephone conversation with author, 5 May 1995.

Wilson, Larry. Interview with author. Des Moines, 2 August 1995.

Government Documents

Iowa. Board of Conservation. Minutes, 1918–1935.

———. *Iowa Parks: Conservation of Iowa Historic, Scenic, and Scientific Areas.* Des Moines, 1919.

———. *Administration of Iowa Parks, Lakes and Streams.* Des Moines, 1931.

Iowa. Department of Health. Division of Public Health Engineering. *Report on the Investigation of Pollution of the Des Moines River From Estherville to Farmington, 1928–1934.* Des Moines, 1934.

Iowa. Drainage, Waterways and Conservation Commission. *Report of the Iowa State Drainage, Waterways and Conservation Commission.* Cedar Rapids: Torch Press, 1911.

Iowa. Executive Council. Minutes as contained with Board of Conservation Minutes.

Iowa. Fish and Game Commission. Minutes, 1931–1935.

———. *Report of the Fish and Game Commission,* 1934.

Iowa. Fish and Game Warden. *Report of the State Fish and Game Warden,* 1914–1918.

Iowa. Interim Flood Control Committee. *Report of the Interim Flood Control Committee,* 1948.

Iowa. Natural Resources Council. *Supplemental Report on Water Resources and Water Problems, Des Moines River, Iowa.* Des Moines, 1955.

Iowa. State Conservation Commission. Minutes, 1935–1986.

———. *Outdoor Recreation in Iowa.* Des Moines, 1966.

———. *Outdoor Recreation Iowa* (First Revision). Des Moines, 1968.

———. *"Rediscovering" Ledges State Park: A Master Plan for Redevelopment.* Des Moines, 1981.

Iowa. State Highway Commission. *Report of the State Highway Commission, Part One.* Des Moines: Robert Henderson, State Printer, 1917.

Iowa. State Planning Board. *A Preliminary Report of Progress.* Des Moines, 1934.

———. *The Second Report.* Des Moines, 1935.

Iowa. Study Committee of Water Rights and Drainage Laws. *Report of the Iowa Study Committee of Water Rights and Drainage Laws.* Des Moines, 1956.

U.S. Army. Corps of Engineers. Rock Island District. *Final Environmental Impact Statement, Saylorville Lake Flood Control Project, Des Moines River, Iowa.* Rock Island, Illinois: 1974.

U.S. Congress. House. Committee on Agriculture. *Hearings on H.R. 4088: A Bill to Establish the Upper Mississippi River Wild Life and Fish Refuge.* 68th Cong., 1st Sess., 1924.

U.S. Department of Interior. National Park Service. *State Park Statistics.* Reports compiled from the annual records on state parks and related recreation areas. Variously issued through the Recreation Planning Division, Cooperative Activities Division, or Division of Recreation Resource Planning. Washington, D.C., 1952–1960.

———. *A Study of the Park and Recreation Problem of the United States.* Washington, D.C., 1941.

U.S. Interagency Floodplain Management Review Committee. *Sharing the Challenge: Floodplain Management Into the 21st Century.* Washington, D.C., 1994.

U.S. Outdoor Recreation Resources Review Commission. *Outdoor Recreation for America*. Washington, D.C., 1962.

Serial Publications, Newspapers

Annals of Iowa
Boone News-Republican
Bulletin: Iowa State Parks, 1923–1927
Christian Science Monitor
Decorah Journal
Des Moines Register
Fort Dodge Messenger
Iowa Clubwoman, 1927– [continues *Iowa Federation News*]
Iowa Conservation, 1917–1923
Iowa Conservationist, 1941–
Iowa Federation News, 1922–1927 [Iowa Federation of Women's Clubs]
Iowa Journal of History and Politics
Iowa Waltonian
Palimpsest
Proceedings of the Iowa Academy of Science
Proceedings of the Iowa Forestry and Conservation Association, 1914–1915. Cedar Rapids: Torch Press, 1916 [continues *Proceedings IPFA*]
Proceedings of the Iowa Park and Forestry Association, 1902–1913
Report of the Iowa State Horticultural Society
Spencer Daily Reporter

Books, Articles, Reports, and Theses

Aikman, J. M. "What an Academy Can Do to Promote the Conservation of Natural Resources," *Proceedings IAS* 52 (1945), 29–37.

Anfinson, John O. "Commerce and Conservation on the Upper Mississippi River." *Annals of Iowa* 52 (1993), 385–417.

Bennett, George. "The American School of Wild Life Protection, McGregor Heights—1923." *Iowa Conservation* 7 (1923), 40–44.

———. "Keeping Iowa's Water Pure." *Proceedings IAS* 31 (1924), 431–435.

———. "The National Park Conference at Des Moines, Iowa, January 10–11–12, 1921." *Iowa Conservation* 5 (1921), 14–25.

———. "The National Park of the Middle West." *Iowa Conservation* 2 (1918), 43–47, 67–70.

Bennett, Henry Arnold. "Fish and Game Legislation." *IJHP* 24 (July 1926), 335–444.

Beyer, Samuel W. "Some Problems in Conservation." *Proceedings IAS* 26 (1919), 37–46.

Blair, Karen J. *The Clubwoman as Feminist: True Womanhood Redefined*. New York: Holmes & Meier, 1980.

Bliss, Robert M. "A Baltimore Piano Maker's Investment Saved Some of Iowa's Native Prairie." *Iowa Conservationist* 8 (1949), 155.

Boulding, Kenneth E. "Commons and Community: The Idea of a Public." In *Managing the Commons*, edited by Garrett Hardin and John Baden. San Francisco: W. H. Freeman, 1977.

Brabham, Bill. "Land and Water Conservation Funds." *Iowa Conservationist* 25 (1966), 31.

———. "What the Land and Water Conservation Fund Means to Iowa." *Iowa Conservationist* 24 (1965), 67.

Brice, Petrides & Associates, Inc. *Brushy Creek State Recreation Area Environmental Impact Study*. Waterloo, Iowa, 1982.

Brice, Petrides-Donohue. *Proposed Dam and Lake, Brushy Creek State Recreation Area, Webster County, Iowa, Final Environmental Impact Statement*. Waterloo, Iowa, 1992.

Briggs, John E. "The Legislation of the Thirty-ninth General Assembly of Iowa," *IJHP* 19 (1921), 489–666.

Briggs, John E., and Cyril B. Upham. "The Legislation of the Thirty-eighth General Assembly," *IJHP* 17 (1919), 471–612.

Briggs, John E., and Jacob Van Ek. "The Legislation of the Fortieth General Assembly of Iowa," *IJHP* 21 (1923), 507–676.

Buckmann, Carol A. *The First 50: The Story of the Izaak Walton League of America, 1923–1973*. Iowa Division, IWLA, 1973.

Buffum, Mrs. Hugh, and Louella Thurston. *History of the Iowa Federation of Women's Clubs, 1893–1968*. Des Moines: Iowa Federation of Women's Clubs [1968].

Bullard, Richard G. "Iowa Water Laws." In *Water Resources of Iowa: Papers Given at a Symposium of the Iowa Academy of Science at the University of Northern Iowa, April 18, 1969*, edited by Paul J. Horick. Iowa City: University of Iowa, 1970.

Burk, Myrle M. "Natural Areas Owned by the Iowa Chapter of the Nature Conservancy." *Proceedings IAS* 80 (1973), 175–177.

Butler, George D. *Municipal and County Parks in the United States, 1935*. Washington, D.C.: National Park Service in cooperation with the National Recreation Association, 1937.

Byington, O. A. "The Passage of the Quail Bill." *Iowa Conservation* 1 (1917), 35–36.

Calvin, Samuel. "The Devil's Backbone." *Midland Monthly* 6 (July 1896), 20–26.

Carhart, A[rthur] R. "A System of Parks—National, State and County." *Report of the Iowa State Horticultural Society* 51 (1917), 79–84.

Carlander, Harriet Bell. "The American School of Wildlife, McGregor, Iowa, 1919–1941." *Proceedings IAS* 68 (1961), 294–300.

Cawley, Edward T. "The History of Prairie Preservation in Iowa." in *Proceedings of the Second Midwest Prairie Conference, 1970*, edited by James H. Zimmerman. Madison: University of Wisconsin, 1972.

———. "The Iowa State Preserves System: A Progress Report," *Proceedings IAS* 76 (1969), 135–141.

Christensen, Thomas P. "The State Parks of Iowa." *IJHP* 26 (1928), 331–414.

Christisen, Donald. "A Vignette of Missouri's Native Prairie." *Missouri Historical Review* 61 (1967), 166–186.

Clawson, Marion, and Jack L. Knetsch. *Economics of Outdoor Recreation*. Baltimore: Johns Hopkins Press for Resources for the Future, 1966.

Clements, Frederic. *Plant Succession: An Analysis of the Development of Vegetation*, Washington, D.C.: Carnegie Institution, 1916.

Clyde, R. W. "Lake Survey of Iowa." *Iowa Conservation* 3 (April–June 1919), 44–45.

Conard, Rebecca. "The Conservation Movement in Iowa, 1857–1942." National Register of Historic Places Documentation, 1991.

———. "National Register of Historic Places Registration: Squirrel Hollow County Park Historic District," 1991. Original copy located at State Historical Society of Iowa, Office of Historic Preservation, Des Moines.

Conservation Study Committee and Ira N. Gabrielson. "Gabrielson, Committee Report" ["Joint Report to Governor Robert D. Blue and the Iowa Conservation Commission"]. *Iowa Conservationist* 7 (1948), 1, 5–8.

Cox, Laurie D. "Use Areas in State Parks." In *American Planning and Civic Annual*, edited by Harlean James. Washington, D.C.: American Planning and Civic Association, 1940.

Cox, Thomas. *The Park Builders: A History of State Parks in the Pacific Northwest*. Seattle: University of Washington Press, 1988.

Crane, Jacob L., Jr. "Preparation of the Iowa Conservation Plan." In *American Civic Annual*, edited by Harlean James. Washington, D.C.: American Civic Association, 1932.

Crane, Jacob L., Jr., and George Wheeler Olcott. *Report on the Iowa Twenty-five Year Conservation Plan*. Des Moines: Meredith, 1933.

Curtis, J[ohn] T. "Scientific Areas in Wisconsin." In *Report on Iowa's Outdoor Resources: Their Conservation and Use in Outdoor Recreation*, edited by Arnold O. Haugen. Des Moines, 1964.

Curtiss, C. F. "Forest and Game Reserves in Iowa." *Proceedings* IFCA (1916), 109–110.

Daly, C. W. "Will Expand Lake Macbride," *Iowa Conservationist* 9 (1950), 20.

Darling, Jay N. "Poverty of Conservation Your National Problem." *Iowa Conservationist* 3 (1944), 89–90, 93–94; 4 (1945), 97–99, 103, 105, 108, 111, 113–115.

Davis, Eugene. "Water Rights in Iowa." *Iowa Law Review* 41 (1956), 216–236.

Deering, Charles C. "The Telephone in Iowa." *Annals of Iowa* 23 (1942), 287–308.

Dubos, Rene. *The Wooing of Earth*. New York: Charles Scribner's Sons, 1980.

Dupree, A. Hunter. *Science in the Federal Government: A History of Policies and Activities to 1940*. Cambridge: Harvard University Press, 1957.

Duram, James C. "Justice William O. Douglas and the Wilderness Mind." In *Great Justices of the U.S. Supreme Court*, edited by William D. Pederson and Norman W. Prozier. New York: Peter Lang, 1994.

Ely, Richard T., et al. *The Foundations of National Prosperity: Studies in the Conservation of Permanent Natural Resources*. New York: Macmillan, 1917.

Engbeck, Joseph H., Jr., and Philip Hyde. *State Parks of California from 1864 to the Present*. Portland, Oregon: C. H. Belding, 1980.

Ennis, Harold J. "The County Conservation Program in Iowa." *Proceedings* IAS 69 (1962), 219–223.

Evison, Herbert. "Plans for State Park Systems." In *American Civic Annual*, edited by Harlean James. Washington, D.C.: American Civic Association, 1932.

Faber, Lester F. "Iowa's 14 Years in the Pittman-Robertson Program." *Iowa Conservationist* 10 (1951), 180.

Flader, Susan L., ed. *Exploring Missouri's Legacy: State Parks and Historic Sites.* Columbia: University of Missouri Press, 1992.

———. *Thinking Like a Mountain: Aldo Leopold and the Evolution of an Ecological Attitude Toward Deer, Wolves, and Forests.* Columbia: University of Missouri Press, 1974.

Fleckenstein, John, comp. *Iowa State Preserves Guide.* Des Moines: Iowa Department of Natural Resources, 1992.

Flickenger, V. W. "Current Land and Waters Development." *Iowa Conservationist* 7 (1948), 17, 22, 24.

———. "Twenty-Five Years of Iowa Parks." *Iowa Conservationist* 6 (1947), 105, 110–111, 113, 118.

———. "Use Areas in Iowa State Parks." In *American Planning and Civic Annual,* edited by Harlean James. Washington, D.C.: American Planning and Civic Association, 1940.

Ford, H. C. "What the State is Doing Toward the Conservation of Iowa Lakes." *Proceedings IFCA* (1916), 176–179.

Freed, H. W. "Iowa's Six New Lakes." *Iowa Conservationist* 6 (1947), 185, 191.

Gabrielson, Ira N. "Report to the Iowa Conservation Commission by the Wildlife Management Institute." *Iowa Conservationist* 14 (March 1955 supplement), 1–7.

———. *Wildlife Conservation.* New York: Macmillan, 1941.

———. *Wildlife Refuges.* New York: Macmillan, 1943.

Gallagher, Annette, C.H.M. "Citizen of the Nation: John Fletcher Lacey, Conservationist." *Annals of Iowa* 46 (1981), 9–22.

Gallaher, Ruth. "The Man." *Palimpsest* 15 (1934), 161–172.

Good, Albert H., ed. *Park and Recreation Structures.* 3 vols. Washington, D.C.: National Park Service, 1938.

———. *Park Structures and Facilities.* Washington, D.C.: National Park Service, 1935.

Gottlieb, Robert. *Forcing the Spring: The Transformation of the American Environmental Movement.* Washington, D.C.: Island Press, 1993.

Greiff, Glory-June. "New Deal Resources in Indiana State Parks." National Register of Historic Places Multiple Property Documentation, 1991.

———. "People, Parks, and Perceptions: Eighty Years of Indiana State Parks." Unpublished manuscript, 1995. Copy in author's possession.

Guralnick, Stanley M. "The American Scientist in Higher Education, 1820–1910." In *The Sciences in the American Context: New Perspectives,* edited by Nathan Reingold. Washington, D.C.: Smithsonian Institution Press, 1979.

Hanson, Robert W. "The Iowa Academy of Science: 1875–1975." *Proceedings IAS* 82 (1975), 2–32.

Hardin, Garrett. "The Tragedy of the Commons." *Science* 162 (1968), 1243–1248.

[Harlan, Edgar R.]. "Iowa State Parks." *Annals of Iowa* 13 (October 1921), 140–144.

Harlan, Edgar R. "Proposed Improvement of the Iowa State Capitol Grounds." *Annals of Iowa* 11 (1913), 96–114.

Harstad, Peter T., and Bonnie Lindemann. *Gilbert N. Haugen: Norwegian-American Farm Politician.* Des Moines and Iowa City: State Historical Society of Iowa, 1992.

Haugen, Arnold O. "History of the Iowa Cooperative Wildlife Research Unit." *Proceedings IAS* 73 (1966), 136–145.

———, ed. *Report on Iowa's Outdoor Resources: Their Conservation and Use in Outdoor Recreation.* Des Moines, 1964.

Hayden, Ada. "Conservation of Prairie." In *Iowa Parks: Conservation of Iowa Historic, Scenic, and Scientific Areas.* Des Moines, 1919.

———. "The Ecologic Foliar Anatomy of Some Plants of a Prairie Province in Central Iowa." *American Journal of Botany* 6 (1919), 69–85.

———. "The Ecologic Subterranean Anatomy of Some Plants in a Prairie Province in Central Iowa." *American Journal of Botany* 6 (1919), 87–104.

———. "Iowa Had a Coat of Many Colors." *Iowa Conservationist* 5 (1946), 25, 31.

———. "Notes on the Floristic Features of a Prairie Province in Central Iowa." *Proceedings IAS* 25 (1918), 368–389.

———. "A Progress Report on the Preservation of Prairie." *Proceedings IAS* 53 (1947), 45–82.

———. "The Selection of Prairie Areas in Iowa Which Should be Preserved." *Proceedings IAS* 52 (1945), 127–148.

Hayden, Ada, and J. M. Aikman. "Considerations Involved in the Management of Prairie Preserves." *Proceedings IAS* 56 (1949), 133–142.

Hays, Samuel P. *Beauty, Health and Permanence: Environmental Politics in the United States, 1955–1985.* New York: Cambridge University Press, 1987.

———. *Conservation and the Gospel of Efficiency: The Progressive Conservation Movement, 1890–1920.* Cambridge: Harvard University Press, 1959.

———. "From Conservation to Environment: Environmental Politics in the United States Since World War II." *Environmental Review* 6 (1982), 14–29.

Henderson, John W. "Legislature Votes Six New Artificial Lakes." *Iowa Conservationist* 6 (1947), 129, 133.

Hershey, H. Garland. "Water Supply." *Proceedings IAS* 52 (1945), 44–47.

Hinshaw, E. C. *Report of the State Fish and Game Warden for the Biennial Period Ending June 30, 1916* (1916).

———. *Report of the State Fish and Game Warden for the Biennial Period Ending June 30, 1918* (1918).

———. *Twenty-first Report ICC of the State Fish and Game Warden, 1913–1914* (1915).

Horack, Frank Edward. "The Legislation of the Thirty-seventh General Assembly," *IJHP* 15 (1917), 503–570.

———. "The Work of the Thirty-fifth General Assembly of Iowa," *IJHP* 11 (1913), 546–600.

Indiana Division of Nature Preserves. *Directory of Indiana's Dedicated Nature Preserves.* Indianapolis: Indiana Department of Natural Resources, 1991.

Isely, Duane. "Ada Hayden: A Tribute." *Journal IAS* 96 (1989), 1–5.

Johnson, Malcolm K. "White Pine Hollow." *Iowa Conservationist* 19 (1960), 93–94.

Johnson, Sandra Jeanne. "Early Conservation by the Arizona Federation of Women's Clubs from 1900 to 1932." M.S. thesis, University of Arizona, 1993.

Keller, Evelyn Fox. *Reflections on Gender and Science*. New Haven: Yale University Press, 1985.

Langenbach, John R. "Restoring a Land Base." In *Restoring America's Wildlife, 1937–1987: The First 50 Years of the Federal Aid in Wildlife Restoration (Pittman-Robertson) Act*. Washington, D.C.: U.S. Department of Interior, Fish and Wildlife Service, 1987.

Lees, James H. "Some Geologic Aspects of Conservation." *Proceedings IAS* 24 (1917), 133–154.

Lendt, David L. *Ding: The Life of Jay Norwood Darling*. 1979. Ames: Iowa State University Press, 1989.

Leopold, Aldo. "Game and Wildlife Conservation." *Condor* 34 (1932), 103–106.

———. "Report of the Iowa Game Survey." *Outdoor America* 11 (August–September 1932), 7–9; (October–November 1932), 11–13, 30–31; (December 1932–January 1933), 10–12, 31; (February–March 1933), 6–8, 21.

———. *A Sand County Almanac*. 1949. New York: Ballantine Books, 1970.

Lovell, Jan. "She Fought to Save Iowa's Prairies." *Iowan* 36 (1987), 22–27, 56–57.

Maass, Arthur. *Muddy Waters: The Army Engineers and the Nation's Rivers*. Cambridge: Harvard University Press, 1951.

Macbride, Thomas H. "The Conservation of Our Lakes and Streams." In *Report of the Iowa State Drainage, Waterways and Conservation Commission*, 1911.

———. "County Parks." *Proceedings IAS* 3 (1896), 91–95.

———. "A Forestry Commission for Iowa." *Report of the Iowa State Horticultural Society for the Year 1908* 43 (1909), 166–170.

———. *On the Campus*. Cedar Rapids: Torch Press, 1925.

———. "Parks and Parks." *Bulletin ISP* 2 (1922), 13–14.

———. "The President's Address." *Proceedings IAS* 5 (1898), 12–23.

———. "President's Address: The Present Status of Iowa Parks." *Proceedings IPFA* 1 (1902), 1–13.

MacDonald, G. B. "The Beginning of a National and State Forestry Program in Iowa." *Ames Forester* 23 (1935), 15–20.

———. "Forestry and the Iowa Farmer." *Ames Forester* 10 (1922), 59–62.

———. "Forestry Progress in Iowa." *Ames Forester* 29 (1941), 7–17.

———. "Multiple Use of State Forests." *Iowa Conservationist* 5 (1946), 9, 11–12.

———. "Present State Forests." *Iowa Conservationist* 4 (1945), 161, 165, 167.

———. "Progress of the Forest Land Acquisition Program in Iowa." *Ames Forester* 25 (1937), 49–57.

———. "Reforestation in Iowa." *Iowa Conservation* 5 (1921), 101–103.

———. "Reforestation Work Under Clarke-McNary Act in Iowa." *Ames Forester* 14 (1926), 21.

———. "Report on the Status of Emergency Conservation Work in Iowa." *Thirty-sixth Annual Year Book*. Des Moines: Iowa Department of Agriculture, 1935.

———. "State Forestry: The Early Period." *Iowa Conservationist* 4 (April 1945), 121, 123, 127.

———. "The Value of Iowa's Lakes for the People." *Proceedings IFCA* (1916), 180–184.

Madson, John. "Iowa's State-Owned Playgrounds." *Iowa Conservationist* 11 (1952), 57, 62.

———. *Where the Sky Began: Land of the Tallgrass Prairie*. Rev. ed. 1982. Ames: Iowa State University Press, 1995.

McClelland, Linda Flint. *Presenting Nature: The Historic Landscape Design of the National Park Service, 1916–1942*. Washington, D.C.: U.S. Department of the Interior, National Park Service, 1993.

McFall, Don, and Jean Karnes, eds. *A Directory of Illinois Nature Preserves*. 2 vols. Springfield: Illinois Department of Natural Resources, 1995.

McKay, Joyce. "Civilian Conservation Corps Properties in Iowa State Parks, 1933–1942. National Register of Historic Places Document, 1989. Original copy at Des Moines: State Historical Society of Iowa, Office of Historic Preservation.

———. "Survey Report: CCC Properties in Iowa State Parks," January 1990. Original copy at Des Moines: State Historical Society of Iowa, Office of Historic Preservation.

Meine, Curt. *Aldo Leopold: His Life and Work*. Madison: University of Wisconsin Press, 1988.

Melosi, Martin V. *Garbage in the Cities: Refuse, Reform, and the Environment, 1880–1980*. College Station: Texas A&M Press, 1981.

Merchant, Carolyn. "Women of the Progressive Conservation Movement, 1900–1916." *Environmental Review* 8 (1984), 57–85.

Merritt, Raymond H. *The Corps, the Environment, and the Upper Mississippi River Basin*. Washington, D.C.: Office of the Chief of Engineers, Historical Division, 1984.

Meyer, Roy W. *Everyone's Country Estate: A History of Minnesota's State Parks*. St. Paul: Minnesota Historical Society Press, 1991.

Myers, Phyllis, and Sharon Green. *State Parks in a New Era*. 3 vols. Washington, D.C.: Conservation Foundation, 1989.

Nash, Jan R. "Plum Grove Historic Site Project: Evaluation of the Interpretation of Plum Grove and Recommendations." Consultant's Report to the National Society of Colonial Dames of America in the State of Iowa. Iowa City, 1993. Copy located at State Historical Society of Iowa, Office of Historic Preservation, Des Moines.

Nash, Roderick. *The Rights of Nature: A History of Environmental Ethics*. Madison: University of Wisconsin Press, 1988.

Natural Heritage Trust. *Fifty Years: New York State Parks, 1924–1974*. Albany, N.Y.: Natural Heritage Trust, 1975.

Nelson, Beatrice Ward. *State Recreation: Parks, Forests and Game Reservations*. Washington, D.C.: National Conference on State Parks, 1928.

Oberholser, Harry C. "The Winneshiek Bottoms Drainage Project." *Iowa Conservation* 7 (1923), 9–10.

O'Bright, Jill York. *The Perpetual March: An Administrative History of Effigy Mounds National Monument*. Omaha: National Park Service, Midwest Regional Office, 1989.

O'Neill, Eugene James. "Parks and Forest Conservation in New York, 1850–1920." Ph.D. dissertation, Columbia University, 1963.

Overfield, Richard A. *Science with Practice: Charles E. Bessey and the Maturing of American Botany*. Ames: Iowa State University Press, 1993.

Pammel, Louis H. "The Arbor Day, Park and Conservation Movements in Iowa." *Annals of Iowa* 17 (1929–30), 83–104, 198–232, 270–313.

————. "Conservation." *Bulletin ISP* 3 (November–December 1925), 66.

————. "The Preservation of Iowa's Lakes from the Standpoint of a Botanist." *Proceedings IFCA* (1916), 173–175.

————. "The Preservation of Natural History Spots in Iowa." *Proceedings IFCA* (1916), 105–108.

————. "Rural Parks." *Iowa State Park Bulletin* 2 (1924), 5.

————. "What the Legislature Did with Reference to State Parks in Iowa." *Iowa Conservation* 3 (1919), 14–15.

————, ed. *Major John F. Lacey Memorial Volume*. Cedar Rapids: Torch Press for the Iowa Park and Forestry Association, 1915.

Parker, Mrs. Addison [Louise]. "Iowa State Parks." In *American Planning and Civic Annual*, edited by Harlean James. Washington, D.C.: American Planning and Civic Association, 1941.

————. "Our Vanishing Prairie." *Iowa Conservationist* 3 (1944), 63.

Persons, Stow. *The University of Iowa in the Twentieth Century: An Institutional History*. Iowa City: University of Iowa Press, 1990.

Peterson, Elmer. *Big Dam Foolishness*. New York: Devin-Adair, 1954.

Pohl, Marjorie Conley. "Louis H. Pammel, Pioneer Botanist: A Biography." *Proceedings IAS* 92 (1986), 1–50.

Prior, Jean C. *Landforms of Iowa*. Iowa City: University of Iowa Press for the Iowa Department of Natural Resources, 1991.

Proceedings of the Second National Conservation Congress at Saint Paul, September 5–8, 1910. Washington, D.C.: National Conservation Congress, 1911.

Proceedings of the Third National Conservation Congress at Kansas City, Missouri, September 25–27, 1911. Kansas City: National Conservation Congress, 1912.

Ramey, Ralph E., and Roger E. Troutman. "Ohio Prairies and the Ohio Prairie Survey Project." In *Fifth Midwest Prairie Conference Proceedings, Iowa State University, Ames, 1976*, edited by David C. Glenn-Lewin and Roger Q. Landers, Jr. Ames: Iowa State University, 1978.

Ranney, Sally Ann Gumaer. "Heroines and Hierarchy: Female Leadership in the Conservation Movement." In *Voices from the Environmental Movement*, edited by Donald Snow, 41–45. Arlington, Va.: The Conservation Fund, 1992.

Riaski, William A. "Concerning State Parks." *Iowa Waltonian* 9 (December 1954), 4–5.

————. "Iowa Conservation Legislation in 1957." *Iowa Waltonian* 11 (July 1957), 2.

————. "Iowa Parks Are Being Neglected." *Iowa Waltonian* 11 (March 1957), 1, 9.

Rieger, John F. *American Sportsmen and the Origins of Conservation*. New York: Winchester Press, 1975.

Robinson, Charles Mulford. *City Planning Report for Des Moines, Ia*. St. Louis, 1909.

Roosa, Dean. "Iowa's Preserves System," *Iowa Conservationist* 35 (1976), 13–14.

Rothman, Hal. *Preserving Different Pasts*. Urbana and Chicago: University of Illinois Press, 1989.

Rudolph, Frederick. *The American College and University: A History*. New York: Vintage Books, 1962.

Runte, Alfred. *National Parks: The American Experience*. Lincoln: University of Nebraska Press, 1987.

Rush, Wilbur A. "Iowa State Parks in 1947." *Iowa Conservationist* 6 (1947), 188.

Sage, Leland. *A History of Iowa*. Ames: Iowa State University Press, 1974.

Salmond, John. *The Civilian Conservation Corps, 1933–1942: A New Deal Case Study*. Durham: Duke University Press, 1967.

Scarpino, Philip V. *Great River: An Environmental History of the Upper Mississippi, 1890–1950*. Columbia: University of Missouri Press, 1985.

Schertz, Mary Winifred Conklin and Walter L. Myers. *Thomas Huston Macbride*. Iowa City: University of Iowa Press, 1947.

Schmidt, G. Perle. "Preservation of Places of Historic Interest in Iowa." In *Iowa Parks*. Des Moines, 1919.

Schmitt, Peter J. *Back to Nature: The Arcadian Myth in Urban America*. Baltimore: Johns Hopkins University Press, 1990.

Schwob, F. T. "Iowa Plans Comprehensive Postwar Conservation Program." *Iowa Conservationist* 3 (1944), 73, 75.

———. "Parks and Recreation from the State's Point of View." In *American Planning and Civic Annual*, edited by Harlean James. Washington, D.C.: American Planning and Civic Association, 1943.

———. "Schwob Makes Recommendation to Postwar Committee." *Iowa Conservationist* 3 (1944), 81–82.

Sears, John F. *Sacred Places: American Tourist Attractions in the Nineteenth Century*. New York: Oxford University Press, 1989.

Seidel, Harris F., ed. *Iowa's Heritage in Water Pollution Control*. Clear Lake: Iowa Water Pollution Control Association, 1974.

Shimek, Bohumil. "Drainage in Iowa." *Proceedings IAS* 31 (1924), 149–155.

———. "In Memoriam." *Proceedings IAS* 41 (1934), 33–37.

———. "Iowa's Natural Parks." *Iowa Conservation* 1 (1967), 16–17.

———. "Papers on the Prairie." *Laboratories of Natural History* [University of Iowa], Bulletin 11 (1925), 1–36.

———. "The Plant Geography of the Lake Okoboji Region." *Laboratories of Natural History*, Bulletin 7 (1915), 1–90.

———. "Prairie Openings in the Forest." *Proceedings IAS* 17 (1911), 16–19.

———. "The Prairies." *Laboratories of Natural History*, Bulletin 6 (1911), 169–240.

———. "The Prairies of the Mississippi River Bluffs." *Proceedings IAS* 31 (1926), 205–212.

———. "Report of the Committee on Conservation." *Proceedings IAS* 34 (1927), 31–33.

Smith, Daryl D. "Iowa Prairie—An Endangered Ecosystem." *Proceedings IAS* 88 (1981), 7–10.

———. "'Mystique' of the Prairie." In *Fifth Midwest Prairie Conference Proceedings, Iowa State University, Ames, 1976*, edited by David C. Glenn-Lewin and Roger Q. Landers, Jr. Ames: Iowa State University, 1978.

Smith, Michael L. *Pacific Visions: California Scientists and the Environment, 1850–1915*. New Haven: Yale University Press, 1987.

Smith, S. Galen. "Natural Area Conservation in Iowa." *Proceedings IAS* 70 (1963), 191–196.

Stearns, Forest, and Cliff Germain. "Natural Areas Preservation Council, 1951–1991." Report prepared for Wisconsin Department of Natural Resources, November 1991.

Stephens, T. C. "Game Protection in Iowa." *Proceedings IFCA* (1916), 61–69.

———. "Needed Changes in Our Game Laws." *Proceedings IFCA* (1916), 14–24.

———. "A Review of Wildlife Protection in Iowa." *Iowa Conservation* 1 (1917), 49–52, 60, 64–68.

Stiles, Bruce F. "Brief Analysis of Iowa Fish and Game Policy." *Iowa Conservationist* 4 (October 1945), 169, 173–174; (November 1945), 178, 184; (December 1945), 185, 190–191; 5 (January 1946), 6–7.

———. "New Laws." *Iowa Conservationist* 6 (1947), 145, 151.

———. "Some Conservation Needs in Iowa as Expressed by a National and State-wide Survey." *Proceedings IAS* 64 (1957), 81–84.

———. "Stiles Gives Brief History of Iowa Wildlife Legislation." *Iowa Conservationist* 2 (September 1943), 65–66; (October 1943), 77–78; continued as "Last Installment of History of Iowa Wildlife Legislation" (November 1943), 86, 88.

Stiles, Bruce F., Everett B. Speaker, Reeve M. Bailey, and George O. Hendrickson. *Wildlife Resources of Iowa*. Des Moines: State of Iowa, 1946.

Swain, Donald C. *Federal Conservation Policy, 1921–1933*. Berkeley and Los Angeles: University of California Press, 1963.

Swisher, J[acob] A. "Historical and Memorial Parks." *Palimpsest* 12 (1931), 201–253.

———. "The Iowa Academy of Science." *IJHP* 29 (1931), 315–374.

———. "The Legislation of the Forty-first General Assembly of Iowa." *IJHP* 23 (1925), 507–625.

———. *The Legislation of the Forty-fourth General Assembly*. Iowa Monograph Series, no. 3, edited by Benjamin F. Shambaugh. Iowa City: State Historical Society of Iowa, 1932.

———. "Plum Grove." *Palimpsest* 29 (1948), 19–32.

Tarr, Joel A., James McCurley, and Terry F. Yosie. "The Development and Impact of Urban Wastewater Technology: Changing Concepts of Water Quality Control, 1850–1930." In *Pollution and Reform in American Cities, 1870–1930*, edited by Martin V. Melosi. Austin: University of Texas Press, 1980.

Taylor, Mrs. H. J. [Rose Schuster]. "Conservation of Life Through City Parks." *Iowa Conservation* 1 (1917), 13.

"Thomas Huston Macbride." *Palimpsest* 15 (May 1934), 161–192.

Thompson, William H. *Transportation in Iowa: A Historical Summary*. Ames: Iowa Department of Transportation, 1989.

Tiffany, Lois Hattery. "Reflections on Women Scientists and the Iowa Academy of Science." *Proceedings IAS* 82 (1975), 94–95.

Tilden, Freeman. *The State Parks: Their Meaning in American Life*. New York: Alfred A. Knopf, 1962.

Tobin, John. "Irrigators' Challenge Is Answered by Waltonian." *Iowa Waltonian* 11 (August–September 1958), 1, 3.

Toney, Thomas E. *Public Prairies of Missouri*. Columbia: Missouri Department of Conservation [1994].

Torrey, Raymond H. *State Parks and Recreational Uses of State Forests in the United States*. Washington, D.C.: National Conference on State Parks, 1926.

Trefethen, James B. *Crusade for Wildlife*. New York: Boone and Crockett Club, 1961.

Van Hise, Charles H. *The Conservation of Natural Resources in the United States*. New York: Macmillan, 1910.

Veysey, Laurence R. *The Emergence of the American University*. Chicago: University of Chicago Press, 1965.

Wells, Mildred White. *Unity in Diversity: The History of the General Federation of Women's Clubs*. Washington, D.C.: General Federation of Women's Clubs, 1953.

Whitley, Mrs. Francis Edmund [Cora]. "Conservation of Natural Resources." *Iowa Federation News* 4 (September–October 1923), 15.

———. "Department of Applied Education, Conservation of Natural Resources." *IFWC Year Book 1925–26* (1926), 251–252.

———. "Our State Parks." *Bulletin ISP* 1 (April 1924), 15–16.

———. "Some Aspects of Conservation in Iowa." *Proceedings of the Iowa Forestry and Conservation Association, 1914–1915*. Cedar Rapids: Torch Press, 1916.

———. "Women's Clubs and Forestry: A Record of Active and Consistent Support of Forest Conservation by the Organized Women of America." *American Forests and Forest Life* 32 (1926), 221–223.

Widney, Stan. "Our Silver Anniversary." *Iowa Conservationist* 19 (1960), 38.

Wieters, A. H. "Status of Stream Pollution in Iowa." *Proceedings IAS* 35 (1928), 63–67.

Wildlife Management Institute. "A Ten Year Program for the Iowa Conservation Commission." Washington, D.C., 1958.

Williams, Michael. *Americans and Their Forests: A Historical Geography*. New York: Cambridge University Press, 1989.

Williamson, Lonnie L. "Evolution of a Landmark Law." In *Restoring America's Wildlife, 1937–1987: The First 50 Years of the Federal Aid in Wildlife Restoration (Pittman-Robertson) Act*. Washington, D.C.: U.S. Department of Interior, Fish and Wildlife Service, 1987.

Wisconsin State Planning Board and Wisconsin Conservation Commission. *A Park, Parkway and Recreational Area Plan*. Madison, 1939.

Wood, Mary I. *The History of the General Federation of Women's Clubs for the First Twenty-Two Years of Its Organization*. New York: General Federation of Women's Clubs, 1912.

Worster, Donald. *Nature's Economy: A History of Ecological Ideas*. New York: Cambridge University Press, 1977.

Wylie, Robert B. "The Scholar." *Palimpsest* 15 (1934), 173–182.

Zieglowsky, Debby J. "Thomas Macbride's Dream: Iowa Lakeside Laboratory." *Palimpsest* 66 (1985), 42–65.

Index

The American Land and Life Series

DATE DUE